Biotechnology of Medicinal Plants

Vitalizer and Therapeutic

Biotechnology of Medicinal Plants

Vitalizer and Therapeutic

Editor

K.G. Ramawat

CRC Press
Taylor & Francis Group
Boca Raton London New York

CRC Press is an imprint of the
Taylor & Francis Group, an **informa** business
A SCIENCE PUBLISHERS BOOK

CRC Press
Taylor & Francis Group
6000 Broken Sound Parkway NW, Suite 300
Boca Raton, FL 33487-2742

First issued in hardback 2019

ISBN-13: 978-1-57808-338-1 (hbk)

Library of Congress Cataloging-in-Publication Data

Biotechnology of medicinal plants : vitalizer and therapeutic / editor, K.G. Ramawat, p. cm.
Includes bibliographical references and index.
ISBN 1-57808-338-9
1. Materia medica, Vegetable. 2. Medicine, Ayurvedic. I. Ramawat,K.G.

RS 164.B488 2004
615'.321--dc22

2004041679

Visit the Taylor & Francis Web site at
http://www.taylorandfrancis.com

and the CRC Press Web site at
http://www.crcpress.com

Preface

In recent years, there has been an increased surge in the demand for plant-based drugs and food additives. Consumers do not prefer the use of synthetic colours and chemicals in food products and drugs. Some countries have even banned the use of synthetic colourants and flavours from use for human consumption. The use of organic fertilizers and natural dyes is the preferred alternative and also fetches higher returns.

The present book provides a comprehensive and useful information on the medicinal plants especially used as food supplement in the form of health vitalizers and invigorators. As such a subjective approach is attempted for the first time, all the information is available in one single volume. It is apparent that the day-to-day stress on the human population as a result of competition for better living, pollution and socio-economic situation is reflected in the form of increased heart ailments, hypertension, diabetes, cancer and decreased vigour and potency. Recent surveys published from China and India show a decrease in the sexual potency of the Chinese males and increased incidences of heart ailments in Indians due to urban pollution and stress, respectively.

In third world countries, markets are flooded with plant-based local formulations for invigorating health. Such formulations are based on folk medicine and an indigenous system of curatives. Though modern medication such as viagra, penegra, etc., has hit the market in recent years, the use of plant-based drugs for health, vigour and aphrodisiac properties is an ancient phenomenon. These drugs are time tested and come without any major side effect, if any at all. In the Indian market, about 75 such food supplements, vitalizers, formulations are available, possessing the essences of *Aloe barbadensis, Asparagus adscendens, Butea monosperma, Chlorophytum borivilianum, Curculigo orchioides, Mucuna pruriens, Saussurea lappa, Smilax officinalis, Terminalia arjuna, Trachyspermum ammi,* and *Withania somnifera* in different percentage and composition. However, *Asparagus, Chlorophytum, Curculigo, Mucuna* and *Withania* are the principal components used most frequently and in high ratio. Some of the preparations also contain ginseng roots, which is much-sought after herb of the East Asian countries. Similar formulations are in demand in the Middle East, Southeast Asia and Europe. Similarly, *Turnera diffusa* var. *aphrodisiaca* is the herb of the New World. Ginkgo biloba extract is widely used in Europe. Humankind is desirous to live long, healthy and young. All these drug-manufacturing firms are encashing on this desire. In this volume on the medicinal values of plants separate chapters are devoted to each herb so as to provide scientific information all at one place in comprehensive chapters written by the experts in their respective fields.

A brief account of the Indian system of medicine, Ayurveda, has been included to give an idea about this curative science. According to Ayurveda, every human being has a different physical and physiological state. Therefore, formulations are prepared according to the state of that particular individual. Immunomodulatory substances of different chemical structures and molecular sizes exist in these plants, which trigger a sequence of various fundamental processes, leading to improved health. It has been shown that certain plant extracts possess significant anti-stress activity, e.g. Indian indigenous drug *Ashwagandha* (*Withania somnifera*) is used as *Medhya Rasaayana* (for the treatment of mental diseases and anxiety states) in Ayurveda. In ancient Ayurvedic texts, *Medhya Rasaayana* materials are supposed to counteract the effect of mental stress by tranquilizing the users and also improving their memory span and intelligence. Plants such as *Centella asiatica, Convovulus pluricaulis, Bacopa monniera, Chlorophytum borivilianum, Curculigo orchioides* and *Nardostachys jatamansi* are prescribed for this therapy. Similarly, Chinese, Korean and Japanese ginseng (*Panax ginseng*), American ginseng (*P. quinquefolius*), Siberian ginseng (*Eleutherococcus senticosus*) and Indian ginseng (*P. pseudoginseng*) are widely used in Oriental traditional medicine. The details about these plants are presented in this book.

The Indian system of medicine also uses several minerals in various formulations. Among the numerous mineral preparations for rejuvenation, *Shilajit* is the most conspicuous mineral. It is a pale brown exudation of variable consistency, extracted from steep rocks (1000-5000 m) of different formations. Found throughout the Himalayas, *Shilajit* is widely used in Oriental medicine to arrest ageing and accelerate the process of rejuvenation—the two major attributes of an Ayurvedic *Rasaayana*. The organic compounds of *Shilajit* can be grouped as humic (80-85% of total organic mass) and non-humic (20-15%) substances. The non-humic component of *Shilajit* contains low molecular weight and organic compounds of various classes such as oxygenated dibenzo (α-pyrones and tirucallane type) triterpene acids. Such compounds are produced by the interaction of the microorganisms with a number of latex- and resin-bearing plants such as *Euphorbia royalena, Trifolium repens, Rhus succedanea* and *Pistasia integerrima*.

Shilajit, apart from being used in diverse clinical conditions in Ayurveda, has also been used to arrest ageing and induce rejuvenation and to improve memory, the major attributes of Ayurvedic *Rasaayana* and *Medhya Rasaayana*. *Shilajit* possesses cell stabilizing and mast cell protecting action and anti-inflammatory and analgesic activity. The neurochemical effects induced by this curative indicate a decrease in rat brain serotin turnover, an after-effect associated with an increase in dopaminergic activity, which explains the nootropic and anxiolytic effects of the drug.

In this book, only works on therapeutic and vitalizer plants have been included. Biotechnology of Medicinal Plants will be a valuable source book for the herbalists, Ayurvedic practitioners, drug manufacturers, botanists, biotechnologists, pharmacologists, agriculturists, phytochemists as well as anyone interested in medicinal plants research and biotechnology.

The editor wishes to thank all the contributors for their help in the preparation of this book.

Udaipur
July 2003

K.G. Ramawat

Contents

List of Contributors

Dr D.K. Arora
Department of Botany, M.L. Sukhadia University, Udaipur 313 002, India

Dr C. Caron
Laboratoire de Pharmacognasie UPRES-A 6013, UFR Pharmacie, Université de Reims, Champagne-Ardenne, CRPB, BP 1039, F-51687 Reims cedex 2, France

Professor J.C. Chenieux
EA 2106, Biomolécules, Biotechnologies Végétales, UFR Pharmacie, Université de Tours, 31 avenue Monge, F-37200 Tours, France

Dr K.T. Choi
Division of Genetics and Physiology, Korea Ginseng and Tobacco Research Institute, Taejon, Korea

Dr E. Couillerot
Laboratoire de Stress, Défenses et Reproduction des Plantes, URVVC EA2069, UFR Sciences, Université de Reims, Champagne-Ardenne, BP 1039, F-51687 Reims cedex 2, France

Dr A. Decendit
Laboratoire de Mycologie et Biotechnologie Végétale, GESUAB EA UFR des Sciences Pharmaceutiques Université victor Segalen Bordeaux, 2, 146 rue Léo Saignat, 33076 Bordeaux Cedex, France

Professor G. Deffieux
Laboratoire de Mycologie et Biotechnologie Végétale, GESUAB EA UFR des Sciences Pharmaceutiques Université victor Segalen Bordeaux-2, 146 rue Léo Saignat, 33076 Bordeaux Cedex, France

Dr C. Dupont
Laboratoire de Pharmacognosie, JE Oncopharmacologie, UFR Pharmacie, Universite de Reims, Champagne-Ardenne, 51 rue Cognacq-Jay, F-51096 Reims cedex, France

Dr. Halina Ekiert
Chair and Department of Pharmaceutical Botany Collegium Medicum, Jagiellonian University, 9 Medyczna Street, 30-688 Kraków. Poland.
e-mail : mfekiert@cyf-kr.edu.pl

Ashok K. Jain
Laboratory of Biomolecular Technology, Department of Botany, M.L. Sukhadia University, Udaipur 313 002, India

Dr J.P. Kamat
Cell Biology and Bioorganic Division, Bhabha Atomic Research Centre, Trombay, Mumbai 400 085, India, e-mail: jkamat@magnum.barc.ernet.in

Dr A. Koulman
Department of Pharmaceutical Biology, GUIDE, University of Groningen, A. Deusinglaan 1, 9713 AV Groningen, The Netherlands.

Professor S. Kintzios
Laboratory of Plant Physiology, Faculty of Agricultural Biotechnology, Agricultural University of Athens, Athens, Greece

Dr S. Krisa
Laboratoire de Mycologie et Biotechnologie Végétale, GESUAB EA UFR des Sciences Pharmacentiques Universite victor Segalen, Bordeaux 2, 146 sue Léo Saignat, 33076, Bordeaux cedex, France.

Dr A.K. Kukreja
Genetic Resources and Biotechnology Division, Central Institute of Medicinal and Aromatic Plants, P.O. CIMAP, Lucknow 226 016, India

Dr O. Makri
Laboratory of Plant Physiology, Faculty of Agricultural Biotechnology, Agricultural University of Athens, Athens, Greece

Dr D.L-Mattar
Laboratory of Pharmacognosy, Faculty of Pharmacy, 1 rue des Louvels, 80037 Amiens Cedex 1, France. e-mail: dominique.mattar@sa.u-picardie.fr

Professor J.M. Mérillon
Laboratoire de Mycologie et Biotechnologie Végétale, GESUAB EA, UFR des Sciences Pharmaceutiques, Universite victor Segalen Bordeaux 2, 146 rue Léo Saignat, 33076, Bordeaux cedex, France.

Dr N. Pras
Department of Pharmaceutical Biology, GUIDE, University of Groningen, A. Deusinglaan 1, 9713 AV Groningen, The Netherlands, e-mail: N.Pras@farm.rug.nl

Dr W.J. Quax
Department of Pharmaceutical Biology, GUIDE, University of Groningen, A. Deusinglaan 1, 9713 AV Groningen, The Netherlands, e-mail: w.j.quax@farm.rug.nl

Dr M. Rajani
Pharmacognosy and Phytochemistry Department, B.V. Patel Pharmaceutical Education and Research Development Centre, Thaltej, Ahmedabad 38004, Gujarat, India, e-mail: perd@wilnetonline.net

Professor K.G. Ramawat
Laboratory of Biomolecular Technology, Department of Botany,
M.L. Sukhadia University, Udaipur 313 001, India.
e-mail: kg _ ramawat@yahoo.com

Professor M.C. Sharma
Department of Materia Medica, National Ayurveda Institute, Amer Road, Jaipur 302 002, India

Dr N. Shrivastava
Pharmacognosy and Phytochemistry Department, B.V. Patel Pharmaceutical Education and Research Development Centre, Thaltej, Ahmedabad 38004, Gujarat, India, e-mail: perd@wilnetonline.net

Dr A.K. Singh
Process and Product Development Division, Central Institute of Medicinal and Aromatic Plants, P.O. CIMAP, Lucknow 226 016, India

Dr K.C. Sonie
Laboratory of Biomolecular Technology, Department of Botany,
M.L. Sukhadia University, Udaipur 313 001, India

Dr C. Trentesaux
JE Oncopharmacologie, UFR Pharmacie, Université de Reims, Champagne-Ardenne, 51 rue Cognacq-Jay, F-51096 Reims cedex, France

Dr. S.R. Venkatachalam
Cell Biology and Bioorganic Division, Bhabha Atomic Research Centre, Trombay, Mumbai 400 085, India

Dr X. Vitrac
Laboratoire de Mycologie et Biotechnologie Végétalé, GESUAB EA, UFR des Sciences Pharmaceutiques, Universite Victor Segalen Bordeaux 2, 146 rue Léo Saignat, 33076 BORDEAUX cedex, France.

CHAPTER 1

Therapeutic Potential of Medicinal Plants: An Introduction

K.G. Ramawat, K.C. Sonie and *M.C. Sharma*

INTRODUCTION

Today, the world population is nearing 5 billion at a rate of growth which is likely to touch 7.5 billion by the year 2020. The World Health Organization (WHO) estimated that 80% of the population of the developing countries rely on traditional medicines, mostly plant-based drugs, for their primary health care needs. Also, modern pharmacopoeia still contains at least 25% drugs derived from plants while many others are synthetic analogues built on prototype compounds isolated from plants. The demand for medicinal plants is steadily increasing in both developing and developed countries due to the growing recognition of drugs based on natural products, food supplements and flavours. Being non-narcotic, having no side-effects and easy availability at affordable prices makes these products sometimes the only source of health care available to the poor. The use of plants in the form of medicine goes back to the stone age. Certainly, the great civilizations of the ancient Chinese, Indians and north Africans have provided written evidence of man's ingenuity in utilizing plants for the treatment of a wide variety of diseases (Phillipson, 2001). Medicinal plants have traditionally occupied an important position in the socio-cultural, spiritual and medicinal arena of rural and tribal lives in India. In this subcontinent, the current herbal drug production is around US$ 1000 million (Anonymous, 2001).

DEPLETION OF NATURAL RESOURCES

Plants are a vast source of numerous valuable drugs of natural origin. The vegetation in Africa and Asia is rich in medicinal plants and biodiversity. Several species are collected from forests in the form of minor forest products. Besides tribals—who are authorized to collect the minor forest produce for their livelihood—a large number of species/materials are illegally collected by traders. This over-exploitation without proper replenishment has led to the removal of certain plant species and large-scale destruction of habitat (Ramawat et al., 1991; Merillon and Ramawat, 1999; Rout et al., 2000). Therefore, plant germplasm resources are being depleted at an alarming rate throughout the third world countries, particularly in nations like India that are burdened with tremendous population pressure. Human greed, particularly commercial need, has been the single factor responsible for such overuse and eventual destruction of natural habitats. The rate of destruction in developing countries has far exceeded their replenishment. Although traditional forest improvement programmes have begun to pay dividend in the industrialized world, efforts towards the same have lagged in the developing countries (Gadgil, 1992). It is estimated that about 1500 plant species in India are threatened—of these about 124 endangered, 81 vulnerable and 100 species have been declared as rare species (Nayar and Sastry, 1987). Gupta and Chadha (1995) enlisted 35 species as medicinally important out of this list of endangered species. The plant species of interest to the Department of Biotechnology, Ministry of Science and Technology, Government of India are listed in Table 1.1. Therefore, it is urgently required to look into reproductive biology, seed biology and germination, as also methods of conservation and multiplication of such species.

While the demand for medicinal plants is growing, some of these curative species are increasingly being threatened in their natural habitat. Plants are still collected from forests and their replenishment in the forests or even cultivation is lacking, e.g. *Commiphora wightii*, used for lipid-lowering properties and *Chlorophytum borivilianum* used as tonic. This makes that particular plant an endangered species. Therefore, information about the demand and supply of such plants should be ascertained in order to develop a proper technology. To meet the future needs of the growing world population, one must encourage the cultivation of medicinal plants.

Table 1.1 Important medicinal and aromatic plant species of the Indian subcontinent

Aconitum species	*Cymbopogon winterianus*	*Podophyllym hexandrum*
Acorus calamus	*Dioscorea deltoidea*	*Pogostemon cablin*
Amomum species	*Ephedra* species	*Psoralea corylifolia*
Andrographis paniculata	*Ferula asafoetida*	*Rauwolfia serpentina*
Azadirachta indica	*Mesua ferrea*	*Rheum emodi*
Cassia angustifolia	*Nardostachys jatamansi*	*Swertia chirata*
Commiphora wightii	*Panax peudo-ginseng*	*Valeriana wallichii*
Coptis teeta	*Phyllanthus fraternus*	*Vetiveria zizanioides*
Crataeva nurvuala	*Picrorhiza kurrooa*	*Withania somnifera*

Source: Department of Biotechnology, Government of India, New Delhi

According to an all-India ethnobiological survey carried out by the Ministry of Environment and Forests, the Government of India, Indians use over 8,000 species of plants for their medicinal properties. Figures 1.1 and 1.2 represent the plants and their various parts used in the Indian system of medicine (Anonymous, 2001).

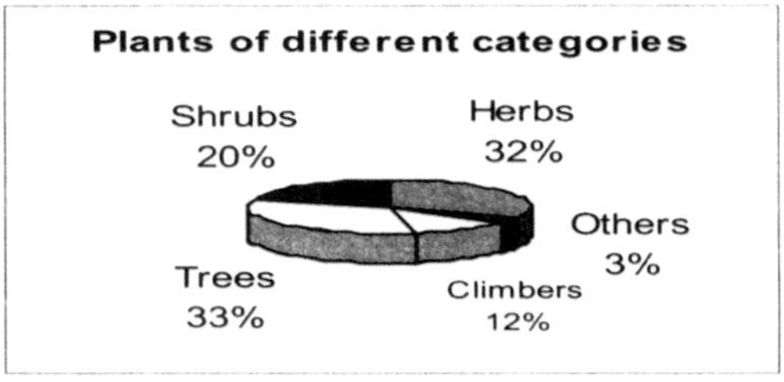

Fig. 1.1 Plants of different categories used as medicinal plants

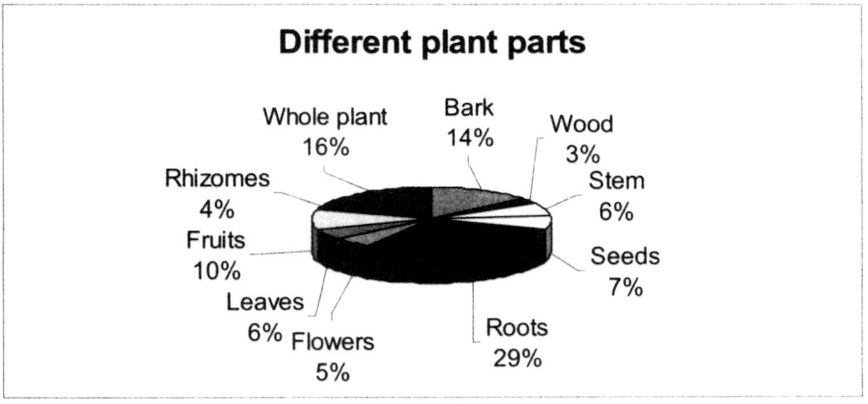

Fig. 1.2 Different plant parts constituting herbal medicine

TRADITIONAL SYSTEM OF MEDICINE

In the countries of Asia and Africa, the traditional system of medicine is practised in the rural and tribal areas where modern medicine is generally inaccessible. It is in these areas that plant-based drugs are frequently used, some of which are really effective. The clinical applications of taxol, etoposide and artemisin have helped to revive an interest in higher plants as sources of new drugs. Several of the traditionally-used plants are not properly investigated phytochemically and pharmacologically. Therefore, much information is not available to the scientific community. It is an established fact that phytochemical investigations result in isolation and identification of compounds present in a major amount in the plant.

However, compounds present in very minute quantities were isolated only when reported in other systems or when the material was reinvestigated with modern sensitive tools. In some cases, the use of in vitro cultures resulted in an isolation of a new unknown compound. Detailed investigation of many such materials by way of sophisticated tools has resulted in the discovery of new molecules hitherto not known in the material, e.g. three stilbene glucosides in cell cultures of *Vitis vinifera* (Waffo Teguo et al., 1998). Despite the belief that the majority of clinical drugs are synthetic in origin, it is interesting to note that 6 out of the top 20 pharmaceutical prescription drugs dispensed in 1996 were natural products and about 50% of them were produced through natural product research (Phillipson, 2001). It is also estimated that 250,000 species of higher plants are yet to be investigated for pharmacological activities. Several natural products are known to possess CNS activity, e.g. caffeine, codeine, morphine, reserpine. Therefore, it is of interest to explore the plants established for such activities in the traditional system for diseases like Alzheimers, Parkinsons, epilepsy, migraine, pain, schizophrenia and sleeping disorders.

Beginning with the mid-sixties, a number of Ayurvedic drugs have been screened for their claimed therapeutic effects. Many new drugs have been developed on the basis of their use in the traditional or folk medicine after a thorough scientific evaluation. The results published so far tend to support Ayurvedic claims (Dev, 1999). Some of the well-established Ayurvedic plant-based drugs are presented in Table 1.2. Most of them have been investigated with modern tools and are observed to be effective on scientific parameters. For example, the establishment of lipid-lowering properties of oleogum-resin of *Commiphora wightii* (Satyavati, 1988), the hypoglycemic effect of *Pterocarpus marsupium* (Ahmed et al., 1991; Manickam et al., 1997; Grover et al., 2002) and memory-enhancing effect of bacosides from *Bacopa monnierie* (Singh et al., 1990) (see Chapter 5 in this book).

The isolation and use of natural products such as digoxin, morphine and quinine have resulted in replacing the plant extracts used with single chemical entities. There is a basic supposition that any plant possessing clinical effectiveness must contain an active principle which can completely replace the plant extract. This may not always be true (Phillipson, 2001). Artimisin is without doubt the potent antimalarial active principle of *Artemisia annua*. Crude extracts *A. annua* contain a plethora of other compounds, including a series of flavonoids, some of which enhance the activity of artemisin against *Plasmodium falciparum* in vitro. These findings lend support to the view that there may be some advantages to the medical use of extracts as opposed to isolated single entities. It is interesting to note that in the Indian system of medicine, a crude plant extract or a mixture of plants is administered for any specific disease. The effect of combined drugs or its extract is different from that of purified compounds. Therefore,

Table 1.2 Some of the ancient Ayurvedic drugs with proven pharmacological claims

S.NO.	Plant		Type of activity
	Botanical terminology	Sanskrit name	
1	*Achyranthes aspera*	Apamrga	Diuretic
2	*Acorus calamus*	Vacha	Tranquilizer
3	*Artemisia vulgaris*	Nagadamni	Cardiac tonic
4	*Asparagus adscendens*	Shatavari	Anabolic and aphrodisiac
5	*Bacopa monnierie*	Brahmi	Memory enhancer
6	*Boerhaavia diffusa*	Punarva	Anti-inflammatory
7	*Butea monosperma*	Palasha	Anthelmintic
8	*Cassia fistula*	Suvarnake	Cathartic
9	*Centella asiatica*	Mandukaparni	Intelligence enhancer
10	*Chlorophytum borivilianum*	Safed musli	Anabolic and aphrodisiac
11	*Commiphora wightii*	Guggul	Hypo-lipidemic and -cholesterolemic
12	*Cuculigo orchioides*	Kali-musli	Anabolic and aphrodisiac
13	*Curcuma longa*	Haridara	Anti-inflammatory
14	*Eugenia jambolana*	Jambu	Hypoglycemic
15	*Nardostachys jatamansi*	Jatamansi	Hepatoprotective
16	*Picrorhiza kurrooa*	Kurua	Hepatoprotective
17	*Rauwolfia serpentina*	Sarpagandha	Hypertension
18	*Sida rhombofolia*	Mahabala	Anabolic
19	*Terminalia arjuna*	Arjun	Cardiotonic
20	*Tinospora cordifolia*	Guduchi, Giloe	Hepatoprotective, immunomodulator
21	*Tylophora indica*	Antamaal	Anti-asthmatic
22	*Withania somnifera*	Ashwagantha	Anabolic

Table 1.3 Plant species whose extracts have been shown to be more effective than isolated pure compounds

Plant species	Active principle	Effective use
Chlorophytum species	Saponins, sterols	Vitalizer
Ephedra gerardiana	Ephedrine	Antispasmodic
Ginkgo biloba	Steroids, flavonoids, etc.	Rejuvenator
Pterocarpus marsupium	Phenolics: epicatechins	Hypoglycemic
Tylophora indica	Tylophorin	Antiallergic, antiasthmatic
Withania somnifera	Steroids	Vitalizer

it is difficult to assign a particular pharmacological effect to the compound in question. Such examples have been listed in Table 1.3, where pure compounds either do not exert the same effect or are less effective.

INDIAN SYSTEM OF MEDICINE

The Indian system of medicine, known as Ayurveda, is one of the oldest systems of medicine prevailing across the world. The word Ayurveda consists of two words— *ayur* meaning life and *veda* meaning knowledge or, in other words, the science or way of living. The practice of this system of medicine is almost 3000-4000 years old. Ayurveda originated through Atharvaveda, an ancient book of knowledge pertaining to the Vedic Age. It has had continuous development upto 1000-500 BC. Its teachings were transmitted orally for thousands of years, then written down for fosterity in melodious Sanskrit poetry. The knowledge was documented in *Charak Samhita* (1000 BC) and *Sushrata Samhita*, which are considered to be authentic books (Tirtha, 1998). *Ras Shastra* was introduced in the twelfth Century AD to beget quick results through the *Ras* (chemical) preparations. After that, the government discontinued patronizing Ayurveda and the developmental activities in the system practically came to a standstill. Ayurveda is considered to be a science of life (Fig. 1.3). This probably makes it an earlier medical science having a positive concept of health to be achieved through a blending of physical, mental, social, moral and spiritual welfare (Tiwari, 1995; Svoboda, 1992). According to ancient books of knowledge, health is considered to be a prerequisite to achieve the supreme end of life. Preventive and curative aspects of diseases are considered as important components of positive health.

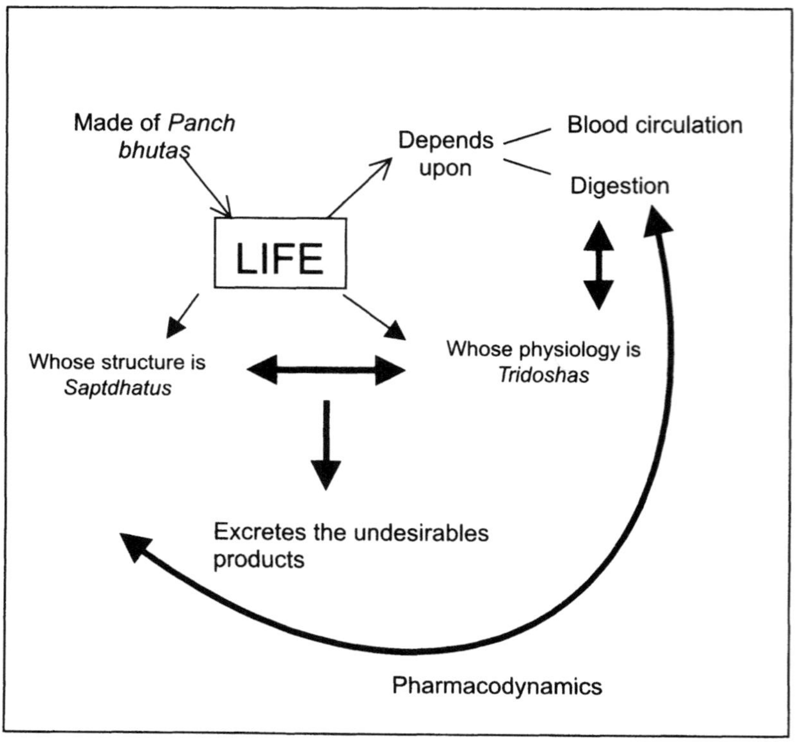

Fig. 1.3 Life, its components and functioning in Ayurveda

CONCEPTS AND PRINCIPLES OF AYURVEDA

The philosophy of Ayurveda is based on the *Panch bhutas*, of which the body is composed of (Fig. 1.3). A healthy person is one in whom there is equilibrium of the humours and body tissues with normal digestive as well as excretory functions, all of which are responses to the gratification of physical sciences and mental as well as spiritual forces. An absence of this equilibrium describes the status of diseases or sickness. The *Vat*, *Pitta*, and *Kapha* are known as the three humours. In a healthy person, these humours are in a state of non-functional equilibrium, and loss of this harmony leads to sickness.

There are four key concepts in Ayurveda (Fig. 1.4). These concepts collectively guide the preventive, promotive and curative aspects of the practice of this alternative remedial science. These concepts are the *Panch bhutas*, *Tridoshas*, *Saptdhatus* and *Malas* (Fig.1.3).

a. *Panch bhutas* (Basic elements)

These *bhutas* refer to the five basic elements of air, water, fire, earth and sky. Creation of all forms of life, including human life, is credited to these elements.

b. *Tridoshas* (Physiological functions)

These concepts together explain all the physiological and biological aspects taking place in a living body. The *Tridoshas* are *Vat*, *Pitta* and *Kapha*. These three terms are broadly comparable to the modern concepts of motion, energy and inertia, respectively (Fig. 1.5).

c. *Dhatus* (Structural components)

Dhatus refer to different vital body organs or parts. These are *Rasa* (body fluid), *Rakta* (blood), *Mansa* (muscular tissues), *Meda* (edipose tissue), *Asthi* (bone tissue), *Majja* (nerve tissue and bone marrow) and *Shukra* (generative tissue including sperm and ova).

d. *Malas* (Excretion)

This concept deals with the waste products of the body. The food consumption by the human body brings into existence and place further the above-mentioned seven *Dhatus* or vital body organs. During this metabolic process, each organ produces a specific waste or *mala* such as stool, urine, sweat, nails, hair, etc.

For promotive and positive aspects of health, Ayurveda has worked out a detailed daily and seasonal patterns of life activities. It emphasizes regulated diet, sleep and sex (Tiwari, 1995).

The diagnosis involves consideration of the entire humanity, including details of the man's temperament, habits, diet, etc. In other words, in Ayurveda, the examination of the patient and the disease both are extremely important.

Treatment in the system generally consists in avoiding the factors responsible for causing this equilibrium of body, matrices or of any of its constituent parts through the use of medicine, suitable diet, activity and regimen for restoring the balance and strengthening the body mechanism in order to prevent or minimize any future occurrence of the diseases. The treatment has to be individualized before being prescribed.

DISEASES AFFLICTING TROPICAL COUNTRIES

In most of the developing countries, insects and pests generate diseases due to the tropical and subtropical climate, while poor hygienic and socioeconomic living conditions are responsible for many diseases like those caused by protozoa such as Amoebiasis caused by *Entamoeba histolytica,* malaria by *Plasmodium falciparum* and filaria by nematodes. Effective plant-based drugs are required to combat these diseases which sometimes become epidemic. In 1996, it was reported that there were between 1.5 and 2.7 million deaths annually due to malaria; a majority

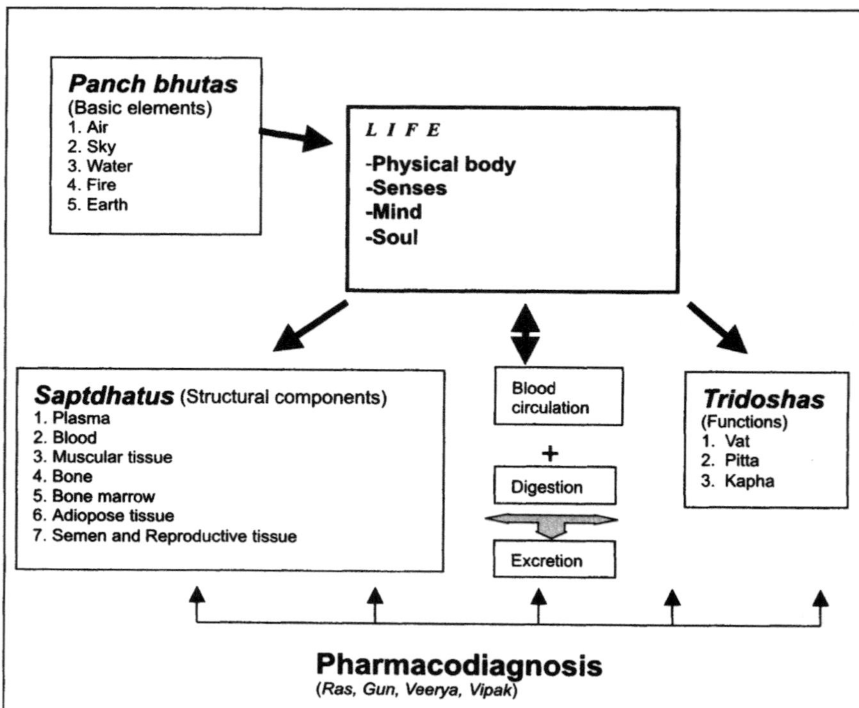

Fig. 1.4 Fundamental concepts of Ayurveda

Tridosha - (Physiological functioning)

1. *Vat*
   - **mental** — enthusiasm, concentration
   - **physical** — respiration, circulation

   Modern: nervous system, cerebrospinal, e.g. hypertension

2. *Pitta*
   - **mental** — intellection
   - **physical** — digestion, assimilation

   Modern: nutritional activity of endocrine, e.g. indigestion, dysentery

3. *Kapha*
   - **mental** — courage, tolerance
   - **physical** — bodily strength, joints function

   Modern: skeletal + anabolic, e.g. muscular pain

Fig. 1.5. Various physiological attributes in Ayurveda

of these were children (Phillipson, 2001). There are about 500 million new incidences of malaria annually. Quinine and artemisin, effective drugs against malaria, are both of plant origin. Several strains of malarial, parasite, *P. falciparum*, developed resistance against quinine and till such time as artemisin was not known, there were no other alternatives. Therefore, we await the discovery of new and effective plant-based drugs against tropical parasitic diseases. A recent survey in India showed that the high current increase in diabetes and heart ailments are due to high fat and low protein dietary habits, luxurious life style (poor physical exercise) and stress to compete. Plant-based drugs are safe and effective but exert pressure on the earth's natural resources. There are many promising plants effectively used in the Indian system of medicine and require a thorough scientific investigation to determine the exact molecule responsible for their proposed activity. Some examples of such plants have been given in Table 1.4.

Table 1.4 Plants with promising therapeutic potentialities used in the Indian system of medicine

Therapeutic potentiality	Plants with promising activity
Immunity enhancer	*Asparagus racemosus, Embelia ribes, Piper longum, Ocimum sanctum*
Hepatoprotective	*Andrographis paniculata, Boerhaavia diffusa, Eclipta alba, Phyllanthus amarus, Picrorhiza kurrooa, Piper longum, Piper nigrum, Ricinus communis, Zingiber officinale*
Atherosclerosis	*Boswellia serrata, Commiphora wightii, Curcuma longa, Garcinia cambogia, Terminalia arjuna, Terminalia chebula*
Cure for Arthritis	*Berberis aristata, Boerhaavia diffusa, Boswellia serrata, Commiphora wightii, Coptis chinensis, Curcuma longa, Glycyrrhiza glabra, Morus alba, Pluchea lanceolata, Ricinus communis*
Diabetes mellitus	*Aegle marmelos, Azadirachta indica, Capparis decidua, Coccinea indica, Eugenia jambolana, Gymnema sylvestrie, Momordica charantia, Pterocarpus marsupium, Trigonella foenum-graecum*
CNS/Brain stimulator	*Bacopa monnierie, Celastrus paniculata, Centella asiatica, Evolvulus alsinoides, Mucuna prurita, Nardostachys jatamansi, Withania somnifera*
Rejuvenation	*Asparagus adscendens, Boerhaavia diffusa, Centella asiatica, Chlorophytum borivilianum, Glycyrrhiza glabra, Phyllanthus emblica, Panax ginseng, Semecarpus anacardium, Withania somnifera*

Modified and updated from Handa (1996).

INDIAN PHARMACEUTICAL INDUSTRY

There are estimated to be over 7800 medicine manufacturing units in India. About 90% of plants used by the industries are collected from the wild. While over 800 species are used in production by this industry, less than 20 species of plants are under commercial cultivation. Over 70% of the plant collections involve destructive harvesting because of the use of parts like roots, bark, wood, stem and the whole plant in case of herbs (Fig. 1.1). This poses a definite threat to the genetic stocks as also to the diversity of medicinal plants. In recent years, the growing demand for herbal products has led to a quantum jump in the volume of plant materials traded within and across the countries. An estimate of the EXIM bank puts the international market of medicinal plants related trade at US$ 60 billion per year, growing at the rate of 7% only.

During the past four decades, bulk production of plant-based modern drugs has become an important segment of the Indian pharmaceutical industry. Some of the pharmaceuticals produced in India presently include morphine, codeine, papaverine, thebaine, emetine, quinine, quinidine, digoxin, caffeine, hyoscine, hyoscyamine, xanthotoxin, psoralen, colchicine, rutin, berberine, vinblastine, vincristine, nicotine, strychnine, brucine, ergot alkaloids, senna glycosides, pyrethroids and podophyllotoxin resin.

Phytopharmaceuticals for which new technology has been developed to undertake large-scale production include L-DOPA from *Mucuna* beans, ajmaline and ajmalicine from *Rauwolfia serpentina* and *Catharanthus roseus,* respectively, and 18β-acetyl glycyrrhetic acid from *Glycyrrhiza glabra* (Handa, 1996).

It is imperative that industries based on plants for obtaining natural product cannot sustain only on wild plants, as a regular supply of uniform material cannot be met in such circumstances. Therefore, it is necessary to bring the plant under cultivation for production of biomass. This can be achieved through conventional and biotechnological methods. Such plants can be reintroduced in nature for conservation as continuous exploitation exert tremendous pressure on the natural resources and make these plants endangered species (Fig. 1.6).

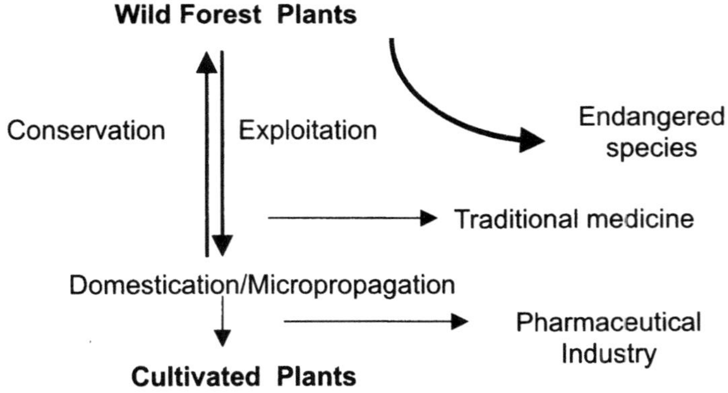

Fig. 1.6 Over-exploitation of medicinal plants leads to endangering their very survival

BIOTECHNOLOGICAL APPROACHES

In comparison to food crops, research efforts to improve upon the medicinal plants are far below the expected requirements. Despite specific constraints imposed by natural characteristic features, successes have been achieved in developing genetically-superior cultivars in some important medicinal plants. Constraints such as more importance to secondary metabolites rather than to primary metabolites; seasonal variation in the rate of biosynthesis of active principles; and variable and complex mode of reproduction are some of the bottlenecks in the way of the improving medicinal plants. Progress has been made with respect to in vitro production of secondary metabolites, micropropagation of selected materials, endangered species and propagule for domestication through biotechnological methods.

The Department of Biotechnology (DBT), under the Ministry of Science and Technology of the Government of India has set up a separate task force to define the thrust areas on medicinal plant research and to monitor the progress of research works in this area. The Indian Council of Agricultural Research under the Ministry of Agriculture is promoting research towards developing agrotechniques for multiplication and domestication of wild medicinal plants of proven therapeutic values (Table 1.5). The application of various approaches to ameliorate the medicinal plants and their active principles are presented in the following pages.

Table 1.5 Medicinal and aromatic plants currently under cultivation/domestication in India and the development of new cultivated varieties through conventional/ biotechnological methods

Plant species	Area (hectares)	Yield (Tonnes/h.)	New varieties released
I—Medicinal plants			
A: Leaf crops			
Atropa belladonna	S	0.75	-
Cassia angustifolia	10000	2.5	1
Digitalis purpurea, D. lanata	S	3	4
Hyoscyamus niger	S	2-3	2*
B: Root crops			
Catharanthus roseus	3000	0.6-0.8	1
Cephaelis ipecacuanha	100	3.0	-
Chlorophytum borivilianum	T	IF	-
Dioscorea deltoidea	-	-	-
D. floribunda	-	30-60F	3*
Glycyrrhiza glabra	-	2-2.5	2*
Rauwolfia serpentina	-	1.5-2.5	1
Withania somnifera	4000	0.3	7
C: Tree bark, gum, oleoresin			
Azadirachta indica	P	-	-
Cinchona species	8000	-	-
Claviceps purpurea	150-200	7.5-10	NA
Commiphora wightii	W	-	NA
D: Flower and fruit crops			
Ammi majus	T	NK	-
Aloe barbadensis	T	NK	-
Datura innoxia	S	1.7 seed	-
Mucuna prurita	T	-	-
Papaver somniferum	18000	0.05 opium	8*
Plantago ovata	50000	1-1.5	3
Psoralea corylifolia	S	-	-
S. lacinatum	S	-	6
Solanum khasianum	S	10F	2
II—Aromatic plants			
A: Leaf crops			
Artemisia pallens	500	15F	-
Cymbopogon flexuosus	20000	0.1 Oil	6*

C. martinii	2000	0.2 Oil	3*
C. winterianus	S	-	4*
Mentha arvensis	S	-	4*
M. citrata	S	-	1*
M. piperita	10000	50-70F	-
M. spicata	S	-	3*
Ocimum basilicum	500	50	1
Ocimum sanctum	-	-	1
Pelargonium graveolens	2000	15-18F	4
Pogostemon patchouli	S	8F	2*
B: Root crops			
Vetiveria zizanioides	-	1-1.5F	1
C: Flower crops			
Crocus sativus	3000	-	-
Humulus lupulus	1000	3.0	-
Jasminum grandiflorum	2000	10F	1
Matricaria chamomilla	-	6.0F	-
Pandanus fascicularis	-	-	-
Polianthes floribunda	-	10-15F	-
Rosa species	2000	3F	1
Tagetes minuta	-	20-25F	-
D: Fruit and seed crops			
Anethum graveolens	-	0.7-0.8F	-
Apium graveolens	5000	0.4-0.5F	-
Abelmoschus moschatus	-	0.7-1.0F	-
Pimpinella anisum	-	0.7-0.8F	-

F = Fresh herbage yield S = Small area under cultivation
NK = Not known T = Trial plant for cultivation
P = Planted W = wild
* All or some varieties developed through somaclonal variation in tissue culture
Data compiled from Sharma and Kumar, 1996; Gupta and Chadha, 1995.

A. Production of active principle

The last four decades have witnessed the efforts to produce useful active principles of medicinal plants. This line of approach has progressed through optimization, selection of cell lines for high yields, cell mutations, biotransformations, manipulations of biosynthetic pathways and lastly, genetic engineering. All strategies were aimed at either increasing the yield of molecule in question or obtaining its desired derivative. However, in most cases, yields of secondary metabolites are too low for commercialization. Growth of differentiated cells and elicitation are some of the other possible approaches to increase production. The technology of large-scale cultivation of organized structures is still being perfected while the elicitation is limited to a specific class of compounds in each plant. Genetic manipulation has great potential in altering the metabolic profile to increase the production of desired products.

The introduction of new genes in plants can nowadays be achieved in various ways. Transformation with *Agrobacterium tumefaciens* and the particle gun are the two most widely used methods. Verpoorte et al. (1999)

discussed this approach at length with *Catharanthus roseus* as model system and involvement of genes of terpenoid metabolism. Even now, problems still remain to master this technology as the biosynthesis of an active principle takes place in many steps involving many enzymes (genes). To learn about the biosynthetic pathway, genes involved, cloning and transfer and expression of genes in a new system is another Herculean task. Our knowledge of many of these aspects is still incomplete. The mechanism involved in the biosynthesis of most of the active principles is as yet poorly understood (Merillon and Ramawat, 1999). Verpoorte et al. (1999) proposed the following approaches to enhance the yield of active principles: (a) increase the flux through the pathway to the desired product, shutting down the side chain biosynthesis of the undesired products; (b) increase the number of producing cells in the cultures, as the cultures are heterogeneous mixture of producer and non-producer cells; and (c) decrease the catabolism of the desired products as the amount of contents is the sum of anabolism and catabolism of the secondary metabolite. The process of achieving industrial level production technology in order to obtain biomass in large amounts through a bioreactor is being evolved simultaneously. We have examples of cultivation of cells in several hundred to several thousand litres vessels, e.g. *Aralia cordata* (Kobayashi et al., 1993) and *Vitis vinifera* (J.M. Merillon personal communication) for anthocyanins and ginseng (Hibino and Ushiyama, 1999).

B. Micropropagation

Micropropagation is the most widely-used application of plant tissue culture. Micropropagation protocols have been developed for a large number of medicinal plants. Micropropagation is used to clone a selected plant material, e.g. in *Mentha* species for high yield of menthol, multiply a wild plant for propagation at desired rate, e.g. *Commiphora wightii*, *Taxus* species and development of planting material for domestication, e.g. *Chlorophytum borivilianum* and *Podophyllum hexandrum*. The ability of plant cells to regenerate through organ or embryo development in liquid cultures can be exploited for large-scale in vitro plant propagation and automation. Shoot formation and propagule singulation still constitute limiting factors in efficacious micropropagation in a liquid medium. The use of biorectors for growing organized cultures needs more technological inputs, which is still employed in limited cases. Control of shoot morphogenesis in liquid media can be attained by culture conditions, which prevent leaf expansion and shoot growth, and simultaneously enhance proliferation. Many forest, food, fodder and ornamental crop plants are micropropagated through somatic embryogenesis using a bioreactor, e.g. potato, banana, poplar, philodendron, *Lilium*, *Narcissus* (Ziv, 1999). Major developments are required to handle the organized structures produced in a bioreactor. There are rare examples where the shoots produced in a bioreactor were directly transferred to soil where they acclimatized within one month without any pretreatment (Hosokawa et al., 1998). It has been observed that the formation of storage organs like tuberous roots (Suri et al., 1999a) or bulbils (Suri et al., 2000) helps in successful establishment of plantlets in soil. Micropropagation has become an important thrust area for medicinal plant research (Rout et al. 2000).

C. Development of new varieties

Increased yield of biomass/active principle can be obtained either by increasing the area under cultivation or by increasing the yield per unit area of a particular crop. There is always a degree limitation in increasing the cultivated area and, hence, to meet the ever-increasing demand of herbal products, one is left with no option but to continuously improve the yield potential of the plant. Therefore, selections through conventional breeding as well as exploitation of somaclonal variation in cultured cells comprise principal methods to improve upon the yield potential. Once selected, the material is multiplied by micropropagation techniques in order to provide seedling material to farmers. New varieties released in India through such efforts are presented in Table 1.5.

D. Genetic mapping

The assessment of genetic diversity and DNA fingerprinting of germplasm are of paramount importance in medicinal plants looking to the large number of patent applications filed every year on therapeutic agents of natural origin. DNA fingerprinting of plant species is just a beginning in developing countries and these sophisticated techniques require sufficient inputs in terms of adequate funds and trained manpower; both of which lacking in these countries. DNA fingerprinting would be helpful in conserving the useful germplasm and also establish its centre of origin to set aside many controversies. In India, the Department of Biotechnology, Government of India, has initiated such a programme and some medicinal plants are being characterized, e.g. *Azadirachta indica* (Singh et al., 1999), *Camellia sinensis* (Paul et al., 1997; S.N. Raina, personal communication), *Commiphora wightii* (M. Lakshmikumaran, personal communication), *Chlorophytum borivilianum* (our unpublished results).

E. Industrial production of biomass/bioactive compounds

Till date, there have been only a few plant cell culture processes in the commercial production of bioactive compounds such as ginseng production by Nitto Denko Company (Osaka, Japan) since 1986, Shikonin production by Mitsui company (Japan) and Paclitaxel production by Samayang Genex Company (Taejon, Korea) since 2001 (Dong and Zhong, 2002). Bioactive secondary metabolites are mainly obtained from biomass produced by in vivo grown plant materials, e.g. ajmalicine from roots of *Catharanthus roseus,* morphine alkaloids from the latex of opium poppy, quinine and taxol from the bark of *Cinchona* and *Taxus,* respectively. When biological/ political problems prevent biomass collection from natural sources, alternative methods like chemical, microbial or plant cell culture are employed. Plant cell cultures grown in a bioreactor are being used to generate the biomass of a number of species in order to produce active principle such as *Taxus, Podophyllum* and *Stizolobium* (Suri et al., 1999b; Hibino and Ushiyama, 1999; Rorrer et al., 1999). However, the number of plant products produced through a bioreactor is not increasing rapidly, on looking at the number of secondary metabolites known to produce through plant tissue cultures. Recent addition to the list of compounds produced at a commercial level through plant cells grown in bioreactor is taxol from *Taxus* species (Dong and Zhong, 2002; Ramawat, 1999; Seki and Furusaki, 1999). Many companies that were once actively involved in developing commercial plant tissue culture, biomass generation through micropropagation or directly producing through cell culture, have discontinued their efforts. The problem lies not only in the difficulties encountered in technological development, but also in the economic factors. Plant Cell Culture Technology, Inc., Japan terminated their commercial production of plant cell culture-derived anthocyanin production primarily because the cost for biomass generation and extraction through in vitro growth was too expensive as compared to anthocyanin extraction from field-grown plant material (Kobayashi, cited by Fu et al., 1999). However, in spite of all odds, plant cells have been grown in bioreactors upto 75000 litres capacity and infrastructure and technologies are being developed to make it a more economically-viable system.

RESEARCH NEED

Medicinal plants comprise the wealth of civilization. They are also involved in social rituals and welfare of human beings. The overexploitation is the impact of increased population pressure. This is more significant in relation to developing countries exploding with population. Therefore, research efforts on the following lines are required in order to meet the increased demand of herbal drugs and, simultaneously, also conserve them.

1. Proper evaluation of medicinal plants for their phytochemical and pharmacological properties based on traditional medicine and ethnobiological information.
2. In situ and ex situ conservation of medicinal plant germplasm, particularly the establishment of herbal gardens.
3. Development of technology for micropagation, reproductive biology

and agricultural practices towards domestication of the selected materials.

4. Improvement in established cultivated species by in vivo and in vitro varietal improvement.
5. Enforcement of regulation regarding more scientific collection, storage and supply of medicinal plant raw materials and their extracts, enabling data generation regarding their future demand.

REFERENCES

Ahmed, F., Khalid, P., Khan, M.M., Rastogi, A.K. and Kidwai, J.R., 1991, Hypoglycemic activity of *Pterocarpus marsupium* wood. J. Ethnopharmacology. 35:71-75.

Anonymous, 2001, Annual Report 1999-2000, The ministry of health and family welfare., Government of India, New Delhi. http://www.mohfw.nic.in

Dev, S., 1999, Ancient-modern concordance in Ayurvedic plants: sum examples. Environ. Health Perspect. 107:783-789.

Dong, H-D. and Zhong J.J., 2002, Enhanced taxane productivity in bioreactor cultivation of *Taxus chilenesis* cells by combining elicitation, sucrose feeding and ethylene incorporation. Enzyme Micriobiol. Technol. 31:116-121.

Fu,T-J., Singh, G. and Curtis, W.R., 1999, Plant cell and tissue culture for food ingredient production. In: Plant Cell and Tissue Culture for the Production of Food Ingredients. Fu, T-J., Singh, G. and Curtis W.R. (eds), Kluwer Academic/Plenum Pub. New York, 1-6.

Gadgil, M., 1992, Biodiversity: Time for bold steps. In: Hindu survey of the Environment 1992, (ed.), Ravi, N. (ed.), National Press, pp. 21-23.

Grover, J.K., Vatsa, V. and Yadav, S.S., 2002, Effect of feeding aqueous extract of *Pterocarpus marsupium* on glycogen content of tissues and key enzymes of carbohydrate metabolism. Mol. Cell. Biochem. 241:53-59.

Gupta, R. and Chadha, K.L., 1995, Medicinal and aromatic plants research in India. In: Advances in Horticulture, Vol. 11, Medicinal and Aromatic Plants. Chadha, K.L. and Gupta, R. (eds), Malhotra Pub. House, New Delhi., pp. 429-451.

Handa, S.S., 1996, Rasaayana Drugs. In: Supplement to cultivation and utilization of medicinal plants. S.S. Handa and M.K. Kaul (eds), RRL, Jammu-Tawi, pp. 509-524.

Hibino, K. and Ushiyama, K., 1999, Commercial production of ginseng by plant tissue culture technology. In: Plant Cell and Tissue Culture for the Production of Food Ingredients. Fu,T-J., Singh, G. and Curtis, W.R. (eds), Kluwer Academic/Plenum Pub. New York, pp. 215-224.

Hosokawa, K., Oikawa, Y. and Yamamura, S., 1998, Mass propagation of ornamental gentian in liquid medium. Plant Cell Rep. 17:747-751.

Kobayashi, Y., Akita, M., Sakamoto,K., Liu, H., Shigheoka, T., Koyano, T., Kawamura, M. and Furuya, T., 1993, Large-scale production of anthocyanin by *Aralia cordata* cell suspension cultures. Appl. Microb. Biotechnol. 40:215-218.

Manickam, M., Ramnathan, M., Jahroomi, F.M.A., Chansouria, J.P. and Ray, A.B., 1997, Antihyperglycemic activity of phenolics from *Pterocarpus marsupium*. J. Nat. Prod. 60:609-610.

Merillon, J.M. and Ramawat, K.G., 1999, Mechanism and Control. In: Biotechnology: Secondary Metabolites. Ramawat, K.G. and Merillon J.M. (eds.), Science Pub. Inc., USA., pp. 241-256.

Nayar, M.P. and Sastry, A.R.K., 1987, Red Data Book of Indian Plants, Vol. II, Nayar M.P., and Sastry, A.R.K., (eds.), Botanical Survey of India, 142. pp.

Paul, S., Wachira, F.N., Powell, W. and Waugh, R., 1997, Diversity and genetic differentiation among population of Indian and Kenyan tea (*Camellia sinensis* (l) O. Kuntze) revealed by AFLP markers. Theor. Appl. Genet. 94:255-263.

Phillipson, J.D., 2001, Phytochemistry and medicinal plants. Phytochem. 56:237-243.

Ramawat, K.G., 1999. Production of anti-tumour compounds. In: Biotechnology: Secondary Metabolites. Ramawat, K.G. and Merillon, J.M. (eds)., Science Pub. Inc., USA., pp. 68-88.

Ramawat, K.G., Bhardwaj, L. and Tewari, M.N., 1991, Exploitation of Indian desert medicinal plants through tissue culture. Indian Review Life Sciences 11:3-27.

Rout, G.R., Samantray, S. and Das, P., 2000, In vitro manipulation and propagation of medicinal plants. Biotechnol. Adv. 18:91-120.

Rorrer, G.L., Mullikin, R., Huang, B., Gerwick, W.H., Maliakal, S., and Cheney, D.P., 1999, Production of bioactive metabolites by cell and tissue cultures of marine macroalgae in bioreactor systems. In: Plant Cell and Tissue Culture for the Production of Food Ingredients. Fu,T-J., Singh, G. and Curtis W.R. (eds), Kluwer Academic/Plenum Pub., New York, pp. 165-184.

Satyavati, G.V., 1988, Gum guggul (*Commiphora mukul*) — the success story of an ancient insight leading to a modern discovery. Indian J. Med. Res. 87:327-335.

Seki, M. and Furusaki, S., 1999, Medium recycling as an operation strategy to increase plant secondary metabolite formation: continuous taxol production. In: Plant Cell and Tissue Culture for the Production of Food Ingredients. Fu,T-J., Singh, G. and Curtis W.R. (eds), Kluwer Academic/Plenum Pub. New York, pp. 157-163.

Sharma J.R. and Kumar, S., 1996, Biotechnology and genetic improvement of medicinal plants. Supplement to cultivation and utilization of medicinal plants. Handa, S.S. and Kaul, M.K., (eds.), RRL, Jammu-Tawi., pp. 753-768.

Singh, A., Negi, M.S., Rajagopal, J. et al., 1999, Assessment of genetic diversity in *Azadirachta indica* using AFLP markers. Theor. Appl. Genet. 99: 272-279.

Singh, H.K., Srimal, R.C., Srivastava, A.K., Garg, N.K. and Dhawan, D.N., 1990, Neurophycopharmaceutical effects of bacosides A & B. Proc. 4th Conf. Neurobiol. Laerning & Memory, California, 79. pp.

Suri, S.S., Arora, D.K. and Ramawat, K.G., 2000, A method for large-scale multiplication of *Curculigo orchioides* through bulbil formation from leaf explants in shake flask cultures. Indian J. Exp. Biol. 38:145-148.

Suri, S.S., Jain, S., Arora, D.K. and Ramawat, K.G., 1999a, In vitro high frequency regeneration of plantlets and tuberous root formation in *Chlorophytum borivilianum*. Gartenbau. 64(3): 106-110.

Suri, S.S., Sharma, R. and Ramawat, K.G., 1999b, Production of secondary metabolites in Bioreactor. In: Role of Biotechnology in Medicinal and Aromatic Plants. Vol. II. I. Khan and Khanum A. (eds.), Ukaz Pub., Hyderabad, pp. 437-461.

Svoboda, R.E., 1992, Ayurveda. Life, Healing and Longevity. Penguin, London, U.K.

Tirtha, S.S., 1998, The Ayurveda Encyclopedia., Ayurvedic Holistic Centre Press, Bayville, NY.

Tiwari, M., 1995, Ayurveda. A life of balance. Healing Art Press, Rochester, VT.

Verpoorte, R., Van der Heijden, R., Ten hoopen, H.J.G. and Memelink, J., 1999, Novel approaches to improve plant secondary metabolite production. In: Plant Cell and Tissue Culture for the Production of Food Ingredients. Fu,T-J., Singh, G. and Curtis W.R. (eds), Kluwer Academic/Plenum Pub. New York, pp. 85-100.

Waffo-Teguo, P., Fauconneau, B., Deffieux, F., Huguet, F., Vercauteren, J. and Merillon, J.M., 1998, Isolation, identification and antioxidant activity of three stilbene glucosides newly extracted from *Vitis vinifera* cell cultures. J. Nat. Prod. 61:655-657.

Ziv, M., 1999, Large-scale plant micropropagation in bioreactor. In vitro Cell Dev. Biol. 35(3): w22.

CHAPTER 2

In vitro Rosmarinic Acid Production: An Update

Olga Makri and *Spiridon Kintzios*

INTRODUCTION

Rosmarinic acid (Figure 2.1), a plant secondary metabolite belonging to the class of hydroxycinnamic esters, is a constituent of the members of Lamiaceae, Boraginaceae and Apiaceae families (De-Eknamkul and Ellis, 1987). Its chemical structure was identified in rosemary extracts (Scarpati and Oriente, 1958). A promising bioactive substance, rosmarinic acid has applications as a food preservative and a part of medication due to its functional properties as an antioxidant and a antimicrobial (Deighton et al., 1993; Frankel et al., 1996; Shetty et al., 1997). It was reported that in bulk corn oil, rosemary extract containing carnosic acid and rosmarinic acid showed significantly more antioxidant activity than carnosol (Frankel et al., 1996). In addition, recent studies show that rosmarinic acid can also

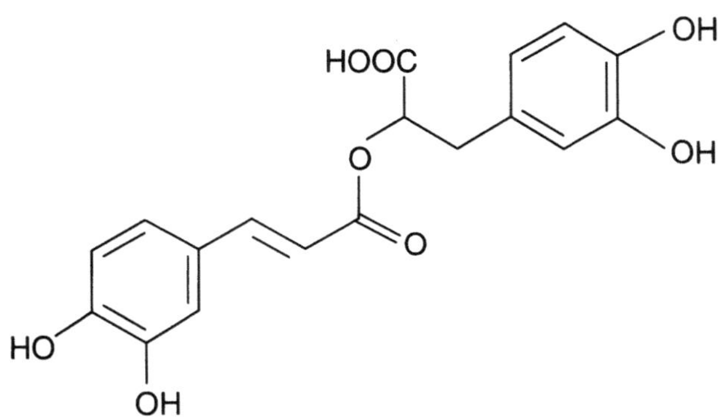

Fig. 2.1 Structure of rosmarinic acid

inhibit inflammatory reactions in vivo (Engleberger et al., 1988; Peake et al., 1991). Zou et al. (1993) reported on the antithrombotic effect of rosmarinic acid, which might be associated with its inhibition of platelet aggregation and promotion of fibrinolytic activity. This plant metabolite also accounts for the antioxidant activity of sage (Baricevic, 2000; Deans and Simpson, 2000). When measuring the radical scavenger effect on 1,1-diphenyl-2-picrylhydrazyl (DPPH) free radical, the antioxidative effect of rosmarinic acid (EC_{50} = 2.7 µg/ml) was comparable to that of ascorbic acid (Lamaison et al., 1991). Rosmarinic acid, as also its methyl ester isolated from EtOAc fraction of the methanolic extract of *Salvia miltiorrhiza*, were proved to have strong inhibitory effects on adenylate cyclase (AC) in both rat brain and in rat erythrocytes (Kohda et al., 1989).

DISTRIBUTION

According to a recent classification, the Lamiaceae consists of 8 subfamilies (Cantino et al., 1992). This hypothesis includes the former results of tests conducted by Wunderlich (1967), El-Gazzar and Watson (1970) and El-Gazzar (1974) and was deduced by taking into account 85 morphological characteristics for a cluster analysis. However, the presence or absence of rosmarinic acid, together with additional chemical markers, suggests a two-subfamily system (Erdtman, 1945; Cantino and Sanders, 1986). Rosmarinic acid is an excellent chemotaxonomic marker, which is present in the subfamily Nepetoideae and absent from the subfamily Lamioideae. In addition to differences in the morphological characters between the two subfamilies, there are chemical features of systematic value such as the high essential oil content of species in the Nepetoideae, the presence of iridoids in the Lamioideae and their absence (with three exceptions) in the Nepetoideae, and the distribution of rosmarinic acid. This compound was absent from all 25 genera of Lamioideae; while 32 out of the 36 genera of Nepetoideae contained it (Janicsak et al., 1999) (see Table 2.1).

BIOSYNTHESIS AND RELATED STUDIES IN VITRO

Elucidation of the biosynthesis pathways of this metabolite was undertaken on callus or cell cultures of mint plants (Ellis and Towers, 1970) and confirmed several years later by Razzaque and Ellis (1977) with cell suspensions of *Coleus blumei*.

Two precursors—tyrosine and phenylalanine—are involved in this biosynthesis procedure (Figure 2.2). Dihydroxyphenylalanine (DOPA) was used as the first intermediary after tyrosine in the first biosynthesis chain. Subsequent studies made it possible to refine the first method proposed and, in 1979, Ellis et al., 1979 proposed a chain in which

Table 2.1. Plant species reported to contain rosmarinic acid

Species	Part	Quantity (ppm)
Borago officinalis L.	Plant	7,000
Glechoma hederacea L.	Leaf	25,000
Hyssopus officinalis L.	Plant	5,000
Lavandula angustifolia L.	Plant	12,000
Lavandula latifolia L.	Plant	7,000
Lavandula × *inermedia* L.	Plant	5,000
Lycopus europeus L.	Plant	37,000
Melissa officinalis L.	Plant	8,400
Menta × *rotundifolia* L.	Plant	32,000
Mentha aquatica L.	Plant	27,000
Mentha arvensis L.	Plant	28,000
Mentha arvensis L.	Leaf	228
Mentha pulegium L.	Plant	30,000
Mentha spicata L.	Plant	43,000
Mentha spicata L.	Shoot	5,900
Mentha x piperita subsp. *Notho* subsp. *piperita* L.	Plant	30,000
Monarda didyma L.	Flower	18,000
Monarda fistulosa L.	Shoot	20,000
Nepeta cataria L.	Plant	2,000
Ocimum basilicum L.	Plant	1,300
Origanum majorana L.	Plant	33,000
Origanum vulgare L.	Plant	55,000
Origanum vulgare L.	Leaf	790
Perilla frutescens L.	Leaf	10,300
Perilla frutescens L.	Stem	2,900
Prunella vulgaris L.	Plant	180,000
Rosmarinus officinalis L.	Tissue culture	38,957
Rosmarinus officinalis L.	Leaf	3,500
Rosmarinus officinalis L.	Shoot	13,500
Salvia officinalis L.	Plant	30,000
Salvia officinalis L.	Shoot	5,800
Salvia sclarea L.	Plant	34,000
Saturaja hortensis L.	Plant	26,000
Satureja montana L.	Plant	12,000
Sympytum officinale L.	Leaf	5,000
Thymus serpyllum L.	Shoot	23,000
Thymus vulgaris L.	Plant	26,000

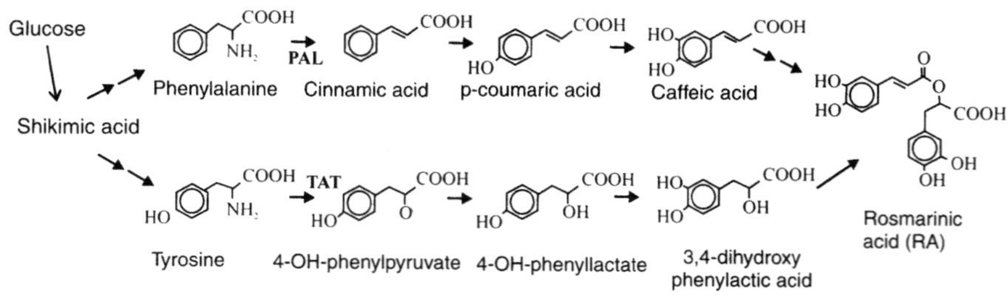

Fig. 2.2 Biosynthesis of rosmarinic acid

4-hydroxyphenylpyruvic acid is the intermediary following tyrosine. The enzymes involved were shown and isolated subsequently. In 1989, the last enzyme in the chain, rosmarinic acid synthase, was isolated and described by Petersen (1989). The mechanisms regulating rosmarinic acid synthesis are still unknown. The chain requires reduction capability in the form of NADPH and NADH. It seems to function in such a way as to keep the release of intermediate substances to a minimum and there is good connection between free precursor amino acids and the rosmarinic acid synthesis chain (De-Eknamkul and Ellis, 1987). This research essentially concerns cell suspensions of *Anchusa officinalis* and *Coleus blumei* (see Hippolyte et al., 2000 for a review). It would seem that the biosynthetic pathway is such as to minimize the release since biosynthetic intermediaries were not detected in the extracts. The state of morphological non-differentiation of cells induced by suspension culture seems to essentially enhance the production of this metabolite.

IN VITRO ROSMARINIC ACID PRODUCTION

Rosmarinic acid has been reported to accumulate in cultured cells of a number of species including *Coleus blumei* (Razzaque and Ellis, 1977; Zenk et al., 1977), *Anchusa officinalis* (De-Eknamkul and Ellis, 1984), *Lithospermum erythrorhizon* (Fukui et al., 1984), *Orthospiron aristatus* (Sumaryono et al., 1991), and *Salvia officinalis* (Hippolyte et al., 1992). Work reported on Lamiaceae mainly concerns undifferentiated cells such as callus and cell suspensions. Cell suspensions of *Salvia officinalis* produce rosmarinic acid in measurable quantities (Hippolyte, 1990). After several subcultures, the callus of *Salvia miltiorrhiza* produces only rosmarinic acid (RA), whereas callus newly initiated from petioles (little dedifferentiated cells) produces RA and lithospermic acid (Morimoto et al., 1994). Cell extracts of other species may contain different phenolic compounds in addition to RA. Cell cultures of *Lavandula vera* produce small quantities of caffeic acid—one of the intermediate substances in the biosynthetic chain (Kovatcheva et al., 1996)—those of *Coleus blumei* synthesize unidentified phenolic compounds but in very small quantities (Razzaque and Ellis, 1977).

At a higher level of morphological differentiation, root cultures of *Ocimum basilicum* induced by *Agrobacterium rhizogenes* produce rosmarinic acid and also lithospermic acid (upto 2% of the dry weight) and lithospermic acid B (0.15% of the dry weight) (Tada et al., 1996).

Production trials have been undertaken on cell suspensions of *Coleus blumei* (Ulbrich et al., 1985) with a view to industrial use.

The hypothesis of Sakuta and Komamine (1987)—according to which secondary metabolites may belong to two categories according to their mode of production (i.e. those whose production is growth-related and those whose synthesis takes place after growth and is dependent on cell differentiation)—does not seem to apply for RA. Here, the connection between primary metabolism and secondary metabolism seems to be caused by pressure in the culture medium in which osmotic pressure may play a regulatory role. Other factors may also be involved, such as the type of hormones. Genetic factors may also determine the mode of production (Kintzios et al., 1996, 1998).

Salvia officinalis

Cell suspensions of *Salvia officinalis* are capable of producing extremely large quantities of rosmarinic acid under stimulation conditions, with 6.4 g l^{-1} representing 36% of the dry weight (Hippolyte, 2000). Hippolyte et al., (1991, 1992) found that the type of culture medium used modified the growth and production of rosmarinic acid by sage cells. Rosmarinic acid production was increased 10-fold in order to attain 6.4 g l^{-1} under optimal conditions. Investigation of cell-growth kinetics showed that a change in the medium caused shifts in peaks of growth and rosmarinic acid production, and modifications of the cell metabolism. By changing the composition of the culture medium, it was possible to manipulate rosmarinic acid production so as to coincide with cell growth or to begin only when the growth had stopped. The mode of production varies from one medium to another. It is difficult to determine the role of phenylalanine as a precursor. Indeed, the increase in the phenylalanine concentration shows that it serves as a limiting factor for production and, to a lesser extent, growth under standard culture conditions. It also has a substantial effect on growth and production mechanisms by considerably shortening the culture cycle. The limiting effect of a precursor has often been observed in studies of the production of secondary metabolites (Margna, 1977). However, in a sucrose-enriched medium, a synergistic effect is observed on production only for a small concentration of this substance; growth and the culture cycle remain unchanged (Hippolyte et al., 1992). The effect of the substance seems to be masked by the sucrose effect. The addition of phenylalanine for other species gives contradictory results. Indeed, it reduces RA production by 35% in suspensions of *Coleus blumei* when added to a standard medium (Razzaque et al., 1977), but improves it by 100% when it is added to a medium containing sucrose at insufficient concentrations (Zenk, 1977). This observation corroborates the results of De-Eknamkul and Ellis (1984): combining the most favorable factors does not necessarily give the best results. Sucrose has a substantial trophic effect as also an osmotic effect. It would seem that both effects were related to the production of rosmarinic acid in our cell suspensions. On the one hand, the trophic factor affects the quantity of rosmarinic acid produced. Very large quantities of RA were assayed; they were similar in MS medium and Heller's medium enriched with sucrose. The effect of sucrose alone on osmotic pressure is difficult to demonstrate. The conditions that were chosen in order to maintain an osmotic pressure identical to that of the standard medium while increasing the sucrose content of the medium (i.e. by decreasing the molarity of the mineral medium) showed an effect of osmotic pressure in sucrose-enriched MS medium on the production quality rather than the quantity. Unfortunately, under these conditions, the possibility that the modifications might be related to the composition of the culture medium cannot be ruled out, even though no phenomena of this type were observed on MS medium and B5 medium, which differ particularly in their nitrogen concentrations.

Salvia miltiorrhiza

Aerial parts of both seed- and stem-regenerated *S. miltiorrhiza* plants demonstrated a higher content of rosmarinic acid (along with lithospermic acid) than their parental plants. The inverse was true for the roots, but, as Morimoto et al. (1994) mentioned, this could simply be due to the age difference of the plant material: parent plants were much older (at least three years old) than the regenerates (only a few weeks old). Accumulation of phenolic acids was essentially the same in shoots either regenerated in vitro from callus cultures or clonally propagated from shoots (Morimoto et al., 1994). Interestingly, accumulation of phenolic acids in the regenerated plants transferred after rooting to vermiculite for 5 weeks, then cultured in soil for another 5 weeks, was found to be higher than in plants cultivated continuously in vermiculite only. Callus cultures themselves, on the other hand, demonstrated a rather poor accumulation potential, which was ca. 3-4 times lower than in regenerated plants (Kintzios, 2000).

In an attempt to produce tanshinones using plant cell culture technology, a Ti-transformed cell culture system of *Salvia miltiorrhiza* was established (Chen et al., 1997). The transformed cells in suspension formed macroscopic clumps of cell aggregates upto 2-3 cm in size rather than homogenous cell suspension. Chen et al. (1999) clearly demonstrated that this transformed cell aggregates retained the ability to produce large amounts of rosmarinic acid after more than two years of subculture in a liquid system. In marked contrast, in the untransformed callus tissues of *Salvia miltiorrhiza*, the content of rosmarinic acid decreased gradually by the fourth subculture (Morimoto et al., 1994). The instability of the undifferentiated tissues in producing secondary metabolites was often observed in other plant cell cultures during prolonged cultivation (Deus-Neuman et al., 1984). The results can be explained in part by the formation of large cell aggregates rather than homogenous cell suspensions in a liquid medium, which mimicked self-immobilized cell culture and thus might encourage differentiation and improve secondary metabolite synthesis (Tsoulpha et al., 1991). In fact, it was found that a number of small leaves with short petioles were differentiated from these macroscopic transformed cell aggregates, which was a typical phenotype, as described by Spencer et al. (1990).

Coleus blumei

Rosmarinic acid accumulation in suspension cultures of *Coleus blumei* Benth. (Lamiaceae) can be stimulated upto 10-fold by raising the sucrose content of the culture medium. Cell cultures in medium with 2% sucrose accumulate only about 2-3% of the dry cell weight as rosmarinic acid, whereas up to 19% rosmarinic acid can be obtained in medium with 4% sucrose. The suspension cultures of *Coleus blumei* synthesize and accumulate all their rosmarinic acid during only a few days at the end of the growth phase. Therefore, this system proved to be very suitable for the investigation of the biosynthetic pathway of rosmarinic acid (Petersen, 1992).

Coleus blumei cell cultures may also display differences in growth and production kinetics. The suspensions maintained by Petersen (1992) synthesized RA at the end of the growth phase with low and high sucrose concentrations, whereas Razzaque and Ellis (1977) observed production starting in the linear growth phase. However, when the latter authors added NAA instead of 2,4-D to the culture medium, RA production began at the beginning rather than the end of the exponential growth phase. Finally, still on the subject of *Coleus blumei* cell suspensions, the production system developed by Ulbrich et al. (1985) induces RA production when the period of growth has been completed.

Ocimum basilicum

Ocimum basilicum L. is a popular lamiaceous plant used as a kitchen herb and as ornamental greenery in house gardens. Tada et al. (1996) derived rosmarinic acid from hairy root cultures of this plant, induced by two types of *Agrobacterium rhizogenes* strains. Five clones of *O. basilicum* L. hairy roots were cultured in three hormone-free liquid media. The hairy roots cultures were able to polymerize the caffeic acid to a trimer (rosmarinic acid) and a tetramer (lithospermic acid B). The results also indicated the importance of the selection of the hairy root clones and the culture conditions (medium, culture period, etc.) for the production and the biosynthetic study of useful secondary metabolites of *Ocimum basilicum* L.

Callus cultures of sweet basil (incubated on solid MS + 2 mg/lt Kin + 2 mg/lt IAA) have been demonstrated in order to accumulate rosmarinic acid at levels almost 2-fold higher than that of the intact plant (Makri and Kintzios, 1999). The pattern of this accumulation varied with time, and production maxima (expressed on a callus fresh weight basis) were observed during the third, fourth and seventh week of culture. When results were expressed per callus dry weight, a maximum production was observed only during the fourth week of culture. Kintzios et al. (manuscript in preparation) investigated the accumulation of rosmarinic acid in different types of sweet basil (*Ocimum basilicum* L.) cell cultures, i.e. callus cultures, cell suspensions and immobilized cell cultures. Leaf-derived suspension cultures of sweet basil, grown in a liquid Murashige and Skoog (MS) medium supplemented with 5% (w/v) sucrose, 10 mg l^{-1} ascorbic acid, 0.1 g l^{-1} meso-inositol, 0.5 g l^{-1} L-phenylalanine and 2 mg l^{-1} 2,4-dichlorophenoxyacetic acid and 2 mg l^{-1} α-napthylacetic acid, were able to accumulate rosmarinic acid at a maximum concentration of 10205.1 µg g^{-1} dry weight, a value 8.5-11 times higher than for callus cultures and donor plants. Cells immobilized in 1.5, 2 or 3 % w/v calcium alginate accumulated RA at a much-reduced rate (<15 µg g^{-1}). RA accumulation in sweet basil did not seem to require the cessation of cell growth, as reported for other species. In every type of culture (callus, cell suspension, immobilized cells), RA usually accounted for approximately 90% of the total phenolic content (Figures 2.3a and 2.3b).

Origanum vulgare

According to Yang and Shetty (1998), it is clear that proline, proline analogues (such as azetidine-2-carboxylate, A2C) and proline precursors (ornithine and arginine) can clearly stimulate rosmarinic acid synthesis in shoot cultures of *Origanum vulgare* clonal line O-1. This strongly suggests that perturbing the proline metabolism can help to redirect metabolites from the pentose phosphate pathway towards total phenolic and rosmarinic acid synthesis. This finding, in turn, indicates that the proline-linked pentose phosphate pahtway may be operational and may also be essential for making precursor intermediates for shikimate and phenylpropanoid pathways.

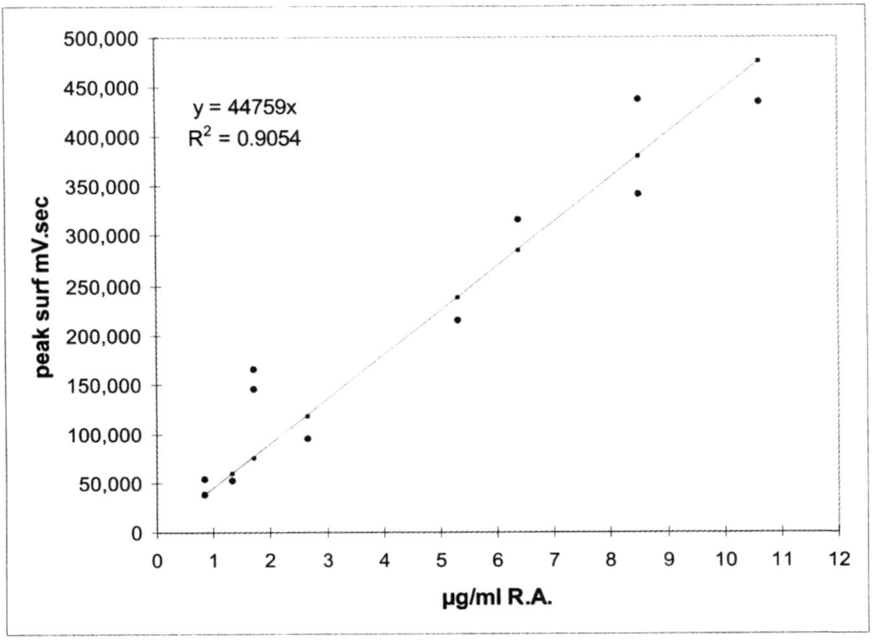

Fig. 2.3a. Regression curve indicating the correlation between total phenolics (HPLC analysis) and rosmarinic acid concentration in *O. basilicum* cell cultures.

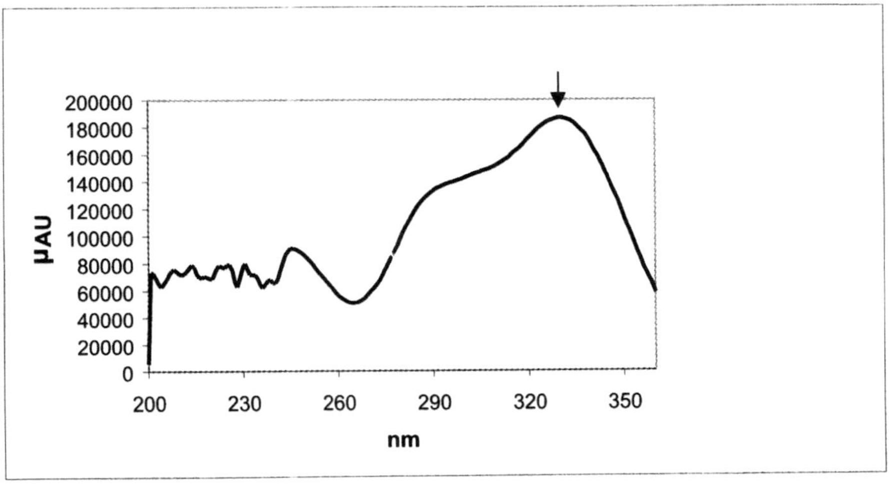

Fig. 2.3b. HPLC-spectrum of total phenolics in *O. basilicum* cell cultures (arrow indicates rosmarinic acid peak absorbance).

Lavandula angustifolia

A methanolic extract from fresh callus cells of *Lavandula angustifolia*, cultured in Linsmaier and Skoog (LS) medium, was tested for the presence of rosmarinic acid, caffeic acid, p-cumaric and ferrulic acids. The extracts contained a comparatively large amount of rosmarinic acid, smaller amounts of caffeic acids and traces of p-coumaric and ferrulic acids (Kovatcheva et al., 1996). Although Banthrope et al. (1985) reported that the main phenolic compound synthesized by *Lavandula vera* callus culture was a blue pigment, which they identified as a complex of Fe^{++} with the isomers of an enol ester formed by condensation of dialdehyde with caffeic acid, Kovatcheva et al. (1996) reported that the compound primarily produced was rosmarinic acid.

Anchusa officinalis

Mizukami and Ellis (1991) observed a sharp increase in both phenylalanine ammonia-lyase (PAL) and tyrosine aminotranferase (TAT), during the lag phase of cell suspension cultures of *Anchusa officinalis*. Such a rapid induction of enzyme activities after cell transfer to a fresh medium has often been reported in case of enzymes involved in secondary metabolism. The rate of rosmarinic acid biosynthesis expressed as net changes of the rosmarinic acid content per gram fresh weight of the cells per day still approximately paralleled the rise and fall of PAL and TAT activities, suggesting that both PAL and TAT may be involved in regulating rosmarinic acid biosynthesis as entry point enzymes.

Hyssopus officinalis

The perennial lamiaceous herb *Hyssopus officinalis* L., originating from the region of south Europe and Asia Minor, is an important medicinal plant used for antispasmodic, stomachic, antifungal and cough treatment. Although the essential oils isolated from this species are popularly available as food and drink additives as well as for cosmetic materials, the bioactive constituents of this plant have been investigated by Murakami et al. (1998). The reports indicated that transformed root cultures of this plant produce high levels of rosmarinic acid and, in addition, related compounds such as lithospermic acid (Kelley et al., 1975) and lithospermic acid B (Tanaka et al., 1989). As *Hyssopus officinalis* transformed root accumulated a high amount of rosmarinic acid and lithospermic acid B, it seems to be one of the most suitable materials for the production of these useful phenolic metabolites.

REFERENCES

Banthrope, D.V., Bilyard, H.J. and Watson D.G., 1985, Pigment formation by callus of *Lavandula angustifolia*. Phytochemistry, 24:2677-2680.

Baricevic, D. and Bartol, T., 2000, The biological/pharmacological activity of the *Salvia* genus. In : Sage — The Genus *Salvia* — Medicinal and Aromatic Plants—Industrial Profiles. Kintzios S., (ed.), Harwood Academic Publishers, pp. 143-182.

Cantino, P.D., Harley, R.M. and Wagstaff, S.J., 1992, Ganera of Labiatae: status and classification. In: Advances in Labiatae Science, pp. 511-522. R.M. Reynolds (ed). Royal Botanic Gardens, Kew.

Cantino, P.D. and Sanders, R.W., 1986, Subfamilial cassification of Labiatae. Systems Bot., 11:163-185.

Chen, H., Chen, F., Zhang, Y.L. and Song J.Y., 1999, Production of rosmarinic acid and lithospermic acid B in Ti transformed *Salvia miltiorrhiza* cell suspension cultures. Process Biochemistry, 34:777-784.

Chen, H., Yuan, J.P., Chen, F., Zhang, Y.L. and Song, J.Y., 1997, Tanshinone production in Ti transformed *Salvia miltiorrhiza* cell suspension cultures. J. Biotechnol., 58:147-156.

De-Eknamkul, W. and Ellis, B.E., 1984, Rosmarinic acid production and growth characterization of *Anchusa officinalis* cell suspention cultutres. Planta Med., 50:346-350.

De-Eknamkul, W. and Ellis, B.E., 1987, Tyrosine aminotransferase: the entrypoint enzyme of the tyrosine-derived pathway in rosmarinic acid biosynthesis. Phytochemistry, 26:1941-1946.

Deans, S.G. and Simpson, J.M., 2000, Antioxidants from *Salvia officinalis*. In : Sage — The Genus *Salvia* — Medicinal and Aromatic Plants—Industrial Profiles. Kintzios S. (ed.), Harwood Academic Publishers, pp. 185-192.

Deighton, N., Glidewell, S.M., Deans, S.G. and Goodman, B.A., 1993, Identification by EPR spectroscopy of carvacrol and thymol as the major sources of free radicals in the oxidation of plant essential oils. J. Food Sci. Agric., 63:221-225.

Deus-Neumann, B. and Zenk, M.H., 1984, Instability of indole alkaloid production in *Catharanthus roseus* cell suspension cultures. Planta Med., 50:427-431.

El-Gazzar, A., 1974, Numerical taxonomy of the Berbenaceae: a reassesement. Egypt. J. Bot., 17:69-83.

El-Gazzar, A. and Watson, L., 1970, Some economic implications of the taxonomy of Lasbiatae: Essential oils and rusts. New Phytol., 69:487-492.

Ellis, B.E. and Towers G.H.N., 1970, Biogenesis of rosmarinic acid in *Mentha*. Biochem. J., 118:291-297.

Ellis, B.E., Remmen, S. and Goeree, G., 1979, Interactions between parallel pathways during biosynthesis of rosmarinic acid in cell suspensions cultures of *Coleus blumei*. Planta, 147:163-167.

Engleberger, W., Hadding, V., Etschenber, E., Graf, E., Leyck, S., Windelmann, J. and Parnham, M., 1988, Rosmarinic acid. A new inhibitor of complement C3-convertase with anti-inflammatory activity. Int. J. Immunopharmacol., 10:729-737.

Erdtman, G., 1945, Pollen morophology and plant taxonomy. IV. Labiatae, Verbenaceae and Avicennaceae. Svensk Bot. Tidskr., 39:279-285.

Frankel, E.N., Huang, S.W., Aeschbach, R. and Prior E., 1996, Antioxidant activity of a rosemary extract and its constituents, carnosic acid, carnosol, and rosmarinic acid, in bulk oil and oil-in-water emulsion. J. Agric. Food Chem., 44:131-135.

Fukui, H., Yazaki, K., and Tabata, M., 1984, Two phenolic acids from *Lithospermum erythrorhizon* cell suspention cultures. Phytochemistry, 23:2398-2399.

Hippolyte, I., 2000, In vitro rosmarinic acid production. In : Sage — The Genus *Salvia* — Medicinal and Aromatic Plants — Industrial Profiles. Kintzios S., (ed.), Harwood Academic Publishers, pp. 233-242.

Hippolyte, I., 1990, Utilisation chez la Sauge (*Salvia officinalis* L.) des techniques de culture de tissus pour la multiplication vegetative par l'isolement d'apex, et la production d'acide rosmarinique par des suspensions cellulaires. PhD thesis, U.S.T.L. Montpellier II.

Hippolyte, I., Mrine, B., Baccu, J.C. and Jonard, R., 1992, Growth and rosmarinic acid production in cell suspention cultures of *Salvia officinalis*. Plant Cell Rep., 11:109-112.

Hippolyte, I., Marin, B., Baccou, J.C. and Jonard, R., 1991, Influence of maintenance medium and sucrose concentration on the production of rosmarinic acid by cell suspensions of sage—*Salvia officinalis* L.). Comptes Rendus de l'Academie des Sciences. Series 3, Sciences de la Vie 313:365-371.

Janicsak, G., Mathe, I., Miklossy-Vari, V. and Blunden, G., 1999, Comparative studies of the rosmarinic and caffeic acid contents of Lamiaceae species. Biochemical Systematics and Ecology, 27:733-738.

Kelley, C.J., Mahajan, J.R., Brooks, L.C., Neubert, L.A., Breneman, W.R. and Carmack, M., 1975, Polyphenolic acids of *Lithospermum ruderale* Dougl. Ex Lehm. (Boraginaceae). 1. Isolation and structure determination of lithospermic acid. J. Org. Chem., 40:1804-1815.

Kintzios, S., 2000, *Salvia* spp.: Tissue culture, somatic embryogenesis, micropropagation and biotransformation. In : Sage — The Genus *Salvia* — Medicinal and Aromatic Plants — Industrial Profiles. Kintzios S. (ed.), Harwood Academic Publishers, pp. 243-250.

Kintzios, S., Nikolaou, A., and Skoula, M., 1998, Somatic embryogenesis and in vitro rosmarinic acid accumulation in *Salvia officinalis* and *S. fruticosa* leaf callus cultures. Plant Cell Rep., 18:462-466.

Kintzios, S., Nikolaou, A., Skoula, M., Drossopoulos, J. and Holevas, C., 1996, Somatic embryogenesis and in vitro rosmarinic acid production from mature leaves of *Salvia officinalis* and *S. fruticosa* biotypes collected in Greece. In: Pank, F. (ed.), Eucarpia Series: Beitr. Z. Zuchtungsforschung, pp. 282-285.

Kohda, H., Takeda, O., Tanaka, S., Yamasaki, K., Yamashita, A., Kurokawa, T. and Ishibashi, S., 1989, Isolation of Inhibitors of Adenylate Cyclase from Dan-shen, the Root of *Salvia miltiorrhiza*. Chem. Pharm. Bull., 37:1287-1290.

Kovatcheva, E., Pavlov, A., Koleva, M., Ilieva, M. and Mihneva, M., 1996, Rosmarinic acid from *Lavandula vera* MM cell culture. Phytochemistry, 43:1243-1244.

Lamaison, J.L., Petitjean-Freytet, C. and Carnat A., 1991, Lamiacées médicinales á propriétés antioxydantes, sources potentielles d'acide rosmarinique. Pharm. Acta Helv., 66:185-188.

Makri, O. and Kintzios., S., 1999, Protein changes associated with in vitro plant regeneration of *Ocimum basilicum* L. ISHS Working Group Quality Management in Micropropagation 'Methods and Markers for Quality Assurance in Micropropagation' 24-27/8/99 University College, Cork/ Ireland.

Margna, U., 1977, Control at the level of substrate supply—an alternative in the regulation of phenylpropanoid accumulation in plant cells. Phytochem., 15:53-57.

Mizukami, H. and Ellis, B., 1991, Rosmarinic acid formation and differential expression of tyrosine aminotransferase isoforms in *Anchusa officinalis* cell suspension cultures. Plant Cell Reports, 10:321-324.

Morimoto, S., Goto, Y. and Shoyama, Y., 1994, Production of lithospermic acid B and rosmarinic acid in callus tissue and regenerated plantlets of *Salvia miltiorhiza*. J. Nat. Pdct., 57:817-823.

Murakami, Y., Omoto, T., Asai, I., Shimomura, K., Yoshihira, K. and Ishimaru, K., 1998, Rosmarinic acid and related phenolics in transformed root cultures of *Hyssopus officinalis*. Plant Cell Tissue and Organ Culture, 53:75-78.

Peake, P.W., Pussel, B.A., Martyn, P., Timmermans, V. and Xcharlesworth, J.A., 1991, The inhibitory effect of rosmarinic acid on complement involves the C5 convertase. Int. J. Immunopharmacol., 13:853-857.

Petersen, M., 1989, Rosmarinic acid synthase from cell cultures of *Coleus blumei*. Planta Med., 55:663-664.

Petersen, M., 1992, New aspects of rosmarinic acid biosynthesis in cell cultures of *Coleus blumei*. Planta Med., 58:578.

Razzaque, A. and Ellis, B.E., 1977, Rosmarinic acid production in *Coleus* cell cultures. Planta, 137:287-291.

Sakuta, M. and Komamine, A., 1987, Cell growth and accumulation of secondary metabolites. In: Cell Culture and Somatic Cell Geneticis of Plants, pp. 97-114. I. Vasil (ed.), Academic Press. Inc., N.Y.

Scarpati, M.J. and Oriente, G., 1958, Isolamento e costituzione dell'acido rosmarinico. Ricerca Sci., 28, 11:2329-2333.

Shetty, K., Ohshima, M., Murakami, T., Oosawa, K. and Ohashi, Y., 1997, Transgenic melon (*Cucumis melo* L.) and potential for expression of novel proteins important to food industry. Food Biotechnol., 11:11-128.

Spencer, A., Hamill, J.D. and Rhodes, J.C., 1990, Production of terpenes by differentiated shoot cultures of *Mentha citrata* transformed with *Agrobacterium tumefacians* T37. Plant Cell Reports, 8:601-604.

Sumaryono, W., Prokasch, P., Hartmann, T., Nimitz, M. and Wray, V., 1991, Induction of rosmarinic acid accumulation in cell suspention cultures of *Orthosiphon aristatus* after treatment with yeast extract. Phytochemistry, 30:3267-3271.

Tada, H., Murakami, Y., Omoto, T., Shimomuta, K., and Ishimaru, K., 1996, Rosmarinic acid and related phenolics in hairy root cultures of *Ocimum basilicum*. Phytochemistry, 42:431-434.

Tanaka, T., Morimoto, S., Nonaka, G., Nishioka, I., Yokozawa, T., Chung, H.Y. and Oura, H., 1989, Magnesium and ammonium-potassium lithospermates B, the active principles having a uremia-preventive effect from *Salvia miltiorrhiza*. Chem. Pharm. Bull., 37:340-344.

Tsoulpha, P. and Doran, P.M., 1991, Solasodine production from self-immobilized *Solanum aviculare* cells. J. Biotechnol., 19:99-110.

Ulbrich, B., Wiesner, W. and Arens, H., 1985, Large scale production of rosmarinic acid from plant cell cultures of *Coleus blumei* Benth. In: Primary and Secondary Metabolism of Plant Cell Cultures, pp. 293-303. K.H. Neumann, W. Barz, E. Reinhart (eds), Springer-Verlag, Berlin.

Yang, R. and Shetty, K., 1998, Stimulation of rosmarinic acid in shoot cultures of oregano (*Origanum vulgare*) clonal line in response to proline, proline analogue, and proline precursors. J. Agric. Food Chem., 46:2888-2893.

Wunderlich, R., 1967, Ein vorschlag zu einer naturlichen gliederung der Labiaten auf grund der pollenkorner, der samenentwicklung und des reifen samens. Oesterr. Bot. Z. 114:383-483.

Zenk, M.H., El-Shagi, H. and Ulbrich, B., 1977, Production of rosmarinic acid by cell suspension cultures of *Coleus blumei*. Naturwissenschaften, 64:585-586.

Zou, Z.W., Xu, L.N. and Tian, J.Y., 1993, Antithrombotic and antiplatelet effects of rosmarinic acid, a water soluble component isolated from Radix *Salviae Miltiorrhizae* (Danshen). Acta Pharmaceutica Sinica, 28, 4:241-245.

CHAPTER 3

Grapevine Polyphenols and their Biological Effects

Vitrac Xavier, Krisa Stéphanie, Decendit Alain, Deffieux Gérard,
and *Mérillon Jean-Michel**

INTRODUCTION

Vitis vinifera L. is a perennial woody vine belonging to the Vitaceae family (Bombardelli and Morazzoni, 1995). This climber is a productive plant considered to be the world's premier fruit with nearly 9 million hectares of viticultural land in 1990 (Coombe, 1989; Fillion et al., 1999). It is used for wine, juice, fresh consumption (table grapes), dried fruit and distilled liquor.

Polyphenolics are important constituents of grapes in determining the colour, taste and body of wines. Unlike other alcoholic beverages, red wine, which is obtained after maceration, contains phenolic compounds in high concentration upto 4 g/l, but relatively low quantities are present in white and rose wines, i.e. about a tenth of those of red wines (Waterhouse and Teissèdre, 1997).

The interest in wine's potential to improve heath is based on numerous epidemiological studies which have shown strong correlations between regular moderate wine consumption and major diseases (cardiovascular and cancer). Haslam (1996) has suggested that phenolic constituents could be responsible, for a large part, for these protective effects due to their physical and chemical properties (complexation, antioxidant).

*To whom correspondence should be addressed Tel (33) 5 57 57 46 89.
Fax: (33) 5 56 57 46 88. E-mail: jean-michel.merillon@phyto.u-bordeaux2.fr.

EPIDEMIOLOGY

There is strong evidence to suggest that vegetables and fruits protect against carcinogenic and coronary heart disease (Willett, 1994). A review published by Block et al. (1992) showed that a statistically significant protective effect of fruit and vegetable consumption against cancer was found in 128 out of 156 studies. This dietary component has an anti-carcinogenic action at a wide range of sites, particularly those of the respiratory and digestive tracts, as also on the hormone-related types of cancer (Hill, 1995). Regarding coronary heart disease, the typical Mediterranean diet (characterised by high consumption of bread, vegetables, fruits, vegetable fat and wine) probably explains the low disease rate observed in these countries such as in southern France (Renaud and de Lorgeril, 1992; de Lorgeril and Salen, 1998).

Some epidemiologic studies in the United States have examined the relationship between wine consumption and the risk of cancer. Contrary to other alcoholic beverages, the moderate wine consumption was associated with a decrease (or no increase) in the risk of oral and pharyngeal cancer (Blot et al., 1988, Macfarlane et al., 1995) and breast cancer (Longnecker et al., 1995). Farchi et al. (1992) using small cohorts in Italy, found a minimum risk of cancer and cardiovascular diseases associated with moderate alcohol consumption (wine being the main beverage). The inverse association between moderate alcohol consumption and coronary heart disease has been well established by several reports, and a portion of this benefit seems to be from alcohol itself (Rimm et al., 1996). Considering the most recent statistics of the World Health Organization (1995), this reduction in the mortality rate from coronary heart disease in France as compared to the USA was 61% for men and 69% for women. As compared to the UK, the reduction was 68% and 71%, respectively (Renaud and de Lorgeril, 1992; Renaud and Guegen, 1998). This finding constitutes the 'French paradox' because saturated fat intakes and serum cholesterol concentrations are similar in these three countries. This paradox could be attributable in part to high wine consumption (Renaud and de Lorgeril, 1992). Renaud's group has evaluated, prospectively, the health risk of grape wine and beer drinking in 36,250 middle-aged men from eastern France (Renaud et al., 1999). Compared to abstainers, a moderate daily intake of wine (22-32 g of alcohol, i.e. 2-3 glasses) was associated with a lower risk of death due to cardiovascular diseases (40%), cancer (22%), other causes (42%), and all causes (33%) (Figure 3.1). In the case of beer drinkers, the risk for cardiovascular diseases was less significant (30%) and there was no reduction in the rate of mortality from cancer and all causes.

Similar results to those mentioned above have been found in several

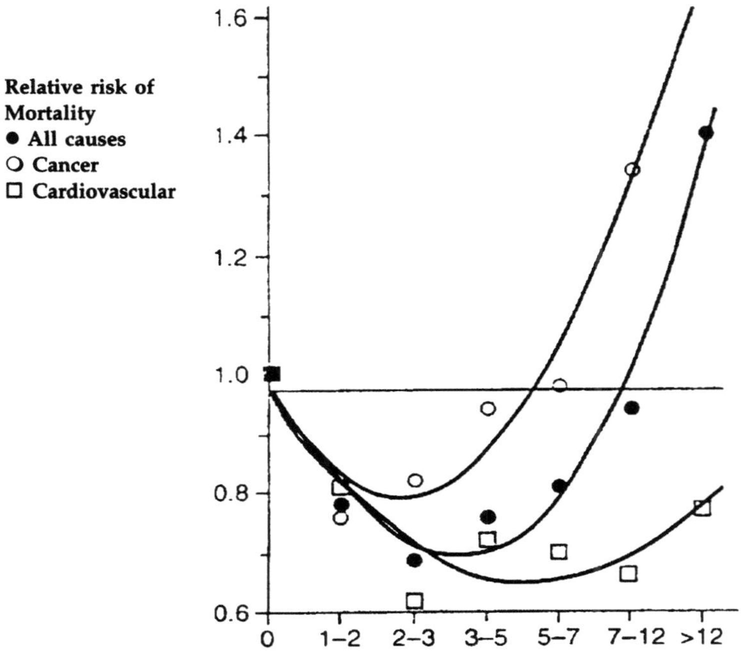

Fig. 3.1 Relative risk of death from all causes, cancer, and cardiovascular diseases in relation to intake of alcohol, 82% being wine. The risk is set at 1.0 for non-drinkers. Reproduced with permission from Renaud and Guegen, (1998).

large Danish cohort studies in which wine intake has a beneficial effect on all-cause mortality attributable to a reduction in death from both coronary heart disease and cancer (Gronbaek et al., 1995; 2000). Other alcoholic beverages (beer, spirits) are positively related to death from cancer. Concerning the risk of upper digestive tract cancer, a considerable increase of the risk was observed from a moderate intake of the latter but not for wine (Gronbaek et al., 1998). Moreover, during the past 15 years, mortality from coronary heart disease has declined by 30% in Denmark, which may be due to the increase in wine drinking during that period (from 17% to 30% of the total alcohol intake) with the opening up of the European market (Gronbaek et al., 1995).

These studies have established the non-linear relationships between alcohol consumption and mortality, which are either U- or J-shaped. This data indicate that drinking wine moderately may have beneficial effects on health, whereas excessive consumption is harmful and associated with increased mortality.

A protective effect of wine against the occurrence of dementia was also found in Aquitaine (France) in a prospective study that involved approximately 4000 subjects aged 65 and over (Orgogozo et al., 1997). The risk of developing a dementia is divided by about a factor of 2 in moderate drinkers (1/4-1/2 L), compared to non-drinkers and light drinkers. The most frequent age-related degenerative disease is Alzheimer's disease, which accounts for approximately 70% of the cases of dementia. Flavonoid intake appears to be inversely related to the risk of incident dementia (Commenges et al., 2000).

CHEMISTRY OF POLYPHENOLS

Polyphenols constitute one of the most numerous and ubiquitous groups of plant metabolites and are an integral part of human diet. A variety of beverages of vegetable origin—including wine—contain several non-flavonoid classes of phenolic compounds, such as phenolic acids (cinnamic and benzoic acid derivatives) and stilbenes (resveratrol and derivatives). Another major group of polyphenols, which is diverse in chemical structures and characteristics, also occurs in plant foods and beverages. This group comprises the flavonoids, in which the different classes are mainly flavan-3-ols (catechins and procyanidins), flavonols and anthocyanins.

NON-FLAVONOIDS

STILBENES

CHARACTERIZATION. Stilbenes are natural compounds occurring in a number of plant families, including Vitaceae. Within this family there is *Vitis vinifera* L., which is the most important species grown worldwide for grape and wine production (Bavaresco et al., 1999). Stilbenes are a family of molecules belonging to the non-flavonoid polyphenol group. Their essential structural skeleton comprises two aromatic rings joined by an ethylene bridge (C_6-C_2-C_6). From this relatively simple structure, there is a large array of compounds varying in the number and position of hydroxyl groups, the substitution with sugars, methyl, methoxy and other residues and the steric configuration of the molecules (Figure 3.2).

Among stilbene monomers, resveratrol (3,5,4'-trihydroxystilbene) has been identified as the major biological active compound, and most of the studies have focused on it. The two isomeric forms of resveratrol (*cis*- and *trans*-) have different chemical characteristics and biological activities. The *trans*-isomer is usually the more stable, and *cis*-*trans*-interconversions can occur in the presence of heat or ultraviolet light. These two compounds, having maximal absorbance at 286 (*cis*-isomer) and 306 nm (*trans*-isomer) (Trela and Waterhouse, 1996), are highly fluorescent at 374 nm when excited at 330 nm (Pezet et al., 1994).

Other simple stilbenes have been isolated in *Vitis vinifera*: *trans*-

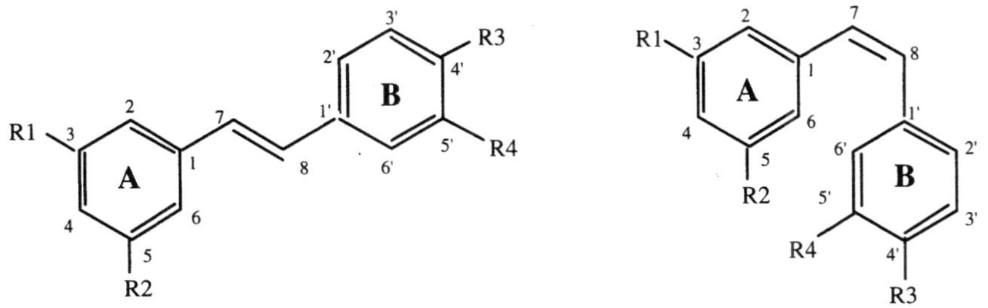

Name	*Structure*			
	R1	*R2*	*R3*	*R4*
trans- and *cis*-resveratrol	OH	OH	OH	H
pterostilbene	OCH3	OCH3	OH	H
trans- and *cis*-piceid	OGlc	OH	OH	H
trans- and *cis*-resveratroloside	OH	OH	OGlc	H
trans-piceatannol	OH	OH	OH	OH
trans- and *cis*-astringin	OGlc	OH	OH	OH

Fig. 3.2 Structures of the main stilbene derivatives from *Vitis vinifera*

pterostilbene, a dimethylated derivative of resveratrol (3,5-dimethoxy-4'-hydroxyresveratrol) (Langcake et al., 1979) and piceatannol (3,5, 4',5' -tetrahydroxystilbene). Furthermore, stilbene monoglucosides have been identified, in the form of piceid and resveratroloside, two β-glucosides of resveratrol (Waterhouse and Lamuela-Raventos, 1994; Waffo-Téguo et al., 1996; 1998) together with astringin (piceatannol 3-O-β-glucoside). These compounds exist in their two isomeric forms, *cis-* and *trans-* (Waffo-Téguo et al., 1998) (Figure 3.2).

Resveratrol oligomeric polyphenols have been isolated primarily from five plant families : Vitaceae, Dipterocarpaceae, Cyperaceae, Gnetaceae and Leguminosae (Sotheeswaran and Pasupathy, 1993). These compounds are either dimers, trimers or tetramers, and fall into two major groups. One group (A) contains one five-membered oxygen heterocyclic ring, usually the *trans*-2-aryl-2,3-dihydrobenzofuran moiety (**X**) (Figure 3.3).

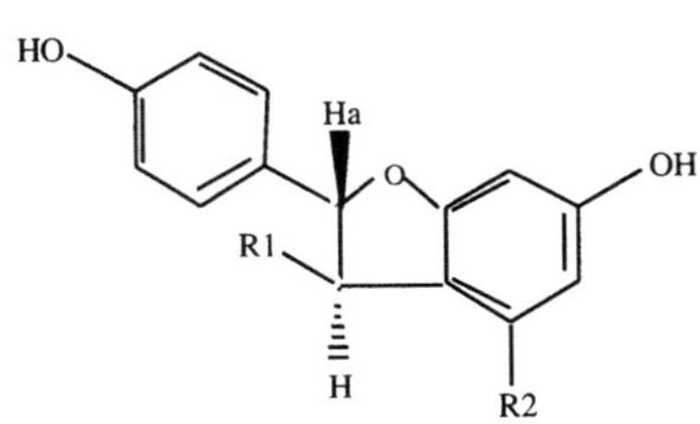

X

Fig. 3.3 General structure of group A resveratrol oligomers

Oligomers belonging to group A are: the dimer ε-viniferin (Figure 3.4) and α-viniferin (a cyclic trimer of resveratrol) (Langcake and Pryce, 1977).

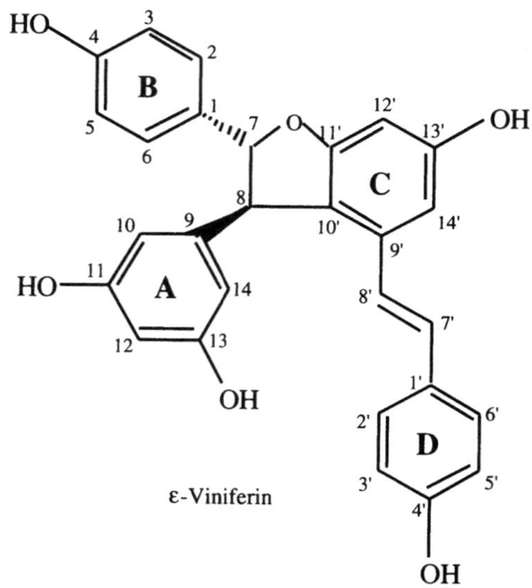

Fig. 3.4 Structure of the dimer ε-viniferin

The other group (group B) does not contain any oxygen heterocyclic ring. An example of a stilbene belonging to this group is pallidol, the symmetrical dimer of resveratrol (Figure 3.5).

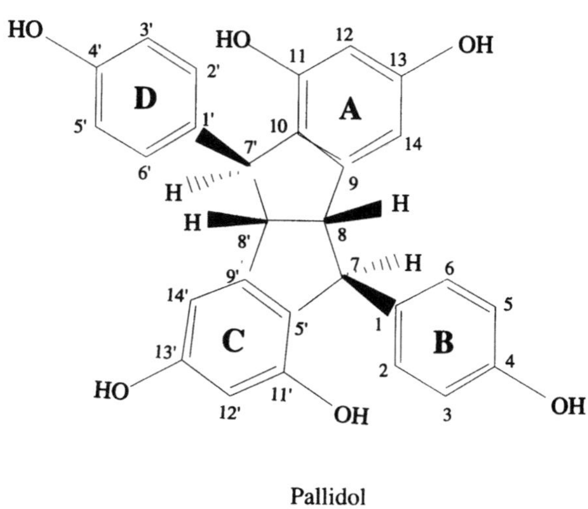

Pallidol

Fig. 3.5 Structure of the resveratrol dimer pallidol

All the oligomers belonging to the group A appear to be formed from resveratrol via the dimer ε-viniferin. The formation of viniferins from resveratrol was shown to involve an oxidative dimerization that could be reproduced by exposing the parent compound to a horseradish peroxidase-hydrogen peroxide system in vitro (Langcake and Pryce, 1977). The group B stilbenes (which do not contain any *trans*-2-aryl-2,3-dihydrobenzofuran moiety) seem to be formed directly from resveratrol monomers without ε-viniferin being an intermediary (Sotheeswaran and Pasupathy, 1993).

Biosynthesis pathway. The immediate precursors of resveratrol are p-coumaroyl CoA and malonyl CoA in a molar ratio of 1 to 3. The latter is derived from elongation of acetyl CoA units and the former from phenylalanine, which can be synthesized from sugars via the shikimate pathway. Following oxidative deamination as catalysed by Phenylalanine Ammonia Lyase (PAL), phenylalanine is converted to cinnamic acid, which in turn, is enzymatically hydroxylated to p-coumaric acid. In the final step, p-coumaroyl CoA is generated from the free coenzyme by a specific CoA ligase. The condensation of p-coumaroyl CoA with 3 molecules of malonyl CoA is accomplished through the activity of the stilbene synthase (STS) (Figure 3.6). However, the exact biosynthetic formation of these derivatives remains unknown.

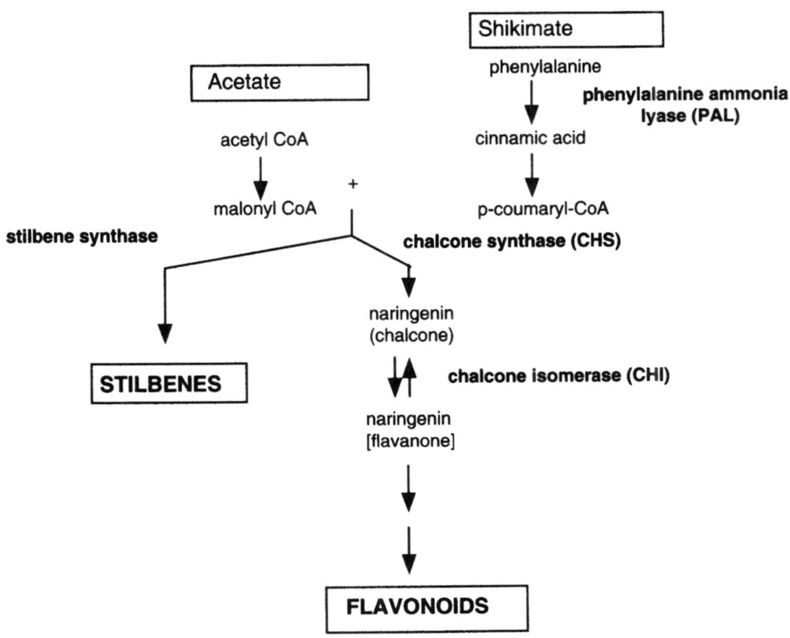

Fig. 3.6 General biosynthesis pathway of stilbenes and flavonoids

Distribution in *Vitis vinifera*. In *Vitis vinifera* varieties, a number of different hydroxystilbenes are present in several parts of the grape plant in the form of constitutive compounds of the lignified organs (roots, canes, seeds, stems, ripe cluster stems), and also as induced substances (in leaves and berries), probably acting as phytoalexins in the mechanisms of grape resistance against pathogens. Both resveratrol and piceid can be found in grape products, with the concentration of the glucoside usually being significantly higher than the aglycone (Waterhouse and Teissedre, 1997).

In grape berries, stilbene synthesis is primarily located at the skin cells and it is either absent or in low quantity in the fruit flesh (Creasy and Coffee, 1988; Jeandet et al., 1995). The greater part of the resveratrol in the skins is in both glycosidic forms (piceid isomers) (Roggero et al., 1997), whereas pterostilbene is detected in very low levels in healthy and immature grape berries (Pezet and Pont, 1988). In a recent study, Romero-Perez et al. (1999) discovered that piceids were the major components of grape juices, averaging a total of 4 mg/L in red grape juices and 0.5 mg/L in white ones. In seeds, only *trans*- and *cis*-resveratrol were detected.

In leaves infected by *Botrytis cinerea*, the main stilbenes detected by Langcake and Pryce (1979) were *trans*-resveratrol (90 μg/g fresh weight), ε-viniferin (30 μg/g) and α-viniferin (20 μg/g). Recently, resveratrol dehydrodimer and piceids were detected in leaves induced by UV irradiation (Jeandet et al., 1997; Douillet-Breuil et al., 1999).

In stems, the stilbenes usually found are *trans*- and *cis*-resveratrol, piceatannol, ε-viniferin (Bourhis et al., 1996) and viniferol A (Yan et al., 2001). A large number of resveratrol derivatives, mainly oligomers, have been detected in roots. These are ε-viniferin, gnetin H, r-viniferin, ampelopsin, hopeaphenol (Mattivi et al., 1998) and r-2-viniferin (Korhammer et al., 1995).

In grape cell suspension cultures, our laboratory has isolated several stilbenes—resveratrol, piceid, resveratroloside and astringin—all in the two isomeric forms *cis*- and *trans*. Furthermore, stilbene oligomers have been identified. Among them, two new stilbene dimer glucosides, resveratrol (E)-dehydrodimer 11- and 11'-O-β-glucoside, have been isolated, together with the known resveratrol (E)-dehydrodimer (Figure 3.7) and pallidol (Figure 3.5) (Waffo-Téguo et al., 2001b).

Determination of stilbenes in wine. *Trans*-resveratrol has been determined in wines, juices, berries and by-products of the grape-juice industry, including pomace, purees and seeds, in order to establish the amount of resveratrol incorporated into the human diet. Several methods have been described. Among them, the method developed by Siemann and Creasy (1992), demonstrated for the first time the presence of resveratrol in wine. This method required organic phase extraction followed by one or more HPLC steps. In this case, large volumes are

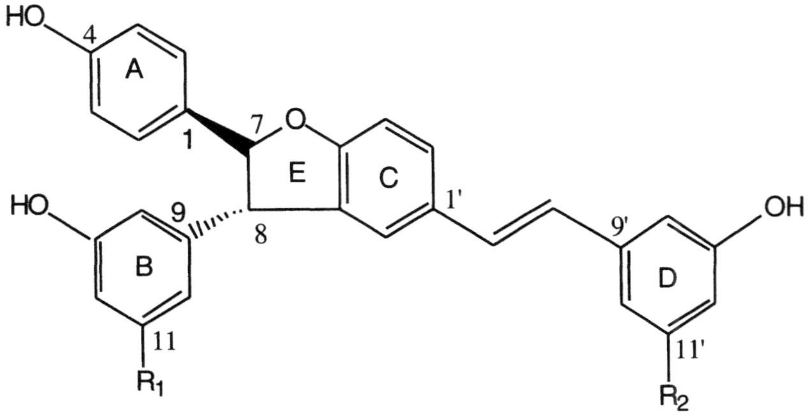

$R_1 = R_2 =$ OH: resveratrol *trans*-dehydrodimer
$R_1 =$ O-Glc; $R_2 =$ OH: resveratrol *trans*-dedhydrodimer-11-O-glucoside
$R_1 =$ OH; $R_2 =$ O-Glc: resveratrol *trans*-dedhydrodimer-11'-O-glucoside

Fig. 3.7 Structures of the stilbene oligomers isolated from grape cell suspension cultures

required and modifications of the phenolic composition and significant losses of compounds occurred due to incomplete extraction or oxidation of resveratrol. The same problems occurred in the method of Lamuela-Raventos and Waterhouse (1993), which required 11 individual steps (extraction, evaporation, washing, filtration) before HPLC. Gas chromatography (GC) has been used with or without MS detection by several authors (Goldberg et al., 1995; Jeandet et al., 1995) but it requires derivatization with TMS in order to enhance the volatility of resveratrol. More recently, a capillary electrophoresis technique has also been proposed (Arce et al., 1998; Gu et al., 2000).

Nevertheless, HPLC has been the most frequently-used method due to its widespread availability. However, there are difficulties in interpreting chromatograms, so detection must be optimized. To this end, fluorescence and electrochemical detections have been used by several authors, and good sensitivity and specificity were obtained in most cases (Pezet et al., 1994; McMurtrey, 1997). Recently, a simple and rapid procedure was proposed that separates and quantifies the four major forms of resveratrol in wine (*cis*- and *trans*-isomers of aglycone and piceids) together with *trans*-astringin and pallidol to analyse French wines (Vitrac et al., 2001a). It uses an HPLC system with direct injection and a fluorescence detector (Figure 3.8).

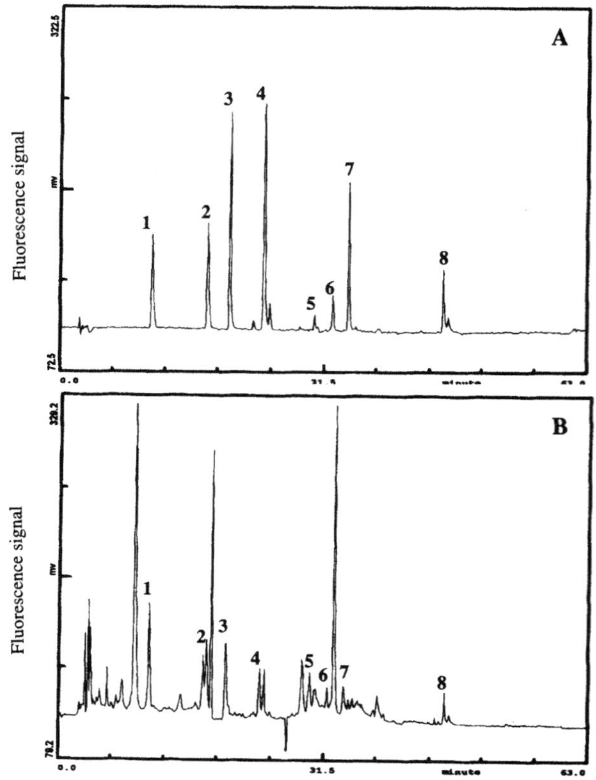

Fig. 3.8 Chromatographic profiles using fluorescence detection for a standard mixture of polyphenols (**A**) and a directly injected sample of red wine (**B**). 1=(+)-Catechin; 2=(–)-Epicatechin; 3=*trans*-Astringin; 4=*trans*-Piceid; 5=*cis*-Piceid; 6=*trans*-Resveratrol; 7=Pallidol; 8=*cis*-Resveratrol.

For many of these stilbenes, the limit of detection is lower than 0.01 mg/L and the limit of quantification is 0.05 mg/L.

In the first paper describing the occurrence of *trans*-resveratrol in wine, Siemann and Creasy (1992) found very low concentrations in white and red wines (<1 mg/L). The highest concentrations were found by Mattivi (1993) in Italian red wines: he recorded the highest concentrations in Cabernet Sauvignon (1.33-7.17 mg/L), Merlot (3.14-6.03 mg/L) and Pinot Noir (3.22-5.93 mg/L).

Resveratrol is now known to occur in wine in free (*cis* and *trans*) and glycosidically-bound form. Free *trans*- and *cis*-resveratrol are present in a concentration range of 0.2-13 mg/L in red wines and 0.1-0.8 mg/L in white wines. For the bound forms of resveratrol (piceids), concentrations are reported to be in a range of 0.3-9 mg/L in red and 0.1-2.2 mg/L in white wines (Lamuela-Raventos et al., 1995; Goldberg et al., 1995). Ribeiro de Lima et al. (1999) even determined piceids in Portuguese red wines in concentrations upto 68 mg/L.

Our study corroborates other researches and experiments which show that red wines from various countries and regions have a low mean concentration of *trans*- and *cis*-resveratrol, inferior to 5 mg/L (Lamuela-Raventos et al., 1995; Goldberg et al., 1995). Concerning *trans*- and *cis*-piceid, we found that their levels exceeded those of the free isomers and reached maximal concentrations of 26 mg/L and 24 mg/L in red wines, respectively (Vitrac et al., 2001a). *Cis*-piceid is typically found at lower concentrations than *trans*-piceid in wines. The maximal value for *cis*-piceid in white wines was found to be 0.9 mg/L, that of the *trans*-isomer being 2.9 mg/L. *Trans*-astringin is sometimes found at higher concentrations than piceids in wines; some wines having concentrations superior to 30 mg/L. However, there were wide variations between the wine samples and *trans*-astringin was not detected in all of them.

In this above-mentioned study, we reported levels of pallidol in red wine for the first time. Concentrations ranged from 0.5 mg/L to 4.8 mg/L for red wines, with a mean value of 2.5 mg/L. On the other hand, pallidol could not be detected in dry or sweet white wines. It has been shown that pallidol and some other resveratrol dimers may be fungal metabolites of resveratrol (Cichewicz et al., 2000). Thus, the occurrence of this compound in wine is certainly due to the oxidation of resveratrol by fungus in infected berries used for vinification. Another resveratrol dimer has been isolated from red wine and identified as Parthenocissin A by our group (Vitrac et al., 2001b) (Figure 3.9). Unfortunately, we were not able to determine its levels in wine because of the insufficient quantity of compound available.

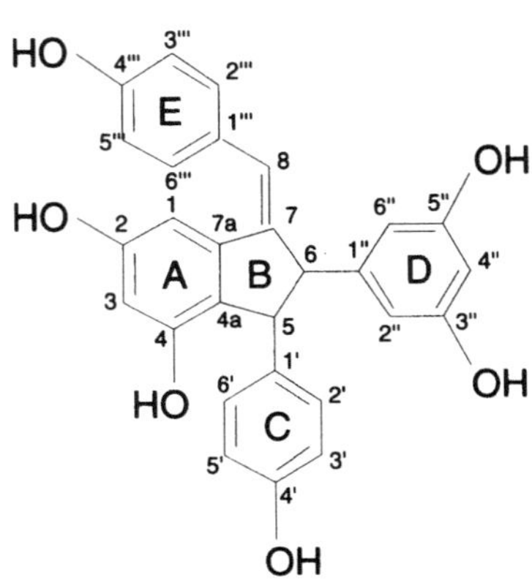

Fig. 3.9 Structure of the resveratrol dimer Parthenocissin A

As already reported (Carando et al., 1999), we found that total stilbene levels reached mean concentrations upto 20 mg/L often with a predominance of the glucoside isomers, depending on multiple factors such as the grape cultivar, fungal pressure and climate.

Red wines contain larger amounts of stilbenes rather than white wines, regardless of the oenological technology applied. The extent of maceration with skins and seeds during fermentation is the main factor determining the concentration of stilbenes in wines. They generally require long maceration on the skins so as to be extracted efficiently (Jeandet et al., 1995; Pezet and Cuenat, 1996).

PHENOLIC ACID DERIVATIVES

Grapes and wine contain several non-flavonoid polyphenols (phenolic acids), which play a primary role in defining the sensorial characteristics of wines and brandies, giving the 'oak wood taste' typical of long-aged products. In hydroalcoholic media, phenolic acids are colorless but they can become yellow after oxidation reactions.

Seven benzoic acids are known to researchers (C6-C1) (Figure 3.10). The most present forms are vanillic acid and gallic acid. Several cinnamic acids (C6-C3) are also present in grape and wine in low levels in their free forms, and are rather esterified, essentially with tartaric acid or glucose moiety. Indeed, we have recently isolated and characterized two compounds whose ^{1}H-NMR spectra are identical: both include a glucopyranose unit linked by an ester linkage with a *p*-coumaroyl part via the C6'', and both show α- and β-configurations in the solution. These compounds were then identified as two isomers of *trans*-*p*-coumaroyl 6''-glucoside (Vitrac et al., 2001b).

Benzoic acid

Cinnamic acid

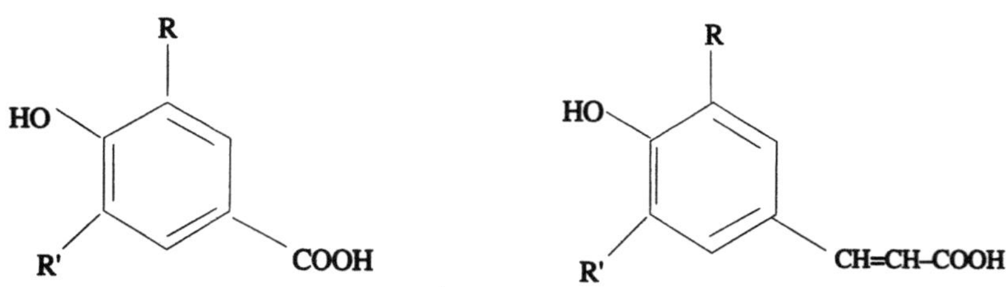

R = R' = H: p-hydroxybenzoic acid
R = OH; R' = H: protocatechuic acid
R = OCH3; R' = H: vanillic acid
R = R' = OH: gallic acid
R = R' = OCH3: syringic acid

R = R' = H: p-coumaric acid
R = OH; R' = H: caffeic acid
R = OCH3; R' = H: ferulic acid
R = R' = OCH3: sinapic acid

Fig. 3.10 Structures of phenolic acids

Phenolic acid concentrations in red wine are 100-200 mg/L, whereas in white wines they are 10-20 mg/L. In a recent study, the mean values of p-coumaric acid for white and red wines were found to be, respectively, 2.2 and 4.1 mg/L, whereas ferulic acid had a mean value of 3.9 and 7.2 in white and red wines, respectively (Gorinstein et al., 2000). In addition, Waterhouse and Teissedre (1997) reported levels of caffeic acid between 7-9 mg/L. Other phenolic acids are present in wines, such as gallic acid (0-10 mg/L in white wines and 65-85 mg/L in red wines). Tyrosol, which is derived from tyrosin after alcoholic fermentation, is always present in red and white wines (20-30 mg/L).

FLAVONOIDS

The flavonoids are a large family of secondary plant metabolites of low molecular weight comprising anthocyanins, flavonols, flavones, catechins and flavanones, many of which are present in plant tissues in relatively high concentrations in the form of sugar conjugates (Harborne and Williams, 2000). The general ring structure for the flavonoids is the flavan nucleus (C_6-C_3-C_6), shown in Figure 3.11. The rings B and C are derived from phenylpropanoid metabolism (shikimate pathway), while the A ring is the result of the condensation of three acetate molecules derived from malonyl-CoA, a reaction catalysed by the chalcone synthase enzyme (CHS) (Figure 3.6).

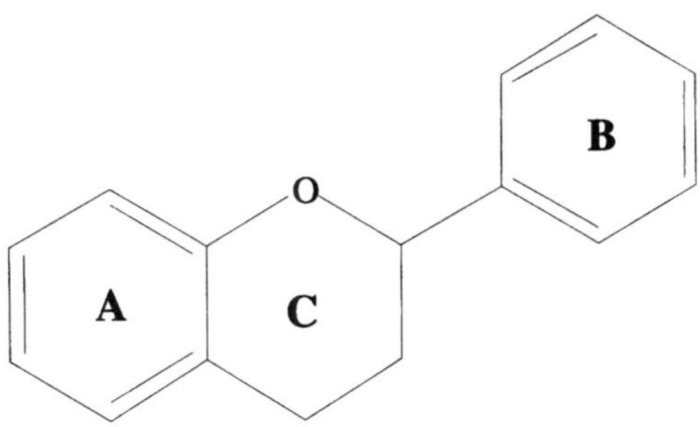

Fig. 3.11 A general structure for flavonoids.

FLAVANOLS

Flavan-3-ols (catechins) are some of the most widely-occurring flavonoids, the most important sources of these compounds in the diet being grapes, wine and tea. Flavan-3-ol monomeric units, (+)-catechin and (-)-epicatechin, as well as several oligomers have been separated and identified in *Vitis vinifera* (Bourzeix et al., 1986; Ricardo da Silva et al., 1992).

Procyanidins are oligomers and polymers of flavan-3-ol units which are (+)-catechin and (–)-epicatechin in the case of procyanidins, which are also known as condensed tannins. The simplest procyanidins are dimeric forms, and the most common of these are the four C4→C8-linked dimers (B1-B4). Procyanidin trimers C1 and C2 are also C4→C8 linked forms. These procyanidin dimers and trimers are sometimes accompanied by lower concentrations of the corresponding C4→C6 isomers (Figure 3.12).

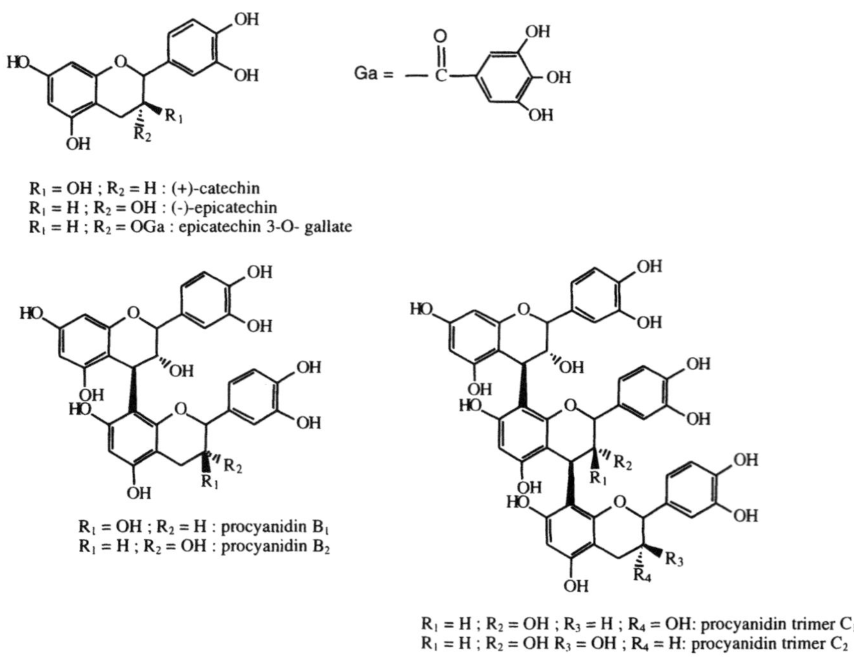

Fig. 3.12 Structures of flavanol monomers and oligomers

The part of the grape containing the highest concentration of catechins is the seeds, from which they are extracted during the wine-making process. Different procyanidins have been isolated and identified in grapes. Many of them are esterified with gallic acid, among which the most common are epicatechin 3-O-gallate, procyanidin B2 3'-O-gallate, (+)-catechin gallate and epigallocatechin 3-O-gallate, which are present in grape skins (Souquet et al., 1996) and seeds (Prieur et al., 1994).

In the studies of Ricardo da Silva et al. (1991a; 1992) in Chardonnay grapes, procyanidin B2 was the major component in seeds (3 mg/g), and procyanidin B1 in stems (2.7 mg/g) and skins (0.04 mg/g). Both B2 and B4 were found mainly in grape seeds for all the varieties studied. The pulp of grape variety was devoid of procyanidins. Trimeric procyanidins with C4→C8 linkages between monomer units (trimers C1 and C2) were also present in large amounts compared to dimeric procyanidins, i.e. procyanidin trimer C2, which is the second major procyanidin quantified in stems and skins of all vine varieties.

Catechin is known to be the most abundant monomeric flavan-3-ol in wine. It appears to be the highest in Pinot noir grapes and wines, averaging 250 mg/L in wines. Generally, it is in the range of 150-220 mg/L. The white varieties contain lower concentrations, averaging 40 mg/L (Waterhouse and Teissèdre, 1997). Epicatechin is typically found at lower concentrations than catechin in grapes (Bourzeix et al., 1986), averaging 80 mg/L. In white wines, procyanidin dimers (B1, B3 and B4) were found to be present at a mean concentration of 5 mg/L, with a higher mean concentration for B1 and B3. The presence of B2 dimer was not detected. B1 is the major component for white and red wines. In red wines, procyanidin dimers (B1-4) were found to be present at mean concentrations of 270 mg/L (Carando et al., 1999).

FLAVONOLS

Vegetables, fruits and beverages constitute the main dietary sources of flavonols, such as conjugates of quercetin, myricetin, kaempferol and isorhamnetin and the corresponding flavones luteolin and apigenin (Figure 3.13).

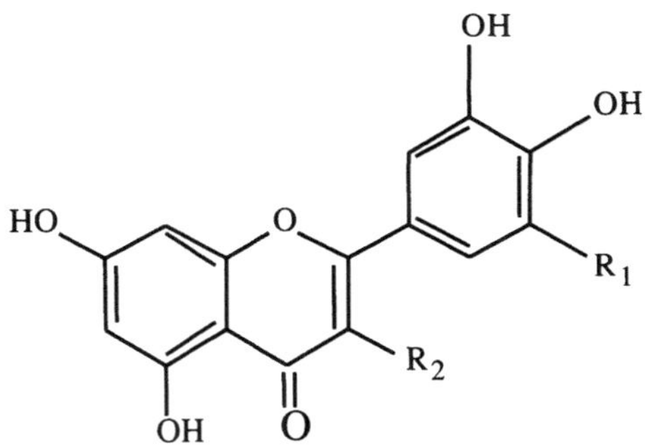

R_1 = H ; R_2 = OH : quercetin
R_1 = H ; R_2 = OGlc : quercitrin
R_1 = OH ; R_2 = OH : myricetin
R_1 = H ; R_2 = OGlc-Rha : rutin
(Glc = glucose; Rha = rhamnose)

Fig. 3.13 Structures of flavonols

In grape, Palomino et al. (2000) found quercetin levels as low as 0.5 mg/100 g. However, it is the most abundant flavonol in wine and its concentration in red varieties ranged from 4 to 16 mg/L (Frankel et al., 1993), while grape juice contained 7-9 mg/L (Hertog et al., 1993). In the white wine varieties, this compound is not detected (Waterhouse and Teissèdre, 1997). Two of its glucosides are also present in wine; one of them is isoquercitrin, and the other is rutin (Figure 3.13), which is found at the level of 9 mg/L (Frankel et al., 1993).

Myricetin was detected in wine by Alonso et al. (1986) and average levels in red wines were found to be between 5 and 13 mg/L (Waterhouse and Teissèdre, 1997). Kaempferol is either absent or in very low quantities.

Quercetin-3-glucuronide (200 mg/kg stem) belongs to an additional group of compounds also characterized in grape stems (Souquet et al., 2000).

Dihydroflavonols : The main dihydroflavonols identified in grape and wine are dihydroquercetol (taxifolin) and dihydromyricetol (Figure 3.14). Recently, astilbin (dihydroquercetin 3-O-rhamnoside) was isolated in pomace (Lu and Foo, 1999) and stems (Souquet et al., 2000). It was previously evidenced in white wine by Trousedale and Singleton (1983), and our laboratory recently isolated this compound in red wine from south-western France together with dihydromyricetin 3-O-rhamnoside, one of its hydroxylated derivatives (Vitrac et al., 2001b). Their levels in red and white wines were also determined. The highest levels of astilbin were found in red wines (15.6 mg/L) with a mean value of 10 mg/L. In white wines, these levels were five times lower, averaging 2.2 mg/L. Concerning its hydroxylated derivative, it was found at higher levels in all the wines, with mean values of 44.7 and 3.0 mg/L in red and white wines, respectively (Vitrac et al., 2001a).

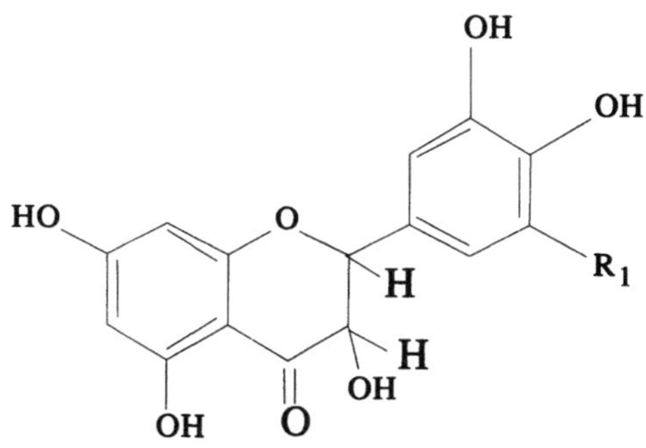

R_1 = H : dihydroquercetol
R_1 = OH : dihydromyricetol

Fig. 3.14 Structures of grape dihydroflavonols

ANTHOCYANINS

Anthocyanins are a group of flavonoids that are related to the color of several aerial and subterranean organs in many plants. In grapevines, anthocyanins are accumulated in leaves during senescence and are responsible for the coloration of grape skins in red cultivars. In grape cultivars named Teinturier, these pigments are also accumulated in the grape pulp. Structurally, they are glycosylated polyhydroxy/methoxy derivatives of 2-phenyl benzopyrilium (flavylium) salts. They can be present as different chemical species (Kennedy and Waterhouse, 2000): at low pH, they are predominantly present in the flavylium cation form, giving a reddish color in aqueous solutions. At higher pH, the flavilium cation is converted into other species, some of them being uncolored. Under certain conditions, this conversion can be virtually irreversible (Figure 3.15).

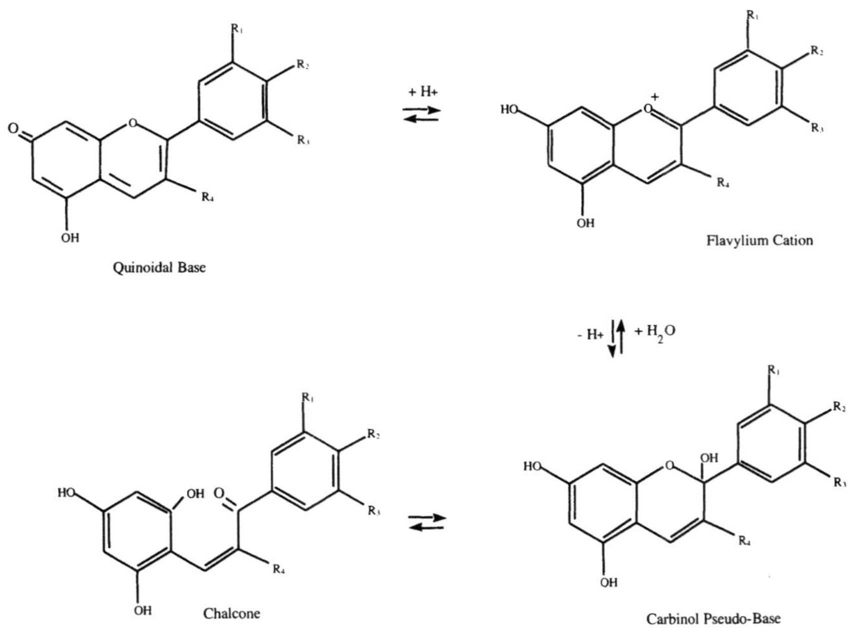

Fig. 3.15 The anthocyanin equilibrium forms

In model solutions, under acidic conditions (pH 4-6) as is typical in cell vacuoles, and in the absence of other substances, most anthocyanins exist in stable colorless forms. Between pH 4 and 6, the four anthocyanin structural types exist in equilibrium (Figure 3.15). However, the co-pigmentation phenomena may occur in the plant cell; in this case, the major form of anthocyanins is the flavilium cation, which gives the red color usually observed in plant vacuoles. As the pH rises, increasing amounts of the anhydrobase are formed. At neutral pH, the anionic form of the quinonoidal base is formed. All anthocyanins have the basic 4'-hydroxyflavillium structure and differ from one another according to the number of hydroxyl groups in the molecule, the degree of methylation of these hydroxyl groups, the nature and number of the sugars attached to the molecule and the position of the attachment. Five aglycones (anthocyanidins) dominate natural anthocyanin structures (Figure 3.16).

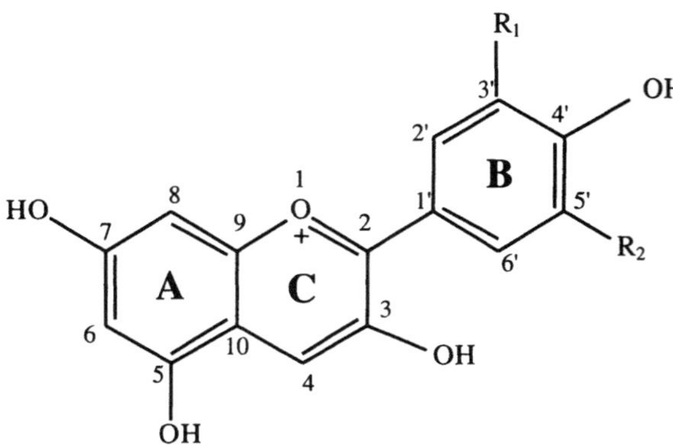

Name	Structure	
	R_1	R_2
Cyanidin	OH	H
Delphinidin	OH	OH
Malvidin	OCH3	OCH3
Peonidin	OCH3	H
Petunidin	OCH3	OH

Fig. 3.16 Basic flavylium structure with substitution pattern for the aglycone of anthocyanins (anthocyanidins) found in grapes

The color of red and black grapes results from the accumulation of anthocyanins that are usually located only in the skin of the berry. The accumulation of anthocyanins in grapes begins only after the phenological stage known as veraison and normally continues throughout the ripening phase of growth. Anthocyanins are hydrophylic in nature and are generally present in the plant cell vacuole. They are affected by weather conditions, especially light intensity and temperature, and by growing regions. When grapes are harvested, the content of anthocyanins in grapes may be as high as 1000 mg/kg of grapes.

Each species or variety of grape has a unique set of anthocyanins. *Vitis vinifera* varieties usually produce 3-monoglucosides, 3-acetylglucosides, and 3-*p*-coumarylglucoside derivatives of the aglycones, with malvidin derivatives often being the major forms present (Boss et al., 1996). Using *Vitis vinifera* cell suspension cultures, our laboratory was able to produce high amounts of anthocyanins (upto 850 mg/L). Moreover, we recently obtained a cell line producing mainly malvidin 3-O-β-glucoside (Krisa et al., 1999a). In most red wines, malvidin-3-O-β-glucoside is also the major anthocyanin.

Total antocyanins occur between 200 and 800 mg/L in young red wines (Ribereau-Gayon, 1982). However, during wine aging, the anthocyanin content of wine becomes progressively more complex. Condensation and disappearance of free anthocyanins occur, which are increasingly replaced by polymeric-colored compounds, thus affecting the wine color characteristics. Anthocyanins are also present in vegetative tissues, in leaves of fall foliage, stems and roots, and in emergent seedlings.

BIOLOGICAL AND PHARMACOLOGICAL ACTIVITIES

Phenolic compounds have a wide range of pharmacological and biological actions in two major domains—cancer and cardiovascular diseases—but the response in humans depends on their absorption and in vivo metabolism.

Bioavailibility and metabolism

Polyphenols are secondary metabolites of those plants which are regularly consumed by humans. The major sources, which differ somewhat between countries, are tea, wine, onions, apples or berries. On the basis of a moderate consumption of red wine, 2 to 3 glasses/day/individual, the daily intake of polyphenols can be estimated at about 300 to 1,200 mg. The advantage of wine with regard to other foods is that these compounds are in a stable solution (Goldberg, 1995). Serafini et al. (1997) suggested that wine ethanol might increase the bioavailability of polyphenols.

Epidemiological and experimental studies support the view that polyphenols may have a positive effect on health. However, less is known about the absorption and metabolism of polyphenols in humans at the usual levels of dietary intake. Xenobiotics, like polyphenols, may undergo modifications such as methylation or hydroxylation, and conjugation with glucuronic acid or sulfate. Conjugation improves the water solubility of polyphenols and consequently favors their elimination. The liver seems to be the main tissue involved in this metabolism, but other tissues that contain conjugative enzymes, such as intestinal mucosa or kidneys, could also be involved.

Non-flavonoids

Stilbenes. Resveratrol occurs naturally in a variety of plants used as traditional medicine in east Asian countries. It is present in wine as two forms of isomers, *cis* and *trans*, mainly in the glycosylated form (piceids). These compounds can be hydrolyzed by glycosidases in the human gastrointestinal tract, leading to *trans*- and *cis*-resveratrol. These substances, which are well known for their potent antioxidant properties, have generated much interest because of their biological activity that may be potentially beneficial to health. However, their metabolism in humans has been poorly studied.

Kuhnle et al. (2000) investigated the absorption of resveratrol across the jejunum and ileum in a rat intestinal model. They showed that upon ingestion, the major product transferred across the intestinal epithelium is a glucuronide of the parent aglycone (> to 90% of the amount absorbed). Soleas et al. (2001) demonstrated the bioavailability of resveratrol with volunteers given 25 mg of this compound. The plasma concentration of conjugated resveratrol was 30-50-fold higher than the free form, and peaked at 30 min. after consumption. However, they did not identify the glucuronidation sites of this stilbene.

In our laboratory, research was conducted on the metabolism of resveratrol through the glucuronidation pathway in humans, and on the identification of some UDP-glucuronosyltransferase isoforms that catalyse the formation of glucuronides. Two glucuronides were formed with *trans*- and *cis*-resveratrol by human liver microsomes, which corresponded to glucuronidation at positions 3 and 4′ (Figure 3.17; Aumont et al., 2001).

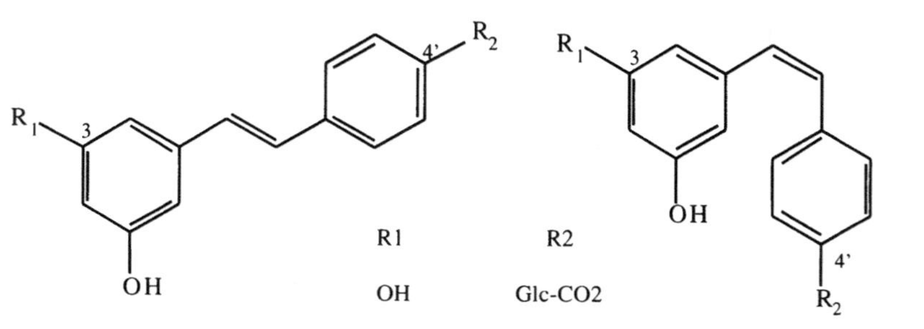

R1	R2
OH	Glc-CO2
Glc-CO2	OH

trans-resveratrol glucuronides　　*cis*-resveratrol glucuronides

Fig. 3.17 *Trans*- and *cis*-resveratrol glucuronides

Phytochemicals such as drugs are detoxified by conjugation with glucuronic acid in different mammalian tissues by UDP glucuronosyltransferase. Glucuronide conjugates have been shown to be less active than the corresponding unconjugated phytochemicals. However, several tissues contain the reverse acting hydrolytic β-glucuronidase, which is responsible for cleavage and reactivation of the glucuronide conjugates (Andlauer et al., 2000). The same authors demonstrated that these enzymes exist in all three segments of the small intestine in rats. However, glucuronides can also have a notable biological activity, as reported for morphine (Abbott and Palmour, 1988) and quercetin (Morand et al., 1998).

Flavonoids

Flavonoids often occur in the form of glycosides in dietary plants such as flavonols and anthocyanins. The linked sugar is often glucose or rhamnose but can also be galactose, arabinose or other sugars. Thus, the question arises as to whether glycosylated forms could be conjugated and absorbed to the same extent as aglycone forms from the digestive tract. On the other hand, flavanols are not usually glycosylated.

FLAVANOLS. The most abundant flavonoids in the diet are flavanols (catechins plus proanthocyanidins) and anthocyanins. Donovan et al. (1999) reported maximum plasma levels of catechin to be 91 nmols/L at 1h, after consumption of red wine containing 35 mg of it. Catechin was present almost exclusively as metabolites in plasma. Twenty percent was methylated metabolites and total catechins (methylated or not) were present as sulfated and/or glucuronidated forms.

This result was supported by an in situ perfusion of the rat intestine which demonstrated the important role of the small intestine in the glucuronidation and methylation of flavonoids. Indeed, mesenteric plasma contained glucuronide conjugates of catechin and 3′-O-methyl catechin and no native catechin could be detected (Donovan et al., 2001).

The same experiment was done with the procyanidin dimers B2 and B5 extracted from cocoa by Spencer et al. (2001), who detected unmetabolized/unconjugated epicatechin monomer (95.8%) and methylated dimer (3.2%) on the serosal side. They suggested that metabolism of monomer and dimer was limited by the dimer cleavage/translocation.

FLAVONOLS. Among flavonoids, quercetin is the major representative of the subclass of flavonols in foods of plant origin. Quercetin is found in wine in the form of aglycon, 3-glucoside and 3-rutinoside. Many studies have examined the beneficial properties of quercetin, namely antioxidant, antiproliferative, anti-inflammatory and anticarcinogenic effects. However, until 2000, the data on the bioavailability of this flavonol after oral intake have been contradictory.

To determine the influence of the sugar moiety on the absorption of quercetin, several authors have compared the bioavailability of different quercetin glycosides in human. They all showed that after consumption of quercetin glycosides, no intact quercetin glucosides and only trace amounts of aglycone were to be found in human plasma, and that the ingested compounds are mostly in the glucuronide form (Graefe et al., 2001, Sesink et al., 2001). The bioavailability was the same for quercetin-3-glucoside and -4′-glucoside (Olthof et al., 2000), and Graefe et al. (2001) concluded that quercetin glucoside is more rapidly absorbed than quercetin rutinoside in humans. In rat, glucosylation improves the absorption of quercetin in the small intestine and quercetin glucoside is hydrolysed before or during its intestinal absorption (Crespy et al., 2001). In view of these studies, it seems that quercetin glycosides are first deglycosylated and then conjugated by glucuronidation. Moreover, structural features of the glucose moiety influence these two reactions.

On the other hand, in an attempt to better understand the mechanism governing the intestinal epithelial membrane transport of flavonoid, Walgren et al. (2000) used the human Caco-2 cell monolayer, a model of intestinal absorption. They demonstrated that quercetin glucoside is transported by sodium-dependent D-glucose transporter (SGLT1) across the apical membrane of enterocytes, and confirmed that Caco-2 is able to hydrolyse absorbed quercetin glucoside within the cytosol (Figure 3.18). However, the glucuronidation of this compound was not determined and the transport across the basolateral membrane was not sought in this study.

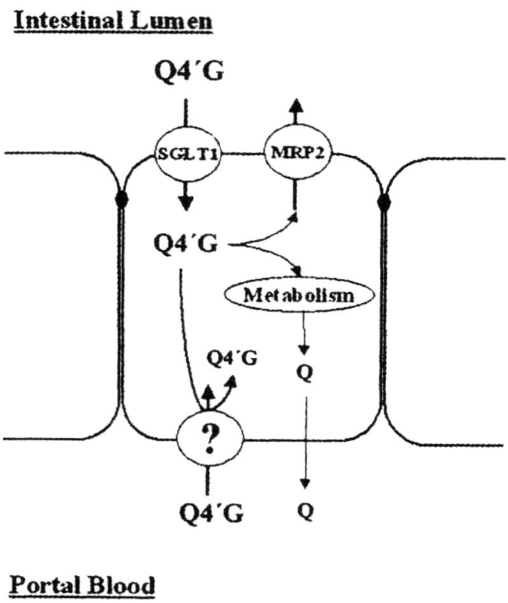

Fig. 3.18 Model of quercetin-4′-β-glucoside absorption in the enterocyte (reproduced with permission from Walgren et al., 2000). Q = quercetin, Q4′G = Quercetin 4′-β-glucoside

ANTHOCYANINS. Anthocyanins are the largest group of water-soluble pigments in the plant kingdom. They are widely distributed in the human diet. The intake of anthocyanins has been estimated to be 180-215 mg/day in the United States (Cao et al., 2001). They mainly exist in wine as glycosides, which were long considered not to be absorbable in human.

Recently, Tsuda et al. (1999) demonstrated that cyanidin-3-O-β-glucoside is present in the plasma of rat after oral administration. The researchers did not observe its aglycone, but showed its metabolism to protocatechuic acid and methyl-cyanidin-3-O-β-glucoside in some tissues.

Other results were observed in humans by Miyazawa et al. (1999) with a red fruit anthocyanin extract, and more recently, by Cao et al. (2001) with elderberry anthocyanins. Interestingly, neither glucuronides nor sulfates of anthocyanins were detected in the plasma, contrary to the other flavonoids present in wine. They only found anthocyanins in the intact glycoside form. The average plasma concentrations of total anthocyanins for these two studies were 30 and 97 nmols/L, and were reached within 60 and 71 min., respectively. Judging from the plasma concentration and urinary recovery, the absorption rate of anthocyanins as a proportion of intake is thought to be very low (< 1%). On the other hand, Miyazawa et al. (1999) found two metabolites identified to be methylated in rat liver.

Therefore, it will be necessary to perform further studies tracing both the intact compounds and their metabolites in blood and digestive tract. For this reason, we investigated the production of ^{13}C anthocyanins using grape cell cultures and ^{13}C-phenylalanine, a precursor of these polyphenols. ^{13}C-phenylalanine feeding experiments were done to optimize its incoporation into anthocyanins (Table 3.1; Krisa et al., 1999b).

Table 3.1 Production of anthocyanins and rate of ^{13}C enrichment of peonidin-3-O-β-glucoside after incorporation of [1-^{13}C] PHE by *Vitis vinifera* cells. Data from Krisa et al. (1999b)

[1-^{13}C] PHE (mM)	Day of addition	Anthocyanins (μmol l^{-1})	Rate of ^{13}C enrichment (E,%)
0	-	296	-
2.5	6	361	55
2.5	8	369	43
2.5	10	302	33
2.5 (×2)	6, 8	298	50
2.5 (×2)	6, 10	341	57
2.5 (×2)	8, 10	374	52
1	6	359	48
1 (×2)	6, 8	469	56
1 (×2)	6, 8, 10	482	65

It was found that the rate of ^{13}C enrichment of anthocyanins such as peonidin-3-O-β-glucoside can reach the level of 65%. Similar results were obtained for ^{13}C enrichment of stilbenes (Krisa et al., 1999c). These compounds will be produced in sufficient quantities in order to examine their absorption and metabolism in humans and also to evaluate their real physiological impact.

On the other hand, the mechanisms underlying the absorption of anthocyanins such as glycosides are not clear. Cao et al. (2001) suggest that the detection of these compounds in plasma in their unchanged glycoside forms may indicate the involvement of a glucose transporter in their absorption in vivo. This hypothesis needs to be investigated.

Cardiovascular protection

ANTIOXIDANT ACTIVITY. Free radicals derived from molecular oxygen, such as superoxide, hydroxyl, hydroperoxyl radicals and nitric oxide, are constantly generated in vivo for specific metabolic purposes (Halliwell, 1994). Free radical concentrations are increased either by their overproduction or by a deficiency in antioxidant defense systems. The reactivity of radicals can cause severe damage to biological molecules, especially to DNA, lipids and proteins (Halliwell, 1994). This damage probably contributes to the development of major chronic diseases including cancer, Parkinson's disease, senile dementia and atherosclerosis.

There is increasing evidence that oxidized low-density lipoproteins (LDL) may be responsible for promoting atherogenesis. Oxidized LDL are a key component in endothelial injury. They may directly injure the endothelium and play an initial role in the increased adherence and migration of monocytes and lymphocytes into the subendothelial space (Ross, 1993). Oxidized LDL may bypass the normal tight control exercized by the classical LDL receptor in the macrophages and be endocytosed via non-regulated scavenger receptors. This rapid accumulation of cholesterol and cholesteryl ester leads to foam cell formation (Esterbauer et al., 1993). Hence, it is conceivable that oxidative stress accelerates atherogenesis by enhancing LDL oxidation and increasing its accumulation in foam cells. The presence of antioxidants can interfere with the peroxidation process by removing the alkoxyl or peroxyl radicals. To be considered as an antioxidant, a compound must satisfy two basic conditions: first, when present at a low concentration relative to the substrate to be oxidized, it can delay or prevent auto-oxidation or free radical-mediated oxidation; secondly, the resulting free radical formed after scavenging must be stable upon further oxidation. Polyphenols may have an important role because of their antioxidant properties by virtue of the free radical scavenging properties of their phenolic groups, allowing them to act as reducing agents, hydrogen- or electron-donating agents or even singlet oxygen scavengers (Paganga et al., 1999).

In in vitro studies with total phenolic compounds from red wine, Frankel et al. (1993) found that red wine diluted 1000-fold containing 10 μmol/L phenolics inhibited human LDL oxidation significantly more than α-tocopherol. In our laboratory, we have also used this relevant test, i.e. Cu^{2+}-mediated lipid peroxidation of LDL lipids, in order to assess the susceptibility of LDL to oxidation, though copper can promote this oxidation by markedly different mechanisms (Ziouzenkova et al., 1998). The results concerning pure compounds belonging to major classes of polyphenols in wine are presented in Table 3.2.

Table 3.2 Antioxidant activities of phenolic compounds extracted from *Vitis vinifera* cell cultures

	LDL EC_{50} (mM)	DPPH EC_{50} (mM)
NON-FLAVONOIDS		
Stilbenes		
Cis-piceid	16.6±1.5	140±12
Trans-piceid	19.3±3.2	200±19
Cis-resveratrol	19.0±2.0	97.0±4.2
Trans-resveratrol	2.6±0.3	74.0±5.3
Astringin	3.1±0.2	30.6±2.1
Astringinin	1.9±0.2	29.0±1.4
FLAVONOIDS		
Catechins		
(+)-Catechin	1.9±0.1	20.2±1.6
(–)-Epicatechin	1.0±0.1	15.7±1.5
Anthocyanins		
Peonidin-3-O-β-glucoside	3.2±0.7	49.1±5.2
Malvidin-3-O-β-glucoside	3.8±0.8	42.1±4.4
DRUG REFERENCE		
Trolox	4.7±0.4	10.1±0.5

Data from Fauconneau et al. (1997)
Each value is the mean of at least three independent experiments ± SD.

In the LDL test, catechins, anthocyanins and some stilbenes are more efficient than trolox, the water-soluble vitamin E analog. Taken together, these results show that the chemical criteria essential for the antioxidant activities of these polyphenolic compounds (flavonoids and stilbenes) are: (i) the presence of the catechol structure or 4′-hydroxy in ring B; and (ii) the presence of the *meta*-hydroxy structure in ring A. In case of flavonoids, glycosylation of the hydroxyl group at C-3 does not seem to alter the antioxidant activity notably, as reported by Teisseidre et al. (1996) and Meyer et al. (1997). Moreover, Rice-Evans et al. (1995) noted the importance of conjugation between rings A and B via a planar C ring. Procyanidin dimers (B2, B8) and trimer (C1) have similar activity to catechin monomers towards inhibiting LDL oxidation (Teissèdre et al., 1996).

In the DPPH test which provides information about the reactivity of the tested compounds with the DPPH stable-free radical, phenolic compounds from wine show scavenger properties (Table 3.2). It has been shown that various catechin monomers, procyanidins, and flavonols have scavenger capacities for superoxide, hydroxyl and peroxy radicals (Torel et al., 1986; Rafat-Husain et al., 1987; Robak and Gryglewski, 1988; Ricardo da Silva et al., 1991b; Maffei Facino et al., 1994). Moreover, anthocyanidins such as pelargonidin can scavenge peroxynitrite (Tsuda et al., 2000).

Polyphenols such as catechin, *trans*-resveratrol and quercetin, can also act by another mechanism, i.e. their complexation with metal ions (iron, copper) which are involved in the generation of free radicals and lipid peroxidation (Afanas'ev et al., 1989; Morel et al., 1993; Frémont, 2000). Moreover, a-tocopherol (the principal form of vitamin E) which functions as a major antioxidant in human LDL can be recycled from its free radical form (α-tocopheryl) by epicatechin itself and derivatives (Zhu et al., 1999).

Red wine polyphenols are also effective in reducing the oxidative damage on normal human red blood cells in vitro (Tedesco et al., 2000). In fact, erythrocytes are particularly exposed to oxidative damage because of the oxygen carrier in the presence of high polyunsaturated fatty acid content on their membranes and high cellular concentration of hemoglobin.

ANTITHROMBOTIC AND VASOPROTECTIVE EFFECTS. Catechins show a protective effect against lipid peroxidation on both unstimulated human platelets and platelets submitted to a peroxidative stress. Interestingly, these compounds also inhibit human platelet aggregation induced by arachidonic acid (Polette et al., 1996). Pace-Asciak et al. (1995) have found that de-alcoholized red wine containing total phenolics inhibits human platelet aggregation induced by ADP and thrombin. This report suggested that *trans*-resveratrol is particularly active by inhibiting the synthesis of certain eicosanoids. We screened various stilbenes obtained from *V. vinifera* cell cultures for their antiplatelet properties on human platelet-rich plasma in which aggregation was induced either with arachidonic acid (AA), ADP, collagen or the TxA2 mimetics U-46619 (Table 3.3; Varache-Lembège et al., 2000). Significant inhibition of AA-induced aggregation was observed for the free hydroxylated compounds (*trans*- and *cis*-resveratrol, and astringinin) with IC_{50} values ranging from 17 to 38 μM, similar to that of acetylsalicylic acid (25 μM). In contrast, the corresponding glycosylated derivatives were weakly active or devoid of any significant activity. These results are in agreement with other reported data (Bertelli et al., 1996; Orsini et al., 1997). On the other hand, only the free hydroxylated stilbenes showed an antiplatelet activity when collagen or ADP was used as stimulating agent, but at much higher concentrations. In addition, all the derivatives were found to be inactive when U-46619 was tested as aggregation inducer.

Table 3.3 In vitro antiplatelet activity of polyhydroxystilbene derivatives from *Vitis vinifera* cell cultures

Compounds	Inhibition of platelet aggregation AA[a]	ADP[b]	Collagen[a]
trans-piceid	450 ± 40	> 800	> 800
trans-resveratrol	17 ± 3	180 ± 16	96 ± 11
cis-piceid	800 ± 50	> 800	> 800
cis-resveratrol	38 ± 6	270 ± 20	192 ± 15
astringin	> 800	> 800	> 800
astringinin	20 ± 4	172 ± 15	80 ± 10
ASAc	25 ± 5	54 ± 7	35 ± 5

a IC_{50}% inhibition concentration obtained from aggregation curve (μM)
Values are the means ± SEM (n = 4)
b Activity on the 2nd wave of ADP-induced aggregation
c Acetylsalicylic acid
Data from Varache-Lembège et al., (2000)

Based on these findings, the anti-aggregating profile observed for stilbenes clearly points to an interaction of the compounds with the platelet arachidonic acid pathway through a mechanism consistent with an inhibition of cyclooxygenase. Moreover, *trans*-resveratrol was found to inhibit calcium influx in thrombin-stimulated human platelets (Dobrydena et al., 1999). Indeed, the increase in free cytosolic calcium concentration is the major intracellular stimulus involved in platelet aggregation.

Several studies have shown that the extent of coronary artery stenosis due to atherosclerotic plaque formation and expansion into the arterial lumen is not sufficient to explain the incidence of clinical events associated with atherosclerosis (Cox and Cohen, 1996). It appears that the generation of clinical events involves plaque rupture, resulting in thrombus formation and arterial occlusion. This rupture is induced by vasomotor disturbances in which oxidized low-density lipoproteins may be involved (Cox and Cohen, 1996). Recently, it was shown that red wine and other grape products which contain polyphenols can induce endothelium-dependent vasorelaxation in the isolated aorta of rat which appears to be mediated by the NO-cGMP pathway (Fitzpatrick et al., 1993; Andriambeloson et al., 1997). Some defined polyphenols from anthocyanins and procyanidin oligomers mimicked this vasorelaxant activity (Freslon et al., 1997; Stoclet et al., 1999). Moreover, red wine polyphenols have been shown to inhibit the proliferation of vascular smooth muscle cells by the downregulation of cyclin A gene, expression (Ijima et al., 2000).

Taken together, these biological activities in vitro of polyphenols (antioxidant, antithrombotic and vasorelaxant) could explain the beneficial effects of wine in the prevention of cardiovascular diseases.

Biological activities after polyphenols or wine ingestion. Some in vivo studies have been carried out in animals and human volunteers in order to show these protective effects after red wine consumption in comparison with white wine or other alcoholized beverages. Klurfeld and Kritchewsky (1981) found that red wine notably reduced coronary atherosclerosis in rabbit. Intravenous and intragastric administration of red wine, grape juice and not white wine inhibited in vivo the platelet activity and thrombosis in canine coronary arteries (Demrow et al., 1995). After 2-4 months of beverage consumption, rats exhibited a reduction of platelet aggregation at the same rate by alcohol, red wine, and white wine (Ruf et al., 1995). By contrast, only red wine did not result in a rebound effect on platelets after deprivation of the alcoholic beverage. This protective effect appears to be essentially associated with polyphenols, which could counteract the known increased lipid peroxidation observed with alcohol drinking. Maffei Facino et al. (1999) found that a three-week supplementation of rat diet with procyanidins from *Vitis vinifera* seeds enhanced the plasma antioxidant capacity and exerted a cardioprotective activity (a reduction of myocardial post-ischaemic damage).

In human experimental studies, the ingestion of red wine led to an increase in the serum antioxidant activity which peaked after 1-2 hours (Maxwell et al., 1994 ; Whitehead et al., 1995). The daily consumption of red wine for two weeks resulted in a reduction in the susceptibility of LDL to oxidation (Kondo et al., 1994; Fuhrman et al., 1995; Nigdikar et al., 1998), as well as a decrease in platelet aggregation and an increase in HDL-cholesterol (Seigneur et al., 1990). In these studies, white wine exhibited no (or negative) activity. However, these results could not be reproduced by other authors (Sharpe et al., 1995; Pace-Asciak et al., 1996; De Rijke et al., 1996; Watkins and Bierenbaum, 1997).

It was also reported that coronary flow-velocity reserve and flow-mediated dilatation of the brachial artery increased specifically after the intake of red wine by volunteers, certainly due to the improvement of endothelial function and the vasorelaxing effects of polyphenols (Shimada et al.,1999; Agewall et al., 2000).

Cancer chemoprevention

The term 'chemoprevention' can be defined as the ingestion of non-toxic quantities of chemical agents (dietary or pharmaceutical) that are capable of preventing, inhibiting or reversing the process of carcinogenesis (Kinghorn et al., 1998). This relatively new concept is based on epidemiological studies which suggest a strong link between the environment (excluding genetic susceptibility) and cancer. There is evidence that 50 to 80% of human cancer is potentially preventable. In the United States, approximately 35% of cancer deaths are attributable to variation in diet. Indeed, carcinogenesis is a multistage process, and tumor development can occupy an important portion of the life span of an individual following long-term exposure to exogenous factors (such as dietary factors) that largely determine the incidence of carcinogenesis (Bailey et al., 1993; Alberts et al., 1999). This treatment offers numerous opportunities for intervention before development of malignant tumors.

The stages of carcinogenesis include (Suschetet et al., 1996; Alberts et al., 1999): (1) initiation (results from the exposure of normal cells to carcinogenic agents which cause a genetic change. Spontaneous mutations also occur through endogenous mechanisms such as free radical production, or through DNA replication errors, even though there may be no exposure to carcinogenic agents); (2) promotion (selective clonal expansion of initiated cells which leads to the appearance of a benign tumor); and (3) progression (genetic change and conversion of benign to malignant tumor).

Chemopreventive agents can act by various mechanisms on this process (Suschetet et al., 1996; Alberts et al., 1999): (1) anti-initiating activities (inhibition of carcinogen formation in the body and of uptake; inhibition of the metabolic activation of carcinogens by Phase I enzymes such as cytochrome P450 enzymes, or increase in their detoxification by Phase II enzymes such as transferases leading to an easier excretion; scavenging of free radicals and trapping ultimate carcinogens preventing their interactions with DNA). These compounds are referred to as 'blocking agents' due to their ability to prevent initiation. Other compounds which inhibit the carcinogenic process after initiation are classified as 'suppressing agents'; (2) anti-promoting and anti-proliferative activities (anti-oxidant activities because free radicals are also involved in the phases of promotion and progression; anti-inflammatory effects by inhibition of arachidonic acid metabolism, the inhibitors of cyclooxygenase and lipoxygenase being considered to be inhibitors of tumor promotion; and inhibition of cell proliferation by modulation of signal transduction, inhibition of polyamine metabolism, inhibition of oncogene activities, induction of differentiation, increase in intercellular communications, increase in apoptosis or inhibition of angiogenesis).

Clifford et al. (1996) reported that a diet with red wine solids rich in phenols delayed the onset of tumors in transgenic mice which spontaneously develop externally visible tumors without carcinogen pretreatment; intact catechin was found in the serum of these mice. In addition, several reports indicate that *trans*-resveratrol inhibits the proliferation of tumor cells (Mgbonyebi et al., 1998; Elattar and Virji, 1999) and also protein kinase activities (Jayatilake et al., 1993). Recently, the cancer chemopreventive activity of *trans*-resveratrol was established in various assays reflecting the three major stages of carcinogenesis (Jang et al., 1997; Jang and Pezzuto, 1998; Jang and Pezzuto, 1999). Thus, anti-initiation activity was demonstrated by its antioxidant and antimutagenic effects, inhibition of carcinogen bioactivation, and induction of phase II drug-metabolizing enzymes. Anti-promotion activity was shown by its anti-inflammatory effects and inhibition of cyclooxygenase activity. Anti-progression activity was also suggested by its induction of human promyelocytic leukemia cell differentiation.

To locate new and as yet undiscovered cancer chemopreventive agents in wine, we have evaluated the inhibitory effect of grape polyphenols on carcinogen-induced preneoplastic lesions in mouse mammary gland organ culture, which is a relevant model to evaluate the efficacy of potential chemopreventive agents (Mehta et al., 1997). Moreover, these polyphenols have been evaluated for their potential to inhibit cyclooxygenase. Indeed, considerable evidence has accumulated to suggest that this enzyme is important for tumorigenesis (Subbaramaiah et al., 1998). Of the two isozymes that lead to the formation of prostaglandins, cyclooxygenase-1, which is constitutively expressed in most tissues, is considered to be involved in physiological cell-cell signaling, whereas cyclooxygenase-2, which is induced by specific events in a limited number of cell types, appears to be involved in inflammation and mitogenesis (Kurumbail et al., 1996). Several mechanisms could explain the causality between cyclooxygenase-2 and cancer, such as a high level of prostaglandins in the tumors, which can favor the growth of malignant cells by increasing cell proliferation, permitting angiogenesis, and inhibiting immune surveillance (Jang and Pezzuto, 1999; Surh et al., 2001). Our results have been presented in Table 3.4 (Waffo Teguo et al., 2001a). *Trans*-astringin, *trans*-resveratrol, *trans*-piceatannol, catechin, and epicatechin 3-O-gallate exhibit significant inhibition of DMBA-induced preneoplastic lesions in mouse mammary gland organ cultures. In this investigation, we showed that two structural criteria appear to be important for this activity of the stibenoids, i.e. the presence of *trans*-geometrical isomerism and the absence of glycosylation. The mechanism by which polyphenolic compounds inhibit carcinogenesis has not been clearly established. As reported previously (Subbaramaiah et al., 1998), *trans*-resveratrol directly inhibits cyclooxygenase activity. However, *cis*-resveratrol is also active against this enzyme, but has exhibited no discernable activity on a carcinogen-induced preneoplastic lesion model. Thus, it seems that activity against the cyclooxygenase-2 target is insufficient to explain the action of *trans*-resveratrol, and so other mechanisms are involved. Moreover, our data show that *trans*-astringin and its aglycone *trans*-piceatannol inhibit the induction of preneoplastic lesions without any apparent activity against the cyclooxygenase-2 enzyme. These two stilbenoids could, thus, act by a novel mechanism. Interestingly, in contrast to *trans*-resveratrol and catechins, they are inactive against physiological cyclooxygenase-1 whose inhibition may lead to side effects such as gastric

Table 3.4 Effects of polyphenols from grape cell cultures on the inhibition of cyclooxygenase 1 and 2, and on the incidence of 7,12-dimethylbenz[a]anthracene-induced lesions in a BALB/c mouse mammary gland culture

Compound	COX-1		COX-2		MMOC[c]
	%Inhibition[a]	IC_{50} (μM)	%Inhibition[b]	IC_{50} (μM)	
crude extract[d]	nd[e]	nd	nd	nd	67.0
trans-astringin	0	nd	0	nd	68.8
trans-resveratroloside	98	4.8	0	nd	58.4
trans-piceid	95	10.6	7	nd	25.0
trans-resveratrol	96	14.9	85	32.2	87.5
trans-piceatannol	0	nd	nd	nd	76.9
cis-resveratroloside	94	17.8	15	nd	nd
cis-piceid	32	nd	13	nd	6.3
cis-resveratrol	86	55.4	70	50.2	53.9
(+)-catechin	95	1.4	36	nd	62.5
(–)-epicatechin	98	3.2	62	nd	50.0
epicatechin 3-*0*-gallate	97	7.5	nd	nd	62.5
procyanidin B2 3′-0-gallate	0	nd	nd	nd	30.8

[a] Compounds were tested at a concentration of 70 μg/mL. Inhibition of >70% was considered significant.
[b] Compounds were tested at a concentration of 50 μg/mL. Inhibition of >90% was considered significant.
[c] Inhibition of 7,12-dimethylbenz[a]anthracene-induced preneoplastic lesions with mouse mammary gland culture. Compounds were tested at a concentration of 10 μg/mL, and the results are expressed as percent inhibition. On the basis of historical controls, inhibition of >60% is considered significant.
[d] Crude extract of stilbenoids and flavans from grape cell cultures.
[e] nd : not determined.
Data from Waffo-Teguo et al. (2001a)

lesions and renal toxicity (Kurumbail et al., 1996). These results suggest that *trans*-astringin and *trans*-piceatannol are attractive new candidates for cancer chemoprevention. Moreover, two stilbene dimers isolated from grape cell culture, resveratrol (*E*)-dehydrodimer and its 11-O-β-D-glucopyranoside, demonstrated strong cyclooxygenase-2 inhibitory activity, but specificity is lacking in both cases (Waffo Teguo et al., 2001b).

The antioxidant properties of these vine polyphenols probably cannot explain their inhibition of carcinogenesis. Indeed, we have previously shown that *trans*-piceatannol (active in the mouse mammary organ culture assay) and epicatechin (inactive in the mouse mammary organ culture assay) exhibit similar marked antioxidant activities, which were assessed by their capacity to prevent Fe^{2+}-induced lipid peroxidation in microsomes; their action on Cu^{2+}-induced lipid peroxidation in LDL, and their capacity to scavenge free radicals (Table 3.2 ; Fauconneau et al., 1997). In addition, procyanidin B2 3′-O-gallate, again inactive in the mouse mammary organ culture model, proved to be the most effective procyanidin from grape seeds in trapping oxygen free radicals (Ricardo da Silva et al., 1991b).

Polyphenols seem to act by more specific mechanisms. *Trans*-resveratrol was found to be an inhibitor of cyclooxygenase-2 gene expression by inhibiting the protein kinase C signal transduction pathway at multiple levels (Subbaramaiah et al., 1998). Moreover, stilbenes, in particular *trans*-resveratrol and *trans*-piceatannol, inhibit several types of cellular kinases, either tyrosine kinases in which the aberrant activities can lead to cancer or serine/threonine kinases including protein kinase D (Jayatilake et al., 1993; Oliver et al., 1994; Wang et al., 1998; Stewart et al., 2000). On the other hand, *trans*-resveratrol has been found to suppress cell transformation and to induce apoptosis through the p53 or CD95 dependent pathways (Huang et al., 1999; Clément et al., 1998), but resveratrol appeared to have no effect on apoptosis in another model, in which it caused a decrease in ornithine decarboxylase activity, a key enzyme of polyamine biosynthesis, which is enhanced in cancer growth (Schneider et al., 2000). In addition, *trans*-resveratrol was shown to be a potent inhibitor of ribonucleotide reductase, a key enzyme in cell proliferation (Fontecave et al., 1998), and to reverse tumor-promoter-induced inhibition of gap-junctional intercellular communication (Nielsen et al., 2000).

In another work, we studied the effects of grapevine stilbenes on cultured human liver myofibroblasts (Godichaud et al., 2000). Indeed, liver myofibroblasts are major actors in the development of liver fibrosis and cancer progression. *Trans*-resveratrol markedly reduced proliferation and migration of myofibroblasts. It can also deactivate human liver myofibroblasts. Other stilbenes such as *trans*-piceid and *cis*-resveratrol were ineffective, but only *trans*-piceatannol exhibited a potent antiproliferative effect. Moreover, *trans*-resveratrol blocks hepatocyte growth factor-induced invasion of hepatocellular carcinoma cells by an unidentified post-receptor mechanism (De Ledinghen et al., 2001).

Based on the structural similarity of *trans*-resveratrol to the synthetic estrogen diethylstilbestrol, some authors have hypothesized that this compound might be a phytoestrogen. Divergent results have been reported on the growth of human breast cancer cells after resveratrol treatment and on estrogen agonism/antagonism of resveratrol (Gehm et al., 1997; Mgbonyebi et al., 1998; Basly et al., 2000).

In the case of flavans, catechins from green tea, in particular the galloylated forms are acknowledged as cancer chemopreventives in Japan (Fujiki, 1999). They had an inhibitory effect on the growth of human lung cancer cells, associated mainly with apoptosis induction and reduction of various cytokine levels, in particular, tumor necrosis factor α. The decreasing order of potency was epicatechin gallate, then epigallocatechin gallate and, finally, epigallocatechin. Epicatechin was not effective, but showed synergistic effects with epigallocatechin gallate. Saeki et al. (2000) showed that a pyrogallol-type structure—either in a B-ring or in 3-O-ester—appears to be a minimum requirement for apoptosis induction by catechins. In vivo in mouse, catechin and epicatechin inhibited lung metastasis (Menon et al., 1995), and exhibited antitumor-promoting effects in epidermis by inhibition of ornithine decarboxylase activity (Gali et al., 1994). The latter effect increased with the degree of polymerization of catechins (trimer > dimer > monomer).

Concerning flavonols, quercetin, which is present in low levels in wine, has been the theme of many studies (Suschetet et al., 1996; Ferry et al., 1996). This compound has antiproliferative activity in various in vitro models and is known to inhibit signal transduction targets such as tyrosine kinase, protein kinase C, and phosphatidyl inositol-3 kinase. Moreover, it induces apoptosis. Ferry et al. (1996) carried out a Phase I clinical trial in cancer patients with quercetin. The i.v. administration led to inhibition of lymphocyte tyrosine kinase activity and evident antitumor activity.

An anthocyanin fraction from red wine was recently found to inhibit the growth of human tumor cells in vitro (Kamei et al., 1998 ; Gomez-Cordoves et al., 2001). The aglycons (anthocyanidins) were more effective than the glycosides (anthocyanins), and cyanidin and delphinidin were most efficient than malvidin (Koide et al., 1996 ; Meiers et al., 2001). These compounds can exhibit a considerable antioxidative activity (Table 3.1; Bohm et al., 1998), and more specific properties such as inhibition of tyrosine kinase activity and antimutagenic activity (Yoshimoto et al., 1999; Meiers et al., 2001).

CONCLUSION

Wine is an abundant source of polyphenolic compounds, including mainly flavonoids (anthocyanins and flavanols) and stilbenes. The moderate consumption of wine—2 to 3 glasses per day—corresponds to about 300 to 1,200 mg of red wine polyphenols. Many of these compounds exhibit potent biological activities in vitro which might be able to favorably influence several physiological and pathological processes, and to provide a protective effect against cardiovascular diseases and cancer, as suggested by epidemiological studies. To assess the protective effects of dietary polyphenols more thoroughly, further investigations of their absorption, in vivo metabolism, and biological activity in humans must be carried out.

REFERENCES

Abbott, F.V. and Palmour, R.M., 1988, Morphine-6-glucuronide: Analgesic effects and receptor binding profile in rats. Life Sci., 43:1685-1695.

Adesanya, S.A., Nia, R., Martin, M.T. and Boukamcha, N., 1999, Stilbene derivatives from *Cissus quadrangularis*. J. Nat. Prod., 62:1994-1995.

Afanas'ev, I.B., Dorozhko, A.I., Brodskii, A.V., Kostyuk, V.A. and Potapovitch, A.I., 1989,

Chelating and free radical scavenging mechanisms of inhibitory actions of rutin and quercetin in lipid peroxidation. Biochem. Pharmacol., 38:1763-1769.

Agewall, S., Wright, S., Douglas, R.N., Whalley, G.A., Duxbury, M. and Sharpe, N., 2000, Does a glass of red wine improve endothelial function? Eur. Heart J., 21:74-78.

Alberts, D.S., Colvin, O.M., Conney, A.H., Ernster, V.L., Garber, J.E. and Greenwald, P., 1999, Prevention of cancer in the next millennium: Report of chemoprevention working group to the American Association for Cancer Research. Cancer Res., 59:4743-4758.

Alonso, E., Estrella, M.I. and Revilla, E., 1986, Chromatogr., **22**, 268-270. In : Waterhouse A.L. and Teissèdre P.L. 9 1997. Levels and phenolics in California varietal wines. In: 'Wine nutritional and therapeutic benefits', T.R. Watkies (ed.) ACS Symp. Set. 661: 12-13.

Andlauer, W., Kolb, J., Siebert, K. and Fürst, P., 2000, Assessment of resveratrol bioavailability in the perfused small intestine of the rat. Drug. Exptl. Clin. Res., 26:47-55.

Andriambeloson, E., Kleschyov, A.L., Muller, B., Beretz, A., Stoclet, J.C. and Andriantsitohaina, R., 1997, Nitric oxide production and endothelium-dependent vasorelaxation induced by wine polyphenols in rat aorta. Brit. J. Pharmacol., 120:1053-1058.

Arce, L., Tena, M.T., Rios, A., and Valcarcel, M., 1998, Determination of *trans*-resveratrol and other polyphenols in wines by a continuous flow sample clean-up system followed by capillary electrophoresis separation. Anal. Chim. Acta, 359:27-28.

Aumont, V., Krisa, S., Battaglia, E., Netter, P., Richard, T., Mérillon, J.M., Magdalou, J., and Sablovic, N., 2001, Regioselective and stereospecific glucuronidation of *trans*- and *cis*-resveratrol in human. Arch. Biochem. Biophys., 393:281-289.

Bailey, G.S., and Williams, D.E., 1993, Potential mechanisms for food-related carcinogens and anticarcinogens. Food Technol., 47:105-118.

Basly, J.P., Marre-Fournier, F., Le Bail, J.C., Habrioux, G., and Chulia, A.J., 2000, Estrogenic/antiestrogenic and scavenging properties of (*E*)- and (*Z*)-resveratrol. Life Sci., 66:769-777.

Bavaresco, L., Fregoni, E., Cantu, E., and Trevisan, M., 1999, Stilbene compounds: From the grapevine to wine. Drug. Exp. Clin. Res., 25:57-63.

Bertelli, A.A.E., Giovannini, L., De Caterina, R., Bernini, W., Migliori, M., Fregoni, M., Bavaresco, L., and Bertelli, A., 1996, Antiplatelet activity of *cis*-resveratrol. Drug. Exp. Clin. Res., 22:61-63.

Block, G., Patterson, B., and Subar, A., 1992, Fruits, vegetables, and cancer prevention: A review of the epidemiological evidence. Nutr. Cancer, 18:1-29.

Blot, W.J., McLaughlin, J.K., Winn, D.M., Austin, D.F., Greenberg, R.S., Preston-Martin, S., Bernstein, L., Schoenberg, J.B., Stemhagen, A., and Fraumeni, J.J.F., 1988, Smoking and drinking in relation to oral and pharyngeal cancer. Cancer Res., 48:3282-3287.

Bohm, H., Boeing, H., Hempel, J., Raab, B., and Kroke, A., 1998, Flavonols, flavone and anthocyanins as natural antioxidants of food and their possible role in the prevention of chronic diseases. Z. Ernährungswiss., 37:147-163.

Bombardelli, E., and Morazzoni, P., 1995, *Vitis vinifera* L. Fitoterapia, 66:291-316.

Boss, P.K., Davies, C., and Robinson, S.P., 1996, Analysis of the expression of anthocyanin pathway genes in developing *Vitis vinifera* L. cv Shiraz grape berries and the implication for pathway regulation. Plant Physiol., 111:1059-1066.

Bourhis, M., Théodore, N., Weber, J.F. and Vercauteren, J., 1996, Isolation and identification or (*Z*) and (*E*)-ε-viniferins form stalks of *Vitis vinifera*. Polyphenols Communications 96, 18th International Conference on Polyphenols, July 15-18, 1996, Bordeaux, 1:43-44.

Bourzeix, M., Weyland, D., Heredia, N. and Desfeux, C., 1986, Etude des catéchines et des procyanidols de la grappe de raisin, du vin et d'autres dérivés de la vigne. Bull. O. I. V., 59:1171-1254.

Cao, G., Muccitelli, H.U., Sanchez-Moreno, C., and Prior, R.L., 2001, Anthocyanins are absorbed in glycated forms in elderly women: A pharmacokinetic study. Am. J. Clin. Nutr., 73:920-926.

Carando, S., Teissedre, P.L., Waffo Teguo, P., Cabanis, J.C., Deffieux, G., and Mérillon, J.M.,

1999, High-performance liquid chromatography coupled with fluorescence detection for the determination of *trans*-astringin in wine. J. Chromatogr. A, 64:748-756.

Cichewicz, R.H., Kouzi, S.A., and Hamann, M.T., 2000, Dimerization of resveratrol by the grapevine pathogen *Botrytis cinerea*. J. Nat. Prod., 36:29-33.

Clément, M.V., Hirpara, J.L., Chawdhury, S.H., and Pervaiz, S., 1998, Chemopreventive agent resveratrol, a natural product derived from grapes, triggers CD95 signaling-dependent apoptosis in human tumor cells. Blood, 92:996-1002.

Clifford, A.J., Ebeler, S.E., Ebeler, J.D., Bills, N.D., Hinrichs, S.H., and Teissèdre, P.L., 1996, Delayed tumor onset in transgenic mice fed an amino acid-based diet supplemented with red wine solids. Am. J. Clin. Nutr., 64:748-756.

Commenges, D., Scotet, V., Renaud, S., Jacqmin-Gadda, H., Barberger-Gateau, P., and Dartigues, J.F., 2000, Intake of flavonoids and risk of dementia. Eur. J. Epidemiol., 16:357-363.

Coombe, B.G., 1989, The grape berry as a sink. Acta Hortic., 239:149-157.

Cox, D.A., and Cohen, M.L., 1996, Effects of oxidized low-density lipoprotein on vascular contraction and relaxation: Clinical and pharmacological implications in atherosclerosis. Pharmacol. Rev., 48:3-19.

Creasy, L.L., and Coffee, M., 1988, Phytoalexin production potential of grape berries. J. Amer. Soc. Hort. Sci., 113:230-234.

Crespy, V., Morand, C., Besson, C., Manach, C., Démigné, C., and Rémésy, C., 2001, Comparison of the intestinal absorption of quercetin, phloretin and their glucosides in rats. J. Nutr., 131:2109-2114.

De Ledinghen, V., Monvoisin, A., Neaud, V., Krisa, S., Payrastre, B., Bedin, C., Desmoulière, A., Bioulac-Sage, P., and Rosenbaum, J., 2001, *Trans*-resveratrol, a grapevin-derived polyphenol, blocks hepatocyte growth factor-induced invasion of hepatocellular carcinoma cells. Int. J. Oncol., 19:83-88.

De Lorgeril, M., and Salen, P., 1998, Lessons from Albania. Lancet, 351:1440.

De Rijke, J., Demacker, P.N.M., Assen, N.A., Sloots, L.M., Katan, M.B. and Stalenhoef, A.F.H., 1996, Red wine consumption does not affect oxidizability of low-density lipoproteins in volunteers. Am. J. Clin. Nutr., 63:329-334.

Demrow, H.S., Slane, P.R., and Folts, J.D., 1995, Administration of wine and grape juice inhibits *in vivo* platelet activity and thrombosis in stenosed canine coronary arteries. Circulation, 91:1182-1188.

Dobrydneva, Y., Williams, R.L., and Blackmore, P.F., 1999, *Trans*-resveratrol inhibits calcium influx in thrombin-stimulated human platelets. Brit. J. Pharmacol., 128:149-157.

Donovan, J.L., Bell, J.R., Kasim-Karakas, S., German, J.B., Walzem, R.L., Hansen, R.J. and Waterhouse, A.L., 1999, Catechin is present as metabolites in human plasma after consumption of red wine. J. Nutr., 129:1662-1668.

Donovan, J.L., Crespy, V., Manach, C., Morand, C., Besson, C., Scalbert, A., and Rémésy, C., 2001, Catechin is metabolized by both the small intestine and liver of rats. J. Nutr., 131:1753-1757.

Douillet-Breuil, A.C., Jeandet, P., Adrian, M., and Bessis, R., 1999, Changes in the phytoalexin content of various *Vitis* spp. in response to ultraviolet C elicitation. J. Agric. Food Chem., 47:4456-4461.

Elattar, T.M.A., and Virji, A.S., 1999, Modulating effect of resveratrol and quercetin on oral cancer cell growth and proliferation. Anti-Cancer Drug., 10:187-193.

Esterbauer, H., Wäg, G., and Puhl, H., 1993, Lipid peroxidation and its role in atherosclerosis. Med. Bull., 49:566-576.

Farchi, G., Fidanza, F., Mariotti, S., and Menotti, A., 1992, Alcohol and mortality in the Italian rural cohorts of the seven countries study. Int. J. Epidemiol., 21:74-82.

Fauconneau, B., Waffo-Téguo, P., Huguet, F., Barrier, L., Decendit, A., and Mérillon, J.M., 1997, Comparative study of radical scavenger and antioxidant properties of phenolic compounds from *Vitis vinifera* cell cultures using *in vitro* tests. Life Sci., 61:2103-2110.

Ferry, D.R., Smith, A., Malkhandi, J., Fyfe, D.W., De Takats, P.G., Anderson, D., Baker, J., and Kerr, D.J., 1996, Phase I Clinical Trial of the flavonoid quercetin: Pharmacokynetics and evidence for *in vivo* tyrosine kinase inhibition. Clin. Cancer Res., 2:659-668.

Fillion, L., Ageorges, A., Picaud, S., Coutos-Thevenot, P., Lemoine, R., Romieu, C., and Delrot, S., 1999, Cloning and expression of hexose transporter gene expressed during the ripening of grape berry. Plant Physiol., 120:1083-1093.

Fitzpatrick, D.F., Hirschfield, S.L., and Coffey, R.G., 1993, Endothelium-dependent vasorelaxing activity of wine and other grape products. Am. J. Physiol., 265:774-778.

Fontecave, M., Lepoivre, M., Elleingand, E., Gerez, C., and Guittet, O., 1998, Resveratrol, a remarkable inhibitor of ribonucleotide reductase. FEBS Lett., 421:277-279.

Frankel, E., Kanner, J., German, J., Parks, E. and Kinsella, J., 1993, Inhibition of oxidation of human low-density lipoprotein by phenolic substances in red wine. Lancet, 341:454-457.

Frémont, L., 2000, Minireview: Biological effects of resveratrol. Life Sci., 66:663-673.

Freslon, J.I., Mendes, A., Desgranges, C., and Vercauteren, J., 1997, Procyanidin oligomers from grape seed induce no-dependent relaxation of rat aortic rings *via* $P2Y_1$ reception activation. Brit. J. Pharmacol., 122:200P.

Fuhrman, B., Lavy, A., and Aviram, M., 1995, Consumption of red wine with meals reduces the susceptibility of human plasma and low-density lipoprotein to lipid peroxidation. Am. J. Clinic. Nutr., 61:549-554.

Fujiki, H., 1999, Two stages of cancer prevention with green tea. J. Cancer Res. Clin. Oncol., 125:589-597.

Gali, H., Perchellet, E., Gao, X., Karchesy, J., and Perchellet, J., 1994, Comparison of the inhibitory effects of monomeric, dimeric, and trimeric procyanidins on the biochemical markers of skin tumor promotion in mouse epidermis in vivo. Planta Med., 60:235-239.

Gehm, B.D., McAndrews, J.M., Chien, P.Y., and Jameson, J.L., 1997, Resveratrol, a polyphenolic compound found in grapes and wine, is an agonist for the estrogen receptor. Proc. Natl. Acad. Sci., 94:14138-14143.

Godichaud, S., Krisa, S., Couronné, B., Dubuisson, L., Mérillon, J.M., Desmoulière, A. and Rosenbaum, J., 2000, Deactivation of cultured human liver myofibroblasts by *trans*-resveratrol, a grapevine-derived polyphenol. Hepatology, 31:922-931.

Goldberg, D.M., Ng, E., Karumanchiri, A., Yan, J., Diamandis, E.P., and Soleas, G.J., 1995, Assay of resveratrol glucosides and isomers in wine by direct-injection high-performance liquid chromatography. J. Chromatogr. A, 708:89-98.

Gomez-Cordovez, C., Bartholome, B., Vieira, W., and Virador, V.M., 2001, Effects of wine phenolics and sorghum tannins on tyrosinase activity and growth of melanoma cells. J. Agric. Food Chem., 49:1620-1624.

Gorinstein, S., Caspi, A., Zemser, M.S., and Trakhtenberg, M.D. 2000, Comparative contents of some phenolics in beer, red and white wines. Nutr. Res., 20:131-139.

Graefe, E.U., Wittig, J., Mueller, S., Riethling, A.K., Uehmleke, B., Drewelow, B., Pforte, H., Jacobasch, G., Derendorf, H. and Veit, M., 2001, Pharmacokinetics and bioavailability of quercetin glycosides in humans. J. Clin Pharmacol., 41:492-499.

Gronbaek, M., Becker, U., Johansen, D., Gottschau, A., Schnohr, P., Hein, H.O., Jensen, G. and Sorensen, T.I.A., 2000, Type of alcohol consumed and mortality from all causes, coronary heart disease, and cancer. Ann. Intern. Med., 133:411-419.

Gronbaek, M., Becker, U., Johansen, D., Tonnesen, H., Jensen, G., and Sorensen, T.I., 1998, Population based cohort study of the association between alcohol intake and cancer of the upper digestive tract. Brit. Med. J., 317:844-847.

Gronbaek, M., Deis, A., Sorensen, T., Becker, U., Schnohr, P., and Jensen, G., 1995, Mortality associated with moderate intakes of wine, beer, or spirits. Brit. Med. J., 310:1165-1169.

Gu, X., Chu, Q., O'Dwyer, M., and Zeece, M., 2000, Analysis of resveratrol in wine by

capillary electrophoresis. J. Chromatogr. A, 881:471-481.
Halliwell, B., and Gutteridge, J.M.C., 1994, Free radicals and antioxidants. Nutr. Rev., 52:253-265.
Harborne, J.B., and Williams, C.A., 2000, Advances in flavonoid research since 1992. Phytochem., 55:481-504.
Haslam, E., 1996, Natural polyphenols (vegetable tannins) as drugs: Possible modes of action. J. Nat. Prod., 59:205-215.
Hertog, M., Hollman, P., Katan, M., and Kromhout, D., 1993, Intake of potentially anticarcinogenic flavonoids and their determinants in adults in The Netherlands. Nutr. Cancer., 20:21-29.
Hill, M.J., 1995, Diet and cancer: A review of scientific evidence. Eur. J. Cancer Prev., 4:3-42.
Huang, C., Ma, W., Goranson, A. and Dong, Z., 1999, Resveratrol suppresses cell transformation and induces apoptosis through a p53-dependent pathway. Carcinogenesis, 20:237-242.
Ijima, K., Yoshizumi, M., Hashimoito, M., Kim, S., Eto, M., Ako, J., Liang, Y.Q., Sudoh, N., Hosoda, K., Nakahara, K., Toba, K. and Ouchi, Y., 2000, Red wine polyphenols inhibit proliferation of vascular smooth muscle cells and downregulate expression of cyclin A gene. Circulation, 101:805-811.
Jang, M. and Pezzuto, J.M., 1999, Cancer chemopreventive activity of resveratrol. Drug. Exp. Clin. Res., 25:65-77.
Jang, M., Cai, L., Udeani, G.O., Slowing, K.V., Thomas, C.F., Beecher, C.W.W., Fong, H.H.S., Farnsworth, N.R., Kinghorn, A.D., Mehta, R.G., Moon, R.C. and Pezzuto, J.M., 1997, Cancer chemopreventive activity of resveratrol, a natural product derived from grapes. Science, 275:218-220.
Jang, M., and Pezzuto, J.M., 1998, Resveratrol blocks eicosanoid production and chemically induced cellular transformation. Pharm. Biol., 326:28-34.
Jayatilake, G.S., Jayasuriya, H., Lee, E.S., Koonchanok, N.M., Geahlen, R.L., Ashendel, C.L., McLaughlin, J.L. and Chang, C.J., 1993, Kinase inhibitors from *Polygonum cuspidatum*. J. Nat. Prod., 56:1805-1810.
Jeandet, P., Bessis, R., Maume, B.F., Meunier, P., Peyron, D., and Trollat, P., 1995, Effect of enological practices on the resveratrol isomer content of wine. J. Agric. Food Chem., 43:316-319.
Jeandet, P., Breuil, A.C., Adrian, M., Weston, L.A., Debord, S., Meunier, P., Maune, G., and Bessis, R., 1997, HPLC analysis of grapevine phytoalexins coupling photodiode array detection and fluorimetry. Anal. Chem., 69:5172-5177.
Kamei, H., Hashimoto, Y., Kide, T., Kojima, T. and Hasegawa, M., 1998, Anti-tumor effect of methanol extracts from red and white wines. Cancer Biother. Radiopharm., 13:447-452.
Kennedy, J.A. and Waterhouse, A.L., 2000, Analysis of pigmented high-molecular-mass grape phenolics using ion-pair, normal-phase high-performance liquid chromatography. J. Chromatogr. A, 866:25-34.
Kinghorn, A.D., Fong, H.H.S., Farnsworth, N.R., Mehta, R.G., Moon, R.C., Moriarty, R.M. and Pezzuto, J.M., 1998, Cancer chemopreventive agents discovered by activity-guided fractionation: A review. Curr. Org. Chem., 2:597-612.
Klurfeld, D.M., and Kritchevsky, D., 1981, Differential effects of alcoholic beverages on experimental atherosclerosis in rabbits. Exp. Mol. Pathol., 34:62-71.
Koide, T., Kamei, H., Hashimoto, Y., Kojima, T., and Hasegawa, M., 1996, Anti-tumor effect of hydrolized anthocyanin from grape rinds and red wine. Cancer Biother. Radiopharm., 11:273-277.
Kondo, K., Matsumoto, A., Kurata, H., Tanahashi, H., Koda, H., Amachi, T. and Itakura, H., 1994, Inhibition of oxidation of low-density lipoprotein with red wine. Lancet, 344:1152.
Korhammer, S., Reniero, F. and Mattivi, F., 1995, An oligostilbene from *Vitis* roots. Phytochem., 38:1501-1504.

Krisa, S., Vitrac, X., Decendit, A., Larronde, F., Deffieux, G. and Mérillon, J.M., 1999a, Obtaining *Vitis vinifera* cell cultures producing higher amounts of malvidin-3-0-β-glucoside. Biotechnol. Lett., 21:497-500.

Krisa, S., Waffo Teguo, P., Decendit, A., Deffieux, G., Vercauteren, J., and Mérillon, J.M., 1999b, Production of ^{13}C-labelled anthocyanins by *Vitis vinifera* cell suspension cultures. Phytochem., 51:651-656.

Krisa, S., Larronde, F., Budzinski, H., Decendit, A., Deffieux, G., and Mérillon, J.M., 1999c, Stilbene production by *Vitis vinifera* cell suspension cultures: methyl jasmonate induction and ^{13}C biolabeling J. Nat. Prod., 62:1688-1690.

Kuhnle, G., Spencer, J.P.E., Chowrimootoo, G., Schroeter, H., Debnam, E.S., Srai, S.K.S., Rice-Evans, C., and Hahn, U., 2000, Resveratrol is absorbed in the small intestine as resveratrol glucuronide. Biochem. Biophys. Res. Co., 272:212-217.

Kurumbail, R.G., Stevens, A.M., Gierse, J.K., McDonald, J.J., Stegeman, R.A. and Pak, J.Y., 1996, Structure basis for selective inhibition of cyclooxygenase-2 by antiinflammatory agents. Nature, 384:644-648.

Lamuela-Raventós, R.M., and Waterhouse, A.L., 1993, Occurrence of resveratrol in selected California wines by a new HPLC method. J. Agric. Food Chem., 41:521-523.

Lamuela-Raventos, R.M., Romero-Pérez, A.N., Waterhouse, A.L. and de la Torre-Boronat, M.C., 1995, Direct HPLC analysis of *cis*- and *trans*-resveratrol and piceid isomers in spanish red *Vitis vinifera* wines. J. Agric. Food Chem., 43:281-283.

Langcake, P., Cornford, C.A., and Pryce, R.J., 1979, Identification of pterostilbene as a phytoalexin from *Vitis vinifera* leaves. Phytochem., 18:1025-1027.

Langcake, P., and Pryce, R.J., 1977, A new class of phytoalexins from grapevines. Experientia, 13:151-152.

Longnecker, M.P., Newcomb, P.A., Mittendorf, R., Greenbreg, E.R., Clapp, R.W., Bogdan, G.F., Baron, J., MacMahon, B., and Willett, C.W., 1995, Risk of breast cancer in relation to lifetime alcohol consumption. J. Natl. Cancer I., 87:923-929.

Lu, Y., and Foo, L.Y., 1999, The polyphenol constituents of grape pomace. Food. Chem., 65:1-8.

Macfarlane, G.J., Zheng, T., Marshall, J.R., Boffetta, P., Niu, S., Brasure, J., Merletti, F. and Boyle, P., 1995, Alcohol, tobacco, diet and the risk of oral cancer: A pooled analysis of three case-control studies. Eur. J. Cancer, 31B:181-187.

Maffei Facino, R., Carini, M., Aldini, G., Bombardelli, E., Morazzoni, P., and Morelli, R., 1994, Free radicals scavenging action and anti-enzyme activities of procyanidines from *Vitis vinifera*. Arzneimitt.-Forsch., 44:592-601.

Maffei Facino, R.M., Carini, M., Aldini, G., Berti, F., Rossoni, G., Bombardelli, E. and Morazzoni, P., 1999, Diet enriched with procyanidins enhances antioxidant activity and reduces myocardial post-ischaemic damage in rats. Life Sci., 64:627-642.

Mattivi, F., 1993, Solid phase extraction of *trans*-resveratrol from wines for HPLC analysis. Z. Lebensm. Unters. F. A., 196:522-525.

Mattivi, F., Reniero, F., Scienza, A. and Farago, S., 1998, Resveratrol oligomers and other stilbenoids are main phenolic constituents of *Vitis* roots. in 'XIXèmes Journées Internationales d'Étude des Polyphenols'. Charbonnier, F., Delacotte J.M. and Rolando C. (eds). Polyphenol Communications 98, Lille, 2:473-474.

Maxwell, S., Cruickshank, A. and Thorpe, G., 1994, Red wine and antioxidant activity in serum. Lancet, 344:193-194.

McMurtreym K.D., 1997, Resveratrol in wine. in 'Wine Nutritional and Therapeutic Benefits', T.R. Watkins (ed.), ACS Symp. Ser. 661:44-55.

Mehta, R.G., Hawthorne, M.E. and Steele, V.E., 1997, Induction of carcinogen-induced precancerous lesions in mouse mammary gland organ culture. Meth. Cell. Sci., 19:19-24.

Meiers, S., Kemeny, M., Weynand, U., Gaspar, R., von Angerer, E. and Marko, D., 2001, The anthocyanidins cyanidin and delphinidin are potent inhibitors of the epidermal growth-factor receptor. J. Agric. Food Chem., 49:958-962.

Menon, L.G., Kuttan, R., and Kuttan, G., 1995, Inhibition of lung metastasis in mice induced by B16F10 melanoma cells by polyphenolic compounds. Cancer Lett., 95:221-225.

Meyer, A.S., Yi, O.S., Pearson, D.A., Waterhouse, A.L., and Frankel, E.N., 1997, Inhibition of human low-density lipoprotein oxidation in relation to composition of phenolic antioxidants in grapes (*Vitis vinifera*). J. Agric. Food Chem., 45:1638-1643.

Mgbonyebi, O.P., Russo, J., and Russo, I.H., 1998, Antiproliferative effect of synthetic resveratrol on human breast epithelial cells. Int. J. Oncol., 12:865-869.

Miyazawa, T., Nakagawa, K., Kudo, M., and Someya, K., 1999, Direct intestinal absorption of red fruit anthocyanins, cyanidin-3-glucoside and cyanidine-3,5-diglucoside, into rats and humans. J. Agric. Food Chem., 47:1083-1091.

Morand, C., Crespy, V., Manach, C., Besson, C., Demigné, C., and Rémésy, C., 1998, Plasma metabolites of quercetin and their antioxidant properties. Am. J. Physiol., 275:212-219.

Morel, I., Lescoat, G., Cogrel, P., Sergent, O., Pasdeloup, N., Brissot, P., Cillard, P. and Cillard, J., 1993, Antioxidant and iron-chelating activities of the flavonoids catechin, quercetin and diosmetin on iron-loaded rat hepatocyte cultures. Biochem. Pharm., 45:13-19.

Nielsen, M., Ruch, R.J. and Vang, O., 2000, Resveratrol reverses tumor-promoter-induced inhibition of gap-junctional intercellular communication. Biochem. Biophys. Res. Comm., 275:804-809.

Nigdikar, S.V., Williams, N.R., Griffin, B.A., and Howard, A.N., 1998, Consumption of red wine polyphenols reduces the susceptibility of low-density lipoproteins to oxidation *in vivo*. Am. J. Clin. Nutr., 68:258-265.

Oliver, J.M., Burg, D.L., Wilson, B.S., McLaughlin, J.L., and Geahlen, R.L., 1994, Inhibition of mast cell Fc ∊ R1-mediated signaling an effector function by the Syk-selective inhibitor, piceatannol. J. Biol. Chem., 269:29697-29703.

Olthof, M.R., Hollman, P.C., Vree, T.B., and Katan, M.B., 2000, Bioavailabilities of quercetin-3-glucoside and quercetin-4'-glucoside do not differ in humans. J. Nutr., 130:1200-1203.

Orgogozo, J.M., Dartigues, J.F., Lafont, S., Letenneur, L., Commenges, D., Salamon, R., Renaud, S. and Breteler, M.B., 1997, Wine consumption and dementia in the elderly: A prospective community study in the Bordeaux area. Rev. Neurol., 153:185-192.

Orsini, F., Pelizzoni, F., Verotta, L. and Aburjai, T., 1997, Isolation, synthesis, and antiplatelet aggregation activity of resveratrol 3-O-β-D glucopyranoside and related compounds. J. Nat. Prod., 60:1082-1087.

Pace-Asciak, C.R., Hahn, S., Diamandis, E.P., Soleas, G. and Goldberg, D.M., 1995, The red wine phenolics *trans*-resveratrol and quercetin block human platelet aggregation and eicosanoid synthesis: Implications for protection against coronary heart disease. Clin. Chim. Acta, 235:207-219.

Pace-Asciak, C.R., Rounova, O., Hahn, S.E., Diamandis, E.P., Elefiherios, P. and Goldberg, D.M., 1996, Wines and grape juices as modulators of platelet aggregation in healthy human subjects. Clin. Chim. Acta, 246:163-182.

Paganga, G., Miller, N., and Rice-Evans, C., 1999, The polyphenolic content of fruit and vegetables and their antioxidant activities. What does a serving constitute? Free Rad. Res., 30:153-162.

Palomino, O., Gomez-Serranillos, M.P., Slowing, K., Carretero, E. and Villar, A., 2000, Study of polyphenols in grape berries by reversed phase high-performance liquid chromatography. J. Chromatogr. A, 870:449-451.

Pezet, R., and Pont, V., 1988, Mise en évidence de ptérostilbène dans les grappes de *Vitis vinifera*. Plant Physiol. Bioch., 26:603-607.

Pezet, R., Pont, V., and Cuenat, P., 1994, Method to determine resveratrol and pterostilbene in grape berries and wines using high-performance liquid chromatography and highly sensitive fluorimetric detection. J. Chromatogr. A, 663:191-197.

Pezet, R., and Cuenat, P., 1996, Resveratrol in wine: Extraction from skin during fermentation and post-fermentation standing of must from Gamay grapes. Am. J. Vitic., 47:287-290.

Polette, A., Lemaitre, D., Lagarde, M. and Véricel, E., 1996, N-3 fatty acid-induced lipid peroxidation in human platelets is prevented by catechins. Thromb. Haemostasis, 75:945-49.

Prieur, C., Rigaud, J., Cheynier, V. and Moutounet, M., 1994, Oligomeric and polymeric procyanidins from grape seeds. Phytochem., 36:781-784.

Rafat-Husain, S., Cillard, J. and Cillard, P., 1987, Hydroxyl radical scavenging activity of flavonoids. Phytochem., 26:2489-2491.

Renaud, S., and De Lorgeril, M., 1992, Wine, alcohol, platelets, and the French paradox for coronary heart disease. Lancet, 339:1523-26.

Renaud, S., Guégen, R., Schenker, J., and D'Houtaud, A., 1998, Alcohol and mortality in middle-aged men from eastern France. Epidemiology, 9:184-188.

Renaud, S., and Guéguen, R., 1998, The French paradox and wine drinking. in 'Alcohol and Cardiovascular diseases'. Novartis Foundation Symposium 216. Wiley, Chichester., pp.208-222.

Renaud, S., Guéguen, R., Siest, G., and Salomon, R., 1999, Wine, beer, and mortality in middle-aged men from eastern France. Arch. Intern. Med., 159:1865-1870.

Ribeiro de Lima, M.T., Waffo-Teguo, P., Teissèdre, P.L., Pujolas, A., Vercauteren, J., Cabanis, J.C. and Mérillon, J.M., 1999, Determination of stilbenes (*trans*-astringin, *cis*- and *trans*-piceid, and *cis*- and *trans*-resveratrol) in Portuguese wines. J. Agr. Food Chem., 47:2666-2670.

Ribereau-Gayon, P., 1982, The anthocyanins of grapes and wines. in: 'Anthocyanins as Food Colors', P. Markakis (ed), Academic Press, New York., pp.209-244.

Ricardo da Silva, J.M., Belchior, A.P., Spranger, M.I. and Bourzeix, M., 1992, Oligomeric procyanidins of three grapevine varieties and wines from Portugal. Sci. Aliment., 12:223-237.

Ricardo da Silva, J.M., Rigaud, J., Cheynier, V., Cheminat, A., and Moutounet, M., 1991a, Procyanidin dimers and trimers from grape seeds. Phytochem., 30:1259-64.

Ricardo da Silva, J.M., Darmon, N., Fernandez, Y. and Mitjavila, S., 1991b, Oxygen free radical scavenger capacity in aqueous models of different procyanidins from grape seeds. J. Agric. Food Chem., 39:1549-1552.

Rice-Evans, C.A., Miller, N.J., Bolwell, P.G., Bramley, P.M. and Pridham, J.B., 1995, The relative antioxidant activities of plant-derived polyphenolic flavonoid. Free Rad. Res., 22:375-383.

Rimm, E.B., Klatsky, A., Grobbee, D., and Stampfer, M.J., 1996, Review of moderate alcohol consumption and reduced risk of coronary heart disease: Is the effect due to beer, wine, or spirits? British Med. J., 312:731-736.

Robak, J. and Gryglewski, R.J., 1988, Flavonoids are scavengers of superoxide anions. Biochem. Pharmacol., 37:837-841.

Roggero, J.P., 1997, Chromatography of phenolics in wine. Biofactors, 6:441-443.

Romero-Pérez, A.I., Ibern-Gomez, M., Lamuela-Raventos, R.M., and De la Torre-Boronat, M.C., 1999, Piceid, the major resveratrol derivative in grape juices. J. Agr. Food Chem., 47:1533-1536.

Ross, R., 1993, The pathogenesis of atherosclerosis: A perspective for the 1990s. Nature, 362:801-809.

Ruf, J.C., Berger, J.L. and Renaud, S., 1995, Platelet rebound effect of alcohol withdrawal and wine drinking in rats. Arterioscler. Thromb. Vasc. Biol., 15:140-144.

Saeki, K., Hayakawa, S., Isemura, M., and Miyase, T., 2000, Importance of a pyrogallol-type structure in catechin compounds for apoptosis-inducing activity. Phytochem., 53:391-394.

Schneider, Y., Vincent, F., Duranton, B., Badolo, L., Gossé, F., Bergmann, C., Seiler, N., and

Raul, F., 2000, Anti-proliferative effect of resveratrol, a natural component of grapes and wine, on human colonic cancer cells. Cancer Lett., 158:85-91.

Seigneur, M., Bonnet, J., Dorian, B., Benchimol, D., Drouillet, F., Gouverneur, G., Larrue, J., Crockett, R., Boisseau, M., Ribéreau-Gayon, P., and Bricaus, H., 1990, Effect of the consumption of alcohol, white wine, and red wine on platelet function and serum lipids. J. Appl. Cardiol., 5:215-222.

Serafini, M., Maiani, G., and Ferro-Luzzi, A., 1997, Effect of ethanol on red wine tannin-protein (BSA) interactions. J. Agr. Food Chem., 45:3148-3151.

Sesink, A.L., O'Leary, K.A., and Hollman, P.C., 2001, Quercetin glucuronides but not glucosides are present in human plasma after consumption of quercetin-3-glucoside or quercetin-4'-glucoside. J. Nutr., 131:1938-1941.

Sharpe, P.C., Mcgrath, L.T., Mcclean, E., Young, I.S. and Archbold, G.P.R., 1995, Effect of red wine consumption on lipoprotein (a) and other risk factors for atherosclerosis. Q. J. Med., 88:101-108.

Siemann, E.H. and Creasy, L.L., 1992, Concentration of the phytoalexin resveratrol in wine. Am. J. Enol. Vitic., 43:49-52.

Shimada, K., Watanabe, H., Hosoda, K., Takeuchi, K. and Yoshikawa, J., 1999, Effect of red wine on coronary flow-velocity reserve. Lancet, 354:1002.

Soleas, G.J., Yan, J., and Goldberg, D.M., 2001, Ultrasensitive assay for three polyphenols (catechin, quercetin and resveratrol) and their conjugates in biological fluids utilizing gas chromatography with mass selective detection. J. Chromatogr. B, 757:161-172.

Sotheeswaran, S. and Pasupathy, V., 1993, Distribution of resveratrol oligomers in plants. Phytochem., 32:1083-1092.

Souquet, J.M., Cheynier, V., Brossaud, F. and Moutounet, M., 1996, Polymeric proanthocyanidins from grape skins. Phytochem., 43:509-512.

Souquet, J.M., Labarbe, B., Cheynier, V. and Moutounet, M., 2000, Phenolic composition of grape stems. J. Agr. Food Chem., 48:1076-1080.

Spencer, J.P., Schroeter, H., Shenoy, B., Srai, S.K., Debnam, E.S. and Rice-Evans, C., 2001, Epicatechin is the primary bioavailable form of the procyanidin dimers B2 and B5 after transfer across the small intestine. Biochem. Biophys. Res. Comm., 285:588-593.

Stewart, J.R., Christman, K.L., and O'Brian, C.A., 2000, Effects of resveratrol on the autophosphorylation of phorbol ester-responsive protein kinases. Biochem. Pharmacol., 60:1355-1359.

Stoclet, J.C., Kleschyov, A., Andriambeloson, E., Diebolt, M., and Andriantsitohaina, R., 1999, Enthelial no release caused by red wine polyphenols. J. Physiol. Pharmacol., 50:535-540.

Subbaramaiah, K., Chang, W.J., Michaluart, P., Telang, N., Tanabe, T. and Inoue, M., 1998, Resveratrol inhibits cyclooxygenase-2 transcription and activity in phorbol. J. Biol. Chem., 273:21875-21882.

Surh, Y.J., Chun, K.S., Cha, H.H., Han, S.S., Keum, Y.S., Park, K.K. and Lee, S.S., 2001, Molecular mechanisms underlying chemopreventive activities of anti-inflammatory phytochemicals: Down-regulation of COX-2 and iNOS through suppression of NF-KB activation. Mutat. Res., 480-481:243-268.

Suschetet, M., Siess, M.H., Le Bon, A.M., and Canivenc-Lavier, M.C., 1996, Anticarcinogenic properties of some flavonoids. *in:* 'Polyphenols 96', Vercauteren J., Chèze C. and Triaud J. (eds), INRA, Paris 1998 *(Les Colloques n° 87)*, Bordeaux:165-204.

Tedesco, I., Russo, M., Russo, P., Iacomino, G., Russo, G.L., Carraturo, A., Faruolo, C., Moio, L., and Palumbo, R., 2000, Antioxidant effect of red wine polyphenols on red blood cells. J. Nutr. Biochem., 11:114-119.

Teissèdre, P.L., Frankel, E.N., Waterhouse, A.L., Peleg, H., and German, J., 1996, Inhibition of *in vitro* human LDL oxidation by phenolic antioxidants from grapes and wines. J. Sci. Food Agric., 70:55-61.

Torel, J., Cillard, J. and Cillard, P., 1986, Antioxidant activity of flavonoids and reactivity with peroxyl radical. Phytochem., 25:383-385.

Trela, B.C., and Waterhouse, A.L., 1996, Resveratrol: Isomeric molar absorptivities and stability. J. Agr. Food. Chem., 44:1253-1257.

Trousdale, E.K. and Singleton, V.L., 1983, Astilbin and engeletin in grapes and wine. Phytochem., 22:619-620.

Tsuda, T., Horio, F. and Osawa, T., 1999, Absorption and metabolism of cyanidin 3-O-β-D-glucoside in rats. FEBS Lett., 449:179-182.

Tsuda, T., Kato, Y. and Osawa, T., 2000, Mechanism for the peroxynitrite scavenging activity by anthocyanins. FEBS Lett., 484:207-210.

Varache-Lembège, M., Waffo Teguo, P. and Richard, T., 2000, Structure-activity relationships of polyhydroxystilbene derivatives extracted from *Vitis vinifera* cell cultures as inhibitors of human platelet aggregation. Med. Chem. Res., 10:253-267.

Vitrac, X., Monti, J.P., Vercauteren, J., Deffieux, G. and Mérillon, J.M., 2001a, Direct high-performance liquid chromatography analysis of resveratrol derivatives and flavonols in wines with absorbance and fluorescence detection. Anal. Chim. Acta (in press).

Vitrac, X., Castagnino, C., Delaunay, J.C., Waffo-Teguo, P., Vercauteren, J., Deffieux, G. and Mérillon, J.M., 2001b, Polyphenols newly extracted from red wine by centrifugal partition chromatography. J. Agric. Food Chem., (in press).

Waffo Teguo, P.W., Decendit, A., Vercauteren, J., Deffieux, G. and Merillon, J.M., 1996, Trans-resveratrol-3-O-β-glucoside (piceid) in cell suspension cultures of *Vitis vinifera*. Phytochem., 42:1591-1593.

Waffo Téguo, P., Fauconneau, B., Deffieux, G., Huguet, F., Vercauteren, J. and Mérillon, J.M., 1998, Isolation, identification, and antioxidant activity of three stilbene glucosides newly extracted from *Vitis vinifera* cell cultures. J. Nat. Prod., 61:655-657.

Waffo-Téguo, P., Hawthorne, M.E., Cuendet, M., Mérillon, J.M., Kinghorn, A.D., Pezzuto, J.M. and Mehta, R.G., 2001a, Potential cancer chemopreventive activities of wine stilbenoids and flavans extracted from grape (*Vitis vinifera*) cell cultures. Nutr. Cancer, 40 (in press).

Waffo Téguo, P., Lee, D., M. M.J., Pezzuto, J.M. and Kinghorn A.D., 2001b, Two new stilbene dimer glucosides from grape (*Vitis vinifera*) cell cultures. J. Nat. Prod., 64:136-138.

Walgren, R.A., Lin, J.T., Kinne, R.K. and Walle, T., 2000, Cellular uptake of dietary flavanoid quercetin 4'-β-glucoside by sodium-dependent glucose transporter SGLT1. J. Pharmacol. Exp. Ther., 294:897-943.

Wang, B.H., Lu Z.X. and Polya, G.M., 1998, Inhibition of eukaryote serine/threonine-specific protein kinases by piceatannol. Planta Med., 64:195-199.

Waterhouse, A.L. and Lamuela-Raventós, R.M., 1994, The occurrence of piceid, a stilbene glucoside, in grape berries. Phytochem., 37:571-573.

Waterhouse, A.L. and Teissèdre, P.L., 1997, Levels and phenolics in California varietal wines. In: 'Wine Nutritional and Therapeutic Benefits', T.R. Watkins (ed.), ACS Symp. Ser. 661:12-23.

Watkins, T.R. and Bierenbaum, M.L., 1997, California wine use leads to improvement of thrombogenic and peroxidation risk factors in hyperlipemic subjects. In: 'Wine Nutritional and Therapeutic Benefits', T.R. Watkins (ed.), ACS Symp. Ser. 661:261-273.

Whitehead, T., Robinson, D., Allaway, S., Syms, J. and Hale, A., 1995, Effect of red wine ingestion on the antioxidant capacity of serum. Clin. Chem., 41:32-35.

Willett, W.C., 1994, Diet and health: What should we eat? Science, 264:532-537.

Yan, K.X., Terashima, K., Takaya, Y. and Niwa, M., 2001, A novel oligostilbene named (+)-viniferol A from the stem of *Vitis vinifera* 'Kyohou'. Tetrahedron Lett., 57:2711-2715.

Yoshimoto, M., Okuno, S., Yoshinaga, M., Yamakawa, O., Yamaguchi, M. and Yamada, J., 1999, Antimutagenicity of sweetpotato (*ιpomoea batatas*) roots. Biosci. Biotechnol. Biochem., 63:537-541.

Zhu, Q.Y., Huang, Y., Tsang, D. and Chen, Z.Y., 1999, Regeneration of α-tocopherol in human low-density lipoprotein by green tea catechin. J. Agric. Food Chem., 47:2020-2025.

Ziouzenkova, O., Sevanian, A., Abuja, P.M., Ramos, P. and Esterbauer, H., 1998, Copper can promote oxidation of LDL by markedly different mechanisms. Free Radical Bio. Med., 24:607-623.

CHAPTER 4

Asparagus racemosus and Radioprotection

Jayashree. P. Kamat and *S.R. Venkatachalam*

INTRODUCTION

Ionizing radiation, with low linear energy transfer (LET) in the biological system, is an important source for the generation of oxygen-derived reactive oxygen species (ROS) and excited states (Von Sonntag, 1987). As such, there is considerable amount of water in all cells apart from the target macromolecules DNA, protein, lipids, etc. Exposure to ionizing radiation results in a radiolytic cleavage of water, giving rise to e^-_{aq}, $\cdot OH^\bullet$ and $H^\bullet$ (Kesavan, 1992; Von Sonntag, 1987). Oxygen, in view of its bi-radical nature, is the most important electron acceptor present in aerobic cells. Oxygen rapidly accepts unpaired electrons and gives rise to a series of partially-reduced ROS such as superoxide (O_2^-), hydroxyl ($\cdot OH^\bullet$), and peroxyl ($ROO^\bullet$) radicals (Kesavan, 1992; Von Sonntag, 1987). The formation of $^1O_2$ takes place by energy transfer reactions as also by the action of hydroperoxyl with superoxide (Kesavan, 1992). Due to the presence of polyunsaturated fatty acids, membranes are highly prone to oxidative damage by ROS (Giusti et al., 1998). Radiation-induced ROS may cause lipid peroxidation and a loss of structural and functional integrity of the membranes, leading directly to cytotoxicity or indirectly to genotoxicity. Oxidative damage is implicated in ageing and several diseases (Burcham, 1998; Shao et al., 1996). In recent years, much attention has been focussed on this subject, especially in the field of clinical medicine, due to its relevance in degenerative diseases (Sies, 1996; Thomas and Kalyanaraman, 1998). It is now a well-established fact that ROS are important in the modification of proteins leading to oxidation of amino acid side chain residues, cleavage of peptide bonds, and formation of covalent protein-

protein cross-linked derivatives (Dean et al., 1997; Stadtman and Berlett, 1997). Oxidized proteins, if they are not degraded by endogenous proteases, accumulate, resulting in ageing and various disorders (Stadtman and Berlett, 1997). Although there are several antioxidant defense mechanisms in the form of chemicals or enzymes (Packer and Ong, 1998), including those present in cells, they are overwhelmed by the high levels of prooxidants generated during radiation. Hence, compounds protecting against ionizing radiation will have potential benefits as radioprotectors. In current years, the information regarding antioxidant defenses by vitamins E and C, protecting membranes against ROS-induced damage, has been well documented (Packer and Ong, 1998). We have also recently shown nicotinamide (vitamin B3) as an antioxidant against oxidative damage in subcellular membranes (Kamat and Devasagayam, 1996).

Several natural compounds from plants exhibit antioxidant and/or radioprotective properties. For instance, bioactive phenolic compounds from *Catharanthus roseus* (Nishibe et al., 1997) have radical scavenging, anticancer and antidiabetic properties. Ayurvedic herbal formulations, Liv 52/Liv 100 exhibited protection against lipid peroxidation in rat liver subcellular fractions (Suja et al., 1997). Chinese herb *Herba epimedii* is known to enhance the levels of several antioxidant enzymes, in addition to lowering the levels of lipid peroxides (Zeng et al., 1997). Various plant products from neem and turmeric showed chemopreventive capacity against oral carcinogenesis (Nagini and Manoharan, 1997). Similarly certain plant extracts used in traditional medicine have therapeutic properties in addition to their antioxidant abilities (De-las-Heras et al., 1998; Lin et al., 1998; Prince et al., 1998). Considerable attention has been focussed on identifying a greater number of naturally-occurring chemopreventive substances capable of inhibiting, retarding or even reversing the multistage carcinogenesis. A wide array of phenolic substances present in dietary and medicinal plants possess substantial anticarcinogenic and antimutagenic activities (Surh, 1999). Capsaicin, a pungent ingredient of hot chilli pepper, protects against experimentally-induced mutagenesis and tumourigenesis. It also induces apoptosis in various malignant cell lines. Hyper-peroxidation of membrane phospolipids reflected well the degree of DNA damage and carcinogenic alteration, and may be a useful intermediate biomarker for initiation of carcinogenesis (Matsumoto et al., 1996). Our studies recently demonstrated the antioxidant effect of chlorophyllin, the sodium copper salt and a water soluble analogue of the ubiquitous green pigment chlorophyll against membrane damage under in vitro and ex vivo condition (Kamat et al., 2000a). *Asparagus racemosus*, a popular Indian plant used as a dietary component, has been found to possess various medicinal and pharmacological properties. However, except for a few reports, its antioxidant properties have not been explored well. Hence, it is of interest to study this aspect. The possible antioxidant properties of the components of this plant may at least partly explain its therapeutic properties. Using rat liver mitochondria as a model system, we have examined the possible membrane protective properties of extracts of *A. racemosus* against γ-radiation.

ASPARAGUS RACEMOSUS

Brief biology of the plant

Our present findings deal with the genus *Asparagus* (Family 'Liliaceae') which comprises under-shrubs or herbs. A total of 22 species have been recorded in India. Among these, *A. racemosus*, commonly known as *Shatavari*, is a perennial, climbing shrub with fleshy roots and stem comprising re-curved spines/cladodes that are actually a modified stem (Abraham, 1998). This plant flowers in the last week of November and the fruit matures 52 days after anthesis. The berries contain two-three seeds, which are spherical or hemispherical. A ten-year-old plant produces about 130 g seeds (Kar and Sen, 1984).

Chemical constituents

The preparation of crude extract and isolation of polysaccharides from *A. racemosus* was carried out in our laboratory, as described previously (Kamat et al., 2000c). In brief, *A. racemosus* was powdered and sequentially extracted in a sohxlet with petroleum ether (60-80°C), chloroform and methanol. The dried residue was extracted with hot water for 5 h, filtered and then an excess of acetone was added. The precipitate was filtered, washed, dried and was re-dissolved in water. A clear supernatant was obtained after centrifugation. Addition of excess acetone to this solution gave a precipitate that was filtered, washed with acetone and dried and dissolved in distilled water, dialyzed overnight and lyophilized. The pale brown product thus obtained was the crude polysaccharide component of *A. racemosus*, with a yield of about 1 gm. This was taken as the crude extract in our experiments.

A portion of this crude extract (700 mg) was purified on a Sephacryl column. Elution was carried out with distilled water, 0.2 M NaCl and finally with 0.2 M NaOH. Analysis of the fractions was performed by HPLC. The fractions (16-25), eluted with distilled water, were pooled and lyophilized to so as to yield the polysaccharide fraction (termed as P3), as a white powder with a yield of 130 mg. Preliminary analysis of the crude extract as well as the purified fractions showed polysaccharide components (Mol. Wt. 2,000 kilodaltons), rich in galactose, glucose and smaller amounts of arabinose. The relative percentage of monosaccharides in crude extract is as follows: galactose — 54%; glucose — 28%; Rhamnose — 4%; xylose —

5% and arabinose — 8%; others — 1% and that of P3 fraction: galactose — 59%; glucose — 38%; others —3%.

Medicinal properties and pharmacology

Asparagus racemosus has significant medicinal properties and forms the essential component of established Ayurvedic drugs. It is used for the treatment of several diseases such as diarrhoea, dysentery, rheumatism, nervous breakdown, and as an aphrodisiac (Chadha, 1985; Nadkarni, 1976). The studies on the total extracts and the isolated principles have revealed a wide range of biological activities, including antitumour (Shao et al., 1996), antifungal (Shimoyamada et al., 1990), antimutagenic (Edenharder, 1990), immunostimulatory (Diwanay and Patwardhan 2003; Dhuley, 1997; Rege et al., 1989; Thatte and Dahanukar, 1988) ulcer protective (Sairam et al., 2003) and diuretic (Balansand and Rayband, 1987) properties. This plant has also been considered to be lactogogue in lactational inadequacy (Sharma et al., 1996). Intraperitoneal administration of *A. racemosus* has resulted in the decrease in adhesion scores attained by animals receiving this extract and thereby prevention and management of post-operative adhesions. These intra-peritoneal adhesions were induced by caecal rubbing in an animal model (Rege et al., 1989). *A. racemosus*, in combination with *Tinospora cordifolia* (TC), inhibits myelosuppressive effects of single and multiple doses of cyclophosphamide in mice (Thatte and Dahanukar, 1988). The crude saponins from the shoots (edible part) of *Asparagus racemosus* were found to possess antitumour activity. They inhibited the growth of human leukemia HL-60 cells in culture and macromolecular synthesis in a dose-and-time-dependent manner (Shao et al., 1996). Collectively, these observations reveal certain beneficial properties of *A. racemosus*.

Biotechnological and tissue culture work

Propagation of *A. racemosus* has been achieved using tissue culture techniques. Some studies on the organogenesis of diploid *A. racemosus* were carried out by callus culture. In these studies, the development of organized structures from the callus of diploid *A. racemosus* was studied. Roots were formed on Murashige-Skoog medium with naphthaleneacetic acid and zeatin. Histological investigations of the organization were made. Cytology of the regenerated plants revealed the diploid constitution and these studies made obvious the importance of organogenesis in the propagation of this medicinal plant (Kar and Sen, 1984). Further, it was also found that in the presence of 2,4-dichlorophenoxyacetic acid and coconut water, the chromosomal variability in callus culture of *A. racemosus* was increased. This study also indicates a gradual polyploidization with increasing age of the callus (Kar and Sen, 1985a). Regeneration of shoot buds and clonal multiplication was achieved by the use of IAA and BAP in the medium. Rooting was achieved with half-strength MS basal medium plus IBA. Complete plants with cladode, crown and root systems were developed in a hormone-free medium. The plants were then successfully transferred to the soil (Kar and Sen, 1985b). γ-irradiation resulted inan inhibition in the growth and morphogenetic potential of the callus culture. The condition and behaviour of nucleus and chromosomes differed markedly from control. Chromosome breaks, clumping, stretching and migration towards the cell wall were the major chromosomal changes noted in the treated calli. The possibility of utilizing genotype from in vitro irradiation for plant improvement has been suggested (Kar and Sen, 1985c).

In our studies, we have measured the oxidative damage in terms of lipid peroxidation and protein oxidation using rat liver mitochondria and examined the possible protective effect of *A. racemosus* against radiation-induced oxidative damage (Kamat et al., 1997; Kamat et al., 2000b). Mitochondria were suspended in a potassium phosphate buffer pH 7.4 with or without *A. racemosus* extracts and exposed to γ-radiation from a ^{60}Co source (dose rate 15 Gy/min., Bhabha Atomic Research Centre). The doses ranged from 45-600 Gy. Radiation-exposed samples without *A. racemosus* served as control. Appropriate sample blanks were taken to ensure that the fractions from *Asparagus* did not interfere with the assay methods.

RESULTS

Our studies showed a significant enhancement in TBARS (marker of lipid peroxidation) and protein carbonyl formation (an index of protein oxidation, $P < 0.001$), at a radiation dose of 450 Gy. Hence, this dose was selected for experimental purposes. Mitochondria in the presence of crude extract of *A. racemosus* showed a marked inhibition in lipid peroxidation at a concentration of 10 μg/ml that amounts to be ~60%. Among the purified fractions of the crude extract, an active fraction consisting of polysaccharides (termed P3) at a similar concentration showed significant inhibition in TBARS (data for other fractions are not included). The P3 fraction, decreased peroxidation by 92% (Fig. 4.1a). The crude as well as P3 fraction showed complete protection against the formation of protein carbonyls at the concentrations of 10 μg/ml, bringing the levels to control values (Fig. 4.1b), ($P< 0.05$). Radiation-induced lipid peroxidation by *A. racemosus* crude and P3 fractions in comparison with established biological antioxidants, ascorbic acid and glutathione are shown in Fig. 4.2. It is clear that, though ascorbate and glutathione showed significant protection, the concentrations required to have effective protection seems to be very high, approximately 20-30 times, compared on the basis of μg/ml.

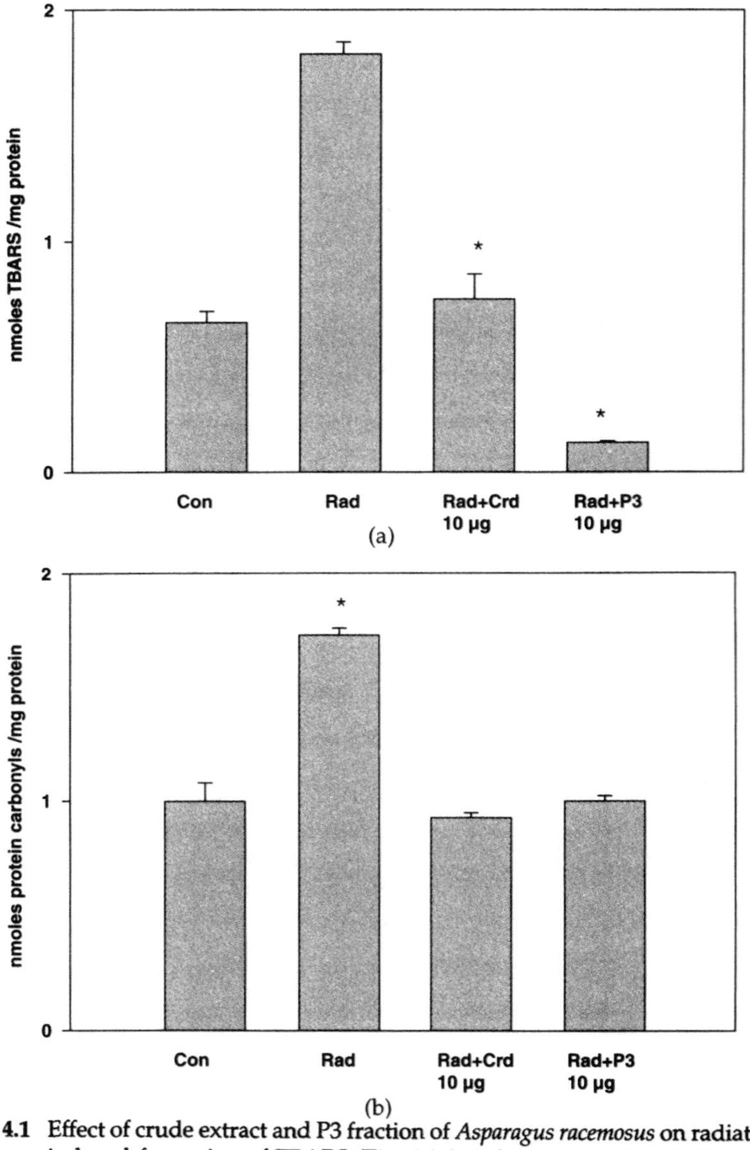

Fig. 4.1 Effect of crude extract and P3 fraction of *Asparagus racemosus* on radiation-induced formation of TBARS (Fig. 4.1a) and protein carbonyl (Fig. 4.1b) in rat liver mitochondria. Mitochondria (2 mg protein/ml) were exposed to radiation at the dose of 450 Gy and the estimation of thiobarbituric acid reactive substances (TBARS) (Fig. 4.1a) and protein carbonyl (Fig. 4.1b) were carried out. Con — control; Rad — radiation; Crd — crude extract 10 µg/ml; P3 — purified P3 fraction; 10 µg/ml . The values are mean ±SEM from 6 experiments. *$P < 0.001$, as compared to radiation.

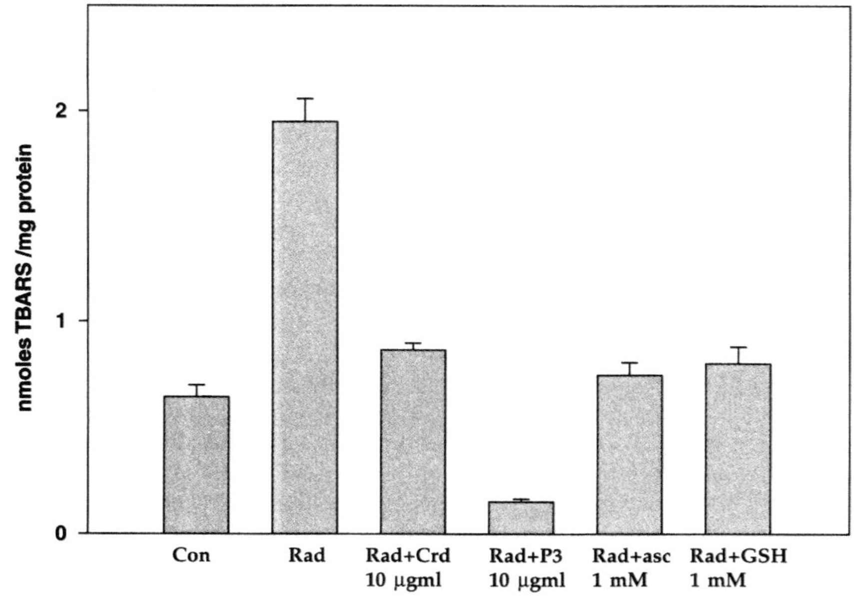

Fig. 4.2 Relative inhibition of radiation-induced lipid peroxidation by *A. racemosus* crude extract and P3 fraction as compared to ascorbic acid and glutathione in rat liver mitochondria. Other legends are same as in Fig. 4.1. The values are mean ±SEM from 6 experiments.

DISCUSSION

Though membranes are crucial components protecting the cells from toxic substances, their interaction with radiation-induced ROS results into deleterious events, such as lipid peroxidation and its cytotoxic products 4-hydroxynonenal that have the ability to interact with DNA, and other macromolecules, possibly leading to diseases (Burcham, 1998; Box and Maccubbin, 1997; Rice-Evans and Burdon, 1993; Von Sonntag, 1987). The biological response to radiation may be modulated by alterations in factors affecting the mechanism of cellular injury. The membrane integrity can be maintained by supplementation of dietary vitamins A, C, E, β-carotene as well as other natural compounds (Kamat and Devasagayam, 1996; Packer and Ong, 1998). The detailed understanding of the oxidative reactions in membranes and their protection by natural compounds will provide therapeutic strategies to minimize such damage in diseases.

The present investigations dealing with *A. racemosus* demonstrated significant protection against radiation-induced oxidative damage in rat liver mitochondria. Radiation-induced TBARS, the stable end product of peroxidation as well as protein oxidation was effectively inhibited by *A. racemosus* even at a low concentration of 10 μg/ml. The results dealing with protein demonstrated that the protection observed by crude extract and P3 fraction of *A. racemosus* was more or less same. Some reports indicated that polysaccharides derived from plants have antioxidant effects against oxidative damage (Gang et al., 1997; Liu et al., 1997; Zen et al., 1997). Antioxidants activity associated with polysaccharides, present in crude extract, may be effective in neutralizing the effect of prooxidants following exposure to γ-radiation.

Asparagus is a popular vegetable consumed in most parts of the world. *Asparagus* shoots, the edible part of the plant, are frequently used in salads, vegetable dishes and soups. In Chinese traditional medicine, this plant has gained a lot of significance. The chemical constituents of this plant have been studied to a certain extent. The compounds so far reported from *Asparagus* include flavonoids, oligosaccharides, aminoacids, sulphur-containing acids and steroidal saponins. The steroidal saponins, methanol soluble derivatives, have been shown to possess significant antitumour activity (Shao et al., 1996). The mechanisms responsible for the observed antioxidant properties of *A. racemosus* are still not clear. Various reports demonstrate that polysaccharides derived from plants exhibit potent antioxidant as well as radioprotective properties (Gang et al., 1997; Liu et al., 1997; Zeng et al., 1997; Zhang et al., 1997). The polysaccharide kreskin has also been shown to have inhibitory effects on the oxidation of LDL (Liu et al., 1997). Additionally, they also have the ability to protect against γ-radiation induced spleen injury (Liu et al., 1996). Therefore, the protective effects observed in the present studies may be explained on the basis of the scavenging ability of polysaccharides present in *A. racemosus* towards different ROS generated during radiation exposure.

The established antioxidants glutathione and vitamin C require 20-30 times more concentration for effective inhibition of TBARS formation comparable to that by *A. racemosus* extracts. It is likely that the effect of these extracts, when compared to that of vitamin E, which is a more potent radio-protector, may be different. Further experiments are needed to confirm this. In this regard, ESR or pulse radiolysis studies may give better understanding of the mechanisms of radioprotection by *A. racemosus*.

In conclusion, our data dealing with *A. racemosus* extracts, suggest that it may act as a potent protector against oxidative damage, including that induced by radiation.

ACKNOWLEDGEMENTS

The authors wish to thank Dr K.B. Sainis, Head, Cell Biology Division and Dr J.P. Mittal, Director, Chemistry Group for their encouragement and

Dr Sipahimalani for his contribution in initiating the studies on this natural product.

REFERENCES

Abraham, V., 1998, Biodiversity at Bhabha Atomic Research Centre, Trombay, p. 119.

Balansand, J. and Rayband M., 1987, Diuretic action of *Asparagus officinalis*. Critical Reviews in Society of Biology 126:954-956.

Burcham, P.C., 1998, Genotoxic lipid peroxidation products: their DNA damaging properties and role in formation of endogenous DNA adducts. Mutagenesis 13:287-305.

Box, H.C. and Maccubbin, A.E., 1997, Lipid peroxidation and DNA damage. Nutrition 13:920-921.

Chadha, Y.R., 1985, The Wealth of India. Publications and Information Directorate, New Delhi, Vol. I., p. 468.

Dean, R.T., Fu, S., Stocker, R. and Davies, M.J., 1997, Biochemistry and pathology of radical mediated protein oxidation. Biochem. J. 324:1-18.

De-las-Heras, B., Slowing, K., Benedi, J., Carretero, E., Ortega, T., Toledo, C., Bermejo, P., Iglesias, I., Abad, M.J., Gomez-Serranillos, P., Liso, P.A. and Villar, A.C., 1998, Antiinflammatory and antioxidant activity of plants used in traditional medicine in Ecuador. J. Ethnopharmacol. 6:161-166.

Dhuley, J.N., 1997, Effect of some Indian herbs on macrophage functions in ochratoxin A treated mice. J. Ethnopharmacol. 58:15-20.

Diwanay, S., and Patwardhan, B., 2003, Immunoprotection by botanical drugs in cancer chemotherapy. J. Ethnopharmacol. (In press).

Edenharder, R., 1990, Antimutagenic activity of vegetable and fruit extracts against in-vitro benzo(a)pyrene. Zeitschrift fur Gesamte. Hygiene 36:144-148.

Gang, Z.Z., Li, L.Z and Xian, L.X., 1997. Study on the isolation, purification and antioxidation properties of polysaccharides from *Spirulina maxima*. Acta Botanica Sinica 39: 77-81.

Giusti, A.M., Raimondi, M., Ravagnan, G., Sapora, O. and Parasassi, ,T. 1998, Human cell membrane oxidative damage induced by single and fractionated doses of ionizing radiation: a fluorescence spectroscopic study. International Journal of Radiation Biology 74:595-605.

Kamat, J.P. and Devasagayam, T.P.A., 1996, Methylene blue plus light induced lipid peroxidation in rat liver microsomes: Inhibition by nicotinamide (vitamin B_3) and other antioxidants. Chem.-Biol. Inter. 99:1-16.

Kamat, J.P., Sarma, H.D., Devasagayam, T.P.A., Nesaretnam, K. and Basiron, Y., 1997, Tocotrienols from palm oil as effective inhibitors of protein oxidation and lipid peroxidation in rat liver microsomes. Mol. Cell. Biochem. 170:131-138.

Kamat, J.P., Boloor, K.K. and Devasagayam, T.P.A., 2000a, Chlorophyllin as an effective antioxidant against membrane damage in vitro and ex vivo. Biochim. Biophys. Acta 1487:113-127.

Kamat, J.P., Boloor, K.K., Devasagayam, T.P.A., Jayashree, B. and Kesavan P.C., 2000b, Differential modification by caffeine of oxygen-dependent and independent effect of γ-radiation on rat liver mitochondria. Int. J. Rad. Biol. (in press).

Kamat, J.P., Boloor K.K., Devasagayam, T.P.A. and Venkatachalam, S.R., 2000c, Antioxidant properties of *Asparagus racemosus* against damage induced by γ-radiation in rat liver mitochondria. J. Ethanopharmacol. 71:425-435.

Kar, D.K. and Sen S., 1984, Organogenesis of diploid *Asparagus racemosus* through callus culture. Proceedings of the Indian National Science Academy, Part B, Biological Sciences 50:113-118.

Kar, D.K. and Sen, S., 1985a, Effect of hormone on chromosome behavior in callus cultures of *Asparagus racemosus*. Biologia Plantarum (Prague) 27:6-9.

Kar, D.K. and Sen S., 1985b, Propagation of *Asparagus racemosus* through tissue culture. Plant Cell Tissue And Organ Culture 5:89-96.

Kar, D.K. and Sen, S., 1985c. Effect of gamma-irradiation on callus of *Asparagus racemosus*. Nucleus 28:52-55.

Kesavan, P.C., 1992, Protection by caffeine against oxic radiation damage and chemical carcinogenesis: Mechanistic consideration. Curr. Sci. 62:791-97.

Lin, C.C., Yen, M.H., Lo, T.S. and Lin, J.M., 1998, Evaluation of the hepatoprotective and antioxidant activity of *Boehmeria nivea* var. *nivea* and *B. nivea* var. *tenacissima*. J. Ethnopharmacol. 60:9-17.

Liu, J., Yeo, H.C., Doniger, S.J. and Ames, B.N., 1997, Assay of aldehydes from lipid peroxidation: gas chromatography-mass spectrometry compared to thioabarbituric acid. Anal. Biochem. 245:161-166.

Liu, S.X., Chen, Y., Zhou, M. and Wan, J., 1997, Protective effect of the polysaccharide kreskin on inhibition of lipopolysaccharide-induced nitric oxide production in macrophages caused by oxidised low-density lipoprotein. Med. Sci. Res. 25:507-509.

Liu, F., Ooi, V.E.C. and Chang, S.T., 1996, Free radical scavenging activities of mushroom polysaccharide extracts. Life Sci. 60:763-771.

Matsumoto, H., Yamane, T., Inagako, M., Iwata, Y., Takahashi, T., Nishimura, H., Nishino, H., Nakagawa, K. and Miyazawa, T., 1996, Inhibition of mucosal lipid hyperoxidation by green tea extract in 1,2–dimethylhydrazine- induced rat colonic carcinogenesis. Cancer Lett. 105:205-209.

Nadkarni, A.K., 1976, Indian Materia Medica, Popular Prakashan, Mumbai, p.151.

Nagini, S. and Manoharan, S., 1997, Biomonitoring the chemopreventive potential of the plant products of neem and turmeric in 4-nitroquinoline-1 oxide induced oral carcinogenesis. J. Clin. Biochem. and Nutrition 23:33-40.

Nishibe, S., Takenaka, T., Fujikawa, T., Yasukawa, K., Takido, M., Morimitsu, Y., Hirota, A., Kawamura, T. and Noro, Y., 1997, Bioactive phenolic compounds from *Catharanthus roseus* and *Vinca minor*. Natural Medicines 50:378-383.

Packer, L. and Ong, A.S.H., 1998, Biological Oxidants and Antioxidants: Molecular Mechanisms and Health Effects. AOCS Press, Champaign, Illinois.

Prince, P.S., Menon, V.P. and Pari, L., 1998, Hypoglycaemic activity of *Syzigium cumini* seeds: effect on lipid peroxidation in alloxan diabetic rats. J. Ethnopharmacol. 61:1-7.

Rege, N.N., Nazareth, H.M., Isaac, A., Karandikar, S.M. and Dahanukar, S.A., 1989, Immunotherapeutic modulation of intraperitoneal adhesions by *Asparagus racemosus*. J. Postgrad. Med. 35:199-203.

Rice-Evans, C. and Burdon, R., 1993, Free radical-lipid interaction and their pathological consequences. Progress in Lipid Res. 32:71-110.

Sairam, K., Priyambada, S., Aryya, N.C., and Goel, R.K., 2003, Gastroduodenal ulcer protective activity of *Asparagus racemosus;* an experimental, biochemical and biological study. J. Ethnopharmacol. 86: 1-10.

Shao, Y., Chin, C-K., Ho, C-T., Ma, W., Garrison, S.A. and Huang, M.T., 1996, Anti-tumour activity of the crude saponins obtained from asparagus. Can. Lett. 104:31-36.

Sharma, S., Ramji, S., Kumari, S. and Bapna, J.S., 1996, Randomized controlled trial of *Asparagus racemosus* (Shatavari) as a lactogogue in lactational inadequacy. Indian Pediatrics 32:675-677.

Shimoyamada, M., Suzuki, M., Sonta, H., Maruyama, M., Okubo, K., 1990, Antifungal activity of the saponin fraction obtained from asparagus and its active principle. Agri. and Biol. Chem. 54: 2553-2557.

Sies, H., 1996, Antioxidants in Disease, Mechanisms and Therapy, Academic Press, New York.

Stadtman, E.R. and Berlett, B.S., 1997, Reactive oxygen mediated protein oxidation in aging and disease. Chem. Res. Toxicol. 10:485-494.

Suja, V., Sharmila, S.L. and Devi, C.S.S., 1997, Protective effect of Liv. 52 and Liv. 100 ayurvedic formulations on lipid peroxidation in rat liver homogenate: an *in vitro* study. Indian J. Exp. Biol. 35:50-52.

Surh, Y., 1999, Molecular mechanisms of chemopreventive effects of selected dietary and medicinal phenolic substances. Mutat. Res. 428:305-327.

Thatte, U.M. and Dahanukar, S.A., 1988, Comparative study of immunomodulating activity of Indian medicinal plants, lithium carbonate and glucan. Methods and Findings in Experimental Clinical Pharmacology 10:639-644.

Thomas, C.E. and Kalyanaraman, B., 1998, Oxygen Radicals and the Disease Process. Harwood Academic Publishers GmbH, Chur, Switzerland.

Von Sonntag, C., 1987, The Chemical Basis of Radiation Biology, Taylor and Francis, London.

Zeng, N., Meng, X., Zhang, Y., Lai, X., Zheng, J., Cheng, L. and Ren, C., 1997, Studies on the antioxidative effect of constituents of herba Epimedii (ESPS). Zhongguo Zhongyao Zazhi 22:46-48 (in Chinese).

Zhang, C., Zheng, S., Zhang, Y., Luo, C. and Guo, C., 1997, The protective effect of polysaccharide and C-phycocyanin from *Spirulina platensis* on acute radiation injury in mice. Acta Nutrimenta Sinica 18: 327-331.

CHAPTER 5

Brahmi (*Bacopa monnieri* (L.) Pennell) — A *Medhya Rasaayana* Drug of Ayurveda

M. Rajani, Neeta Shrivastava and M.N. Ravishankara*

INTRODUCTION

Bacopa monnieri (L.) Pennell, commonly known as Brahmi, is one of the important *medhya rasaayana (medhya* = that which promotes memory and intellect; *rasaayana* = rejuvenative) drugs of Ayurveda. In traditional medicine, the entire plant is used for treatment of asthma, epilepsy and insanity (Nadkarni, 1954; Chopra et al., 1956b; Anonymous 1, 1999). Brahmi is also considered to be astringent, bitter, cooling and act as a laxative. Further, it is indicated in cases of dermatosis, anaemia, diabetes, cough, dropsy, fever, arthritis, anorexia, dyspepsia and insanity. It has been claimed to improve intelligence and memory and revitalize the sensory organs (Anonymous 2, 1994). Brahmi is included in several Ayurvedic formulations prescribed for memory enhancement such as *Brahmighrtam, Saravataristam, Brahmitailam, Misrakasneham* (Anonymous 3, 1978). In a study conducted on Indian medicinal plants (Anonymous 4, 1997), *B. monnieri* was placed second in a priority list of the most important medicinal plants, evaluated on the basis of their medicinal importance, commercial value and potential for further research and development. A monograph on *B. monnieri* is available in the Ayurvedic Pharmacopoeia of India (Anonymous 1, 1999) and Indian herbal pharmacopoeia (Anonymous 5, 1998).

The name brahmi is commonly applied to two different plants, viz., *Bacopa monnieri* and *Centella asiatica* (Linn.) Urban (Family Apiaceae), in

**Author for correspondence*
Dr. M. Rajani
Assistant Director & Head
Pharmacognosy & Phytochemistry Department
B.V. Patel Pharmaceutical Education & Research Development Centre, Thaltej, Ahmedabad-380054 (India)
Phone # : 91-079-27439375; Fax # 91-079-27450449,
Email: perd@perdcentre.com; rajanivenkat@hotmail.com

different regions of India. However, the true brahmi of Ayurveda is claimed to be *B. monnieri,* which is used extensively in the Indian Systems of Medicine for various nervous disorders (Chopra et al., 1956a; Anonymous 1, 1999).

Vernacular names

Hindi	:	Brahmi, Jal brahmi, Jal neem
Sanskrit	:	Brahmi, Nira-brahmi

Brahmi in Ayurveda

In Ayurveda, the properties of brahmi are described as follows (Anonymous 1, 1999; Anonymous 6, 1976) :

Rasa (taste)	:	*Tikta* (bitter), *Kasaya* (astringent)
Gunna (properties)	:	*Laghu* (light, easy to digest), *snigdha* (unctous), *sara* (mild laxative)
Veerya (potency)	:	*Sita* (cooling)
Vipaaka (properties after digestion)	:	*Madhura* (sweet in the first phase of *vipaaka*, i.e. before the drug reaches the stomach)

Brahmi's actions and uses include *deepana* (stomachic), *paachana* (digestive), *rasaayani* (rejuvenative), *swarya* (for voice improvement), *aayushkara* (promotes longevity), *medhya* (promotes memory and intellect), *kushtakandoohara* (for skin disorders and itching), *jwaraghna* (antipyretic), *vishaghna* (antidote for poisons), *apasmaarahara* (antiepileptic), *vedanasthapana* (analgesic), *mehaghna* (relieves polyuria), *shothahara* (for edema).

Plant description

Bacopa monnieri (synonyms: *Bramia monnieri* Pennell, *Moniera cuneifolia* Michx., *Herpestis monniera* (Linn.) H. B. & K., *H. spathulata* Blume., *Gratiola monniera* Linn., *Lysimachia monniera* Linn.) (Family: Scrophulariaceae), is a weak, creeping and somewhat succulent herb that can be found in damp and marshy areas in plains throughout India (Fig. 5.1). The herb is bright green in colour and the creeping stem sends out roots at nodes touching the soil; the leaves are simple opposite, sub-sessile, obovate-oblong, obtuse, glabrous, entire; flowers light violet in colour, axillary, solitary; 5-membered calyx with unequal lobes; corolla gamopetalous, short, funnel-shaped; stamens 4, didynamous; ovary bicarpellary, syncarpus, ovules on swollen axile placenta; and the fruit is a capsule, calyx persistent (Anonymous 2, 1994).

CHEMISTRY

Chemical constituents

B. monnieri plant contains a complex mixture of dammarane type triterpenoidal saponins (Fig. 5.2a), to which the activity of the plant as a nervine tonic—in traditional systems of medicine— is attributed. Of these

Bacopa monnieri (L.) Pennell plant

Fig. 5.1 A twig with flower

saponins, bacoside A and bacoside B have been tested and found to be responsible for the efficacy of the drug. Bacoside A is laevo-rotatory, while bacoside B is dextro-rotatory in nature. The haemolytic action of bacoside B was double that of bacoside A, which could be due to differences in the configuration of carbohydrate part of the molecules (Basu et al., 1967). Bacoside A, which initially appeared to be a single compound, is a mixture of saponins from which bacoside A_2 and A_3 were separated. Bacoside A_3 is the main constituent of the mixture (Rastogi et al., 1994; Rastogi and Kulshreshtha, 1998). The other saponins include bacoside A_1 (Jain and Kulshreshtha, 1993) and bacopasaponins A, B, C, D (Garai et al., 1996a, b), E, F (Mahato et al., 2000), bacopasaponin G (Hou et al., 2002), bacopaside I, II (Chakravarty et al., 2001), bacopasides III, IV, V, bacopasides A, B, C (Hou et al., 2002; Chakravarty et al., 2003), monnierasides I – III, and plantainoside B (Chakravarty et al., 2002). While Chakravarty et al (2001) named a saponin as bcopaside III that has jujubogenin as aglycone, Hou et al (2003) named a saponin as bacopaside III which has pseudojujubogenin as aglycone and a sulfonyl group at 6-*O*- of the glucose. Bacopasaponin A, E and F are jujubogenin bisdesmosides, with the sugar units attached at two positions of the aglycone (Garai et al., 1996a, b; Mahato et al., 2000) (Fig. 5.2).

On hydrolysis, both bacoside A and bacoside B yield a mixture of aglycones, bacogenin A_1–A_4 (Fig. 5.1). The structure of Bacogenin A_1 (m.p. 242°C) was proposed as trihydroxy dammar-en-one (Chatterjee et al., 1965), but later on, based on detailed physico-chemical studies, its structure has been established as 3,18-dihydroxy-20→25-epoxy-22(or 23)-methyl-24-nor-dammar-22-en-16-one (Kulshreshtha and Rastogi, 1973a). This is the most polar component of the aglycone mixture, obtained after hydrolysis of bacoside A and bacoside B.

Bacogenin A_1 is an artifact derived from the genuine sapogenin, pseudojujubogenin, during acid hydrolysis (Kulshreshtha and Rastogi, 1973a, 1974; Kawai et al., 1973; Garai et al., 1996a). Bacogenin A_2 (m.p. 220-2°C) is an isomer of bacogenin A_1, differing either in configuration at C-20 or disposition of *vinylic Me* in a side-chain (Kulshreshtha and Rastogi, 1974). Bacogenin A_3 (m.p. 190°C) is an artifact formed during acid hydrolysis (Chandel et al., 1977). Bacogenin A_4 is identified as ebelin lactone (m.p. 170-5°C), which has also been found to be an artifact, derived from the genuine sapogenin, jujubogenin, during acid hydrolysis (Kulshreshtha and Rastogi, 1973b). On hydrolysis, bacopasaponin A, E and F yield ebelin lactone, while bacopasaponins B, C and D yield bacogenin A_1 as aglycone (Garai et al., 1996a, b; Mahato et al., 2000). Of the two genuine sapogenins, *viz.*, jujubogenin and pseudojujubogenin, the structure of pseudojujubogenin has not been completely elucidated yet, as the configurations at C-20 and C-22 have not

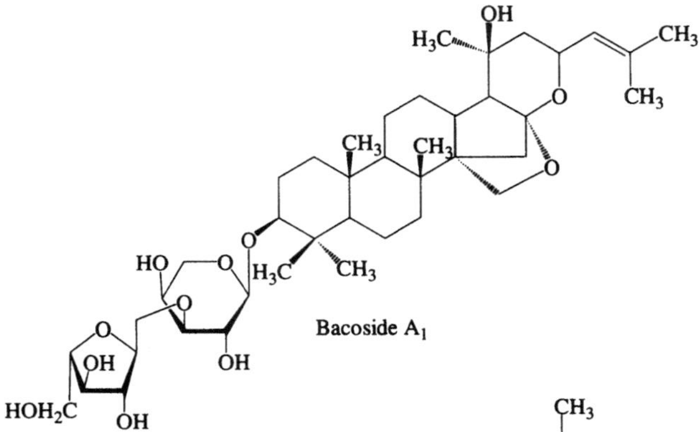

Bacoside A_1

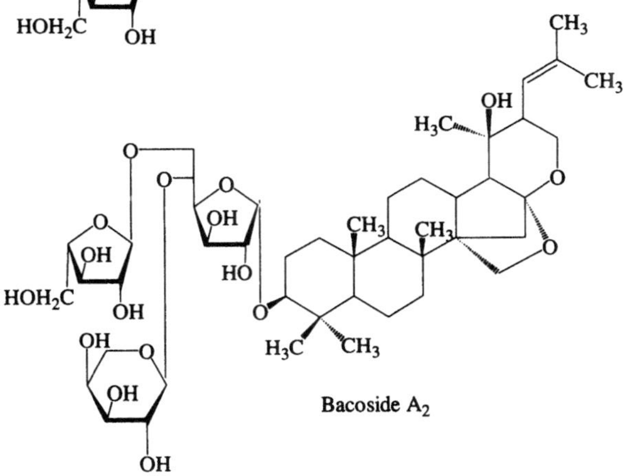

Bacoside A_2

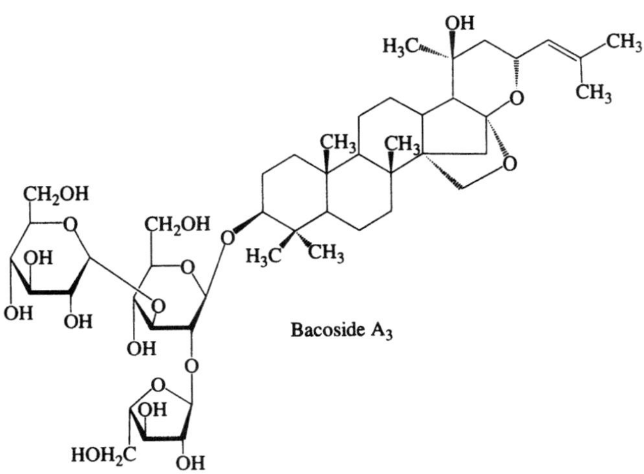

Bacoside A_3

Fig. 5.2 Some chemical constituents of *Bacopa monnieri* - saponins (continued.....)

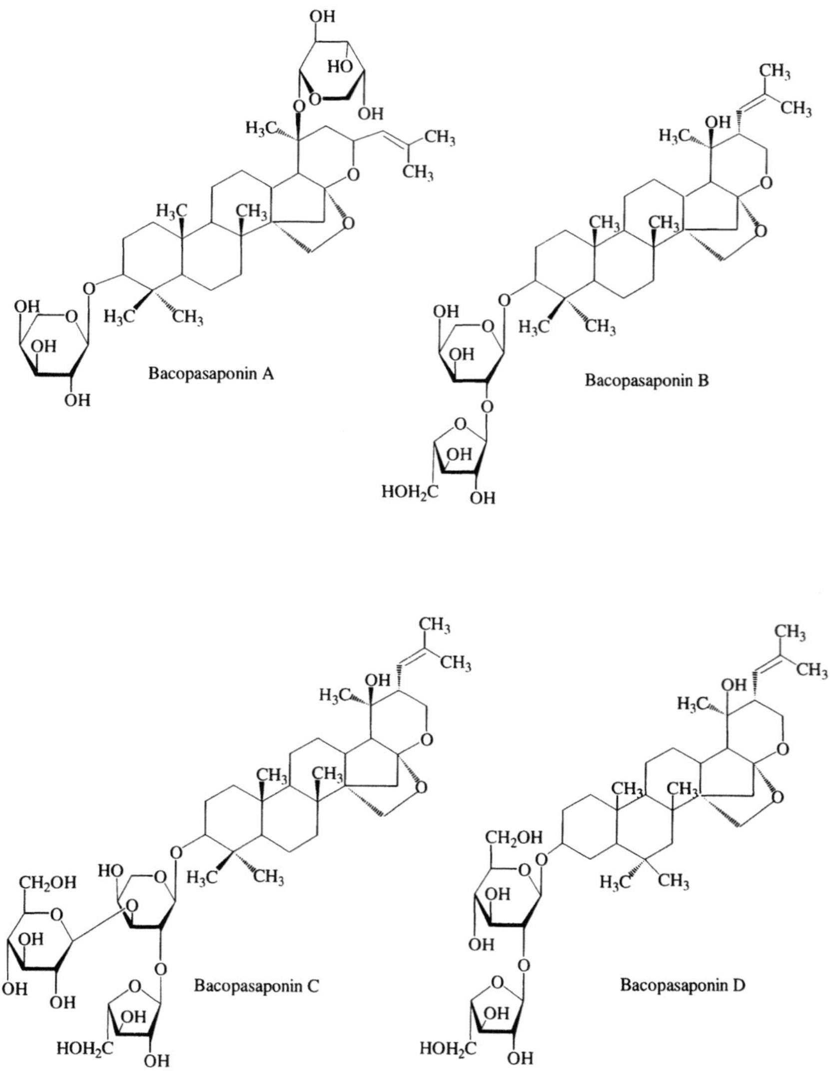

Fig. 5.2 Some chemical constituents of *Bacopa monnieri* - saponins (continued.....)

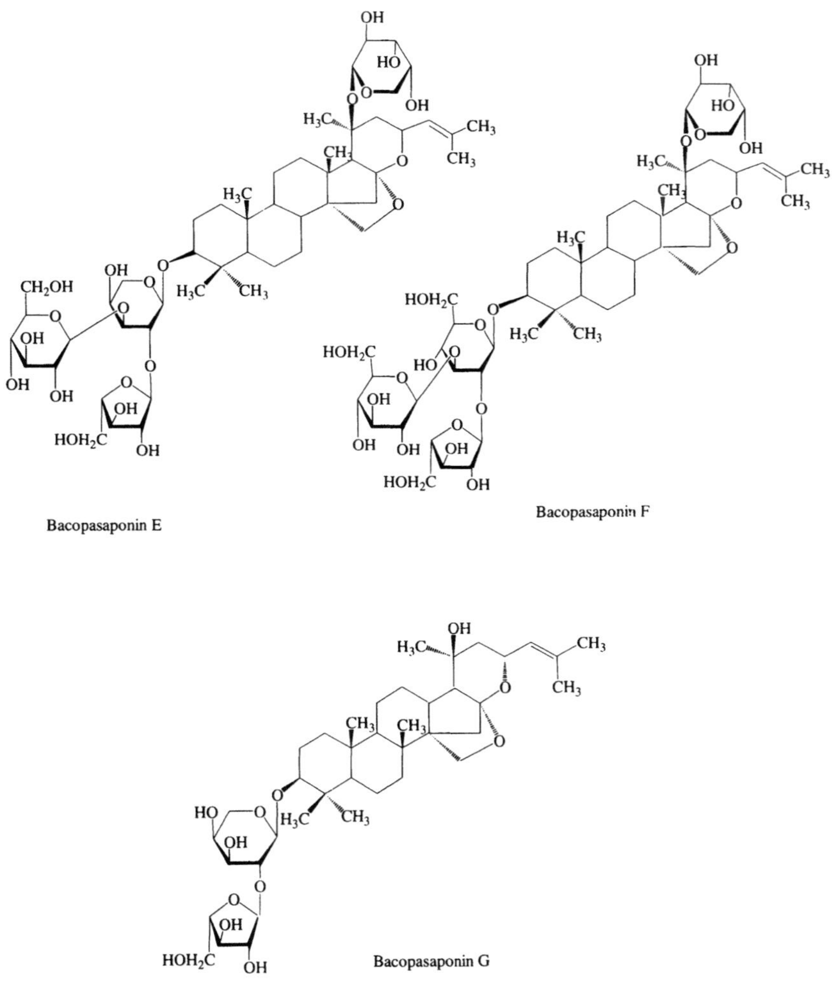

Bacopasaponin E

Bacopasaponin F

Bacopasaponin G

Fig. 5.2 Some chemical constituents of *Bacopa monnieri* - saponins (continued....)

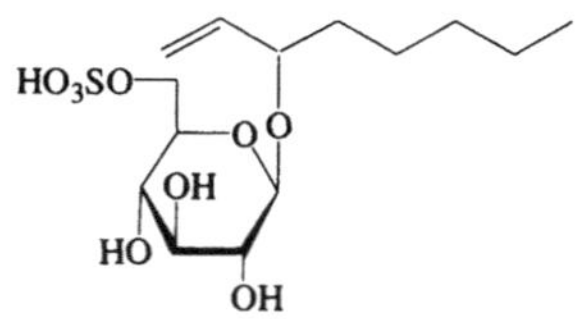

Bacopaside A

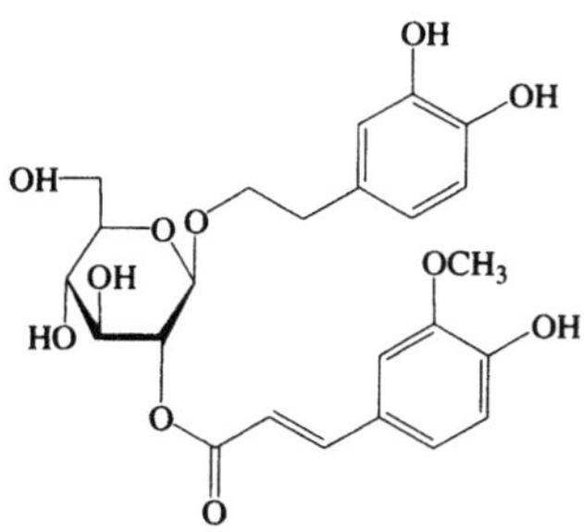

Bacopaside B

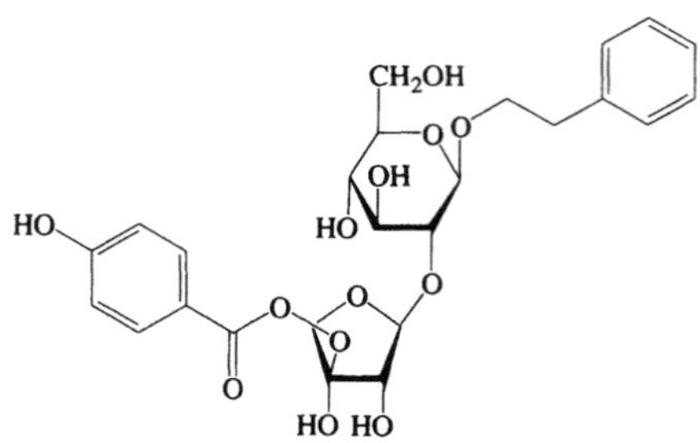

Bacopaside C

Fig. 5.2 Some chemical constituents of *Bacopa monnieri* - saponins.

been determined (Garai et al., 1996a). The details of the saponins, their aglycones and sugar moieties obtained after acid hydrolysis are given in Table 5.1. The saponins differ mainly in the sugar units, in the glycosidic chain and the position of the olefinic side chain in the aglycone (Table 5.1, Figure 5.2; Renukappa et al., 1999). Apart from these, a triterpene bacosine was isolated from the aerial parts of the plant and was identified as lup-20(29)-ene-3α-ol-27oic acid (Vohora et al., 1997).

Other chemical constituents of the plant include D-mannitol (Sastry et al., 1959), hersaponin (m.p. 232-234°C), betulic acid (Chatterji et al., 1963; Rastogi and Dhar, 1960; Jain and Kulshreshtha, 1993), alkaloids, viz., brahmine and herpestine and a mixture of three alkaloids (Bose and Bose 1931; Basu and Walia, 1944; Basu and Pabrai, 1947), flavonoids, viz., luteolin-7-glucoside, glucoronyl-7-apigenin and glucoronyl-7-luteolin (Schulte et al., 1972; Proliac et al., 1991) and phytosterols like stigmasterol, *β*-sitosterol, stigmastanol (Chatterji et al., 1963).

Methods of analysis of saponins of *B. monnieri*

A simple UV-spectrophotometric method was reported by Pal and Sarin (1992) for the estimation of total bacosides as bacoside A. The method involves extraction, acid hydrolysis and estimation of the aglycone thus formed by measuring at 278 nm (Anonymous 4, 1998). Renukappa and co-workers (1999) reported a HPLC coupled to NMR spectrometry, mass spectrometry and bioassay of anthelmintic activity against *Caenorhabditis elegans* as test organism, for rapid and direct determination of active saponins having close structural similarities from crude fraction of the plant. The analysis showed the presence of two pairs of saponins with the same sugar part and two pairs of saponins with similar aglycones. A HPTLC method was reported to determine bacoside A where methanolic extract of the plant was resolved using the solvent system of ethyl acetate-methanol-water (60:14:10), derivatization with vanillin-sulphuric acid reagent and scanning the plate at 620 and 430 nm using dual wavelength absorption-reflection mode (Gupta et al., 1998). In this study, the method of drying of the plant material was shown to affect the bacoside A content, which varied from 1.28 to 3.78%, where the highest yield was obtained with 30 min. of quick drying at 80°C + 4 days at 37°C.

PHARMACOLOGICAL ACTIVITY

Animal experiments

To establish the traditional claims as also to ascertain the active principles, the plant, its alcoholic and aqueous extracts and bacosides have been subjected to pharmacological evaluation in specially-designed animal models.

The ethanolic extract of *B. monnieri* improved the performance of rats in motor learning capabilities (Prakash and Sirsi, 1962). It augmented both the cognitive function and mental retention capacity, as revealed by the studies involving acquisition, consolidation and retention of three newly-acquired behavioural responses in albino rats, viz., foot-shock motivated brightness discrimination response, active conditioned avoidance response and Sidman continuous avoidance response and the two saponins, viz., bacoside A and bacoside B have been found to be responsible for the activity (Singh and Dhawan, 1978, 1982, 1994, 1997; Singh et al., 1988). Bacosides also improved long-term memory (Singh et al., 1990). The plant elevated the cerebral glutamic acid levels and caused a transient increase in GABA levels, which may help in the process of learning (Shukla et al., 1987; Handa, 1996). Further, a mixture of bacosides A and B has been shown to increase protein kinase and bring about an increase in protein and serotonin levels and lower norepinephrine level in hippocampus. However, they did not affect the receptors of biogenic amines and acetylcholine. On the other hand, the bacosides have been shown to decrease the activities of protein kinase and acid and alkaline phosphatase activity

Table 5.1 Certain details of important saponins of *Bacopa monnieri*

Saponin	Description	Aglycone	Sugar	Reference
Bacoside A m.p. 250-1°C (decomp.)	3-(α-L-arabinopyranosyl)-O-β-D-glucopyranosid-10,20-dihydroxy-16-keto-dammarene-(24)-laevo-rotatory	Mixture of bacogenins A_1, A_2, A_3, A_4	Arabinosyl-glucose, with arabinose unit as terminal sugar	Chatterji et al., 1965
Bacoside A_1 m.p. 240°C	3-*O*-[α-L-arabinopyranosyl (1→ 3)-α-L-arabinopyranosyl]-jujubogenin	Ebelin lactone	L-Arabinose	Jain and Kulshreshtha, 1993
Bacoside A_2	3β-*O*-[α-L-arabinofuranosyl (1→6)-*O*-[α-L-arabinopyranosyl(1→5)-*O*-α-D-glucofuranosyl)oxy]pseudojujubogenin] pseudojujubogenin	Bacogenin A_1 and A_2	Glucose and arabinose	Rastogi and Kulshreshtha, 1998
Bacoside A_3	3-β-[*O*-β-D-glucopyranosyl (1→3)-*O*-[α-L-arabinofuranosyl(1→2)]*O*-β-D-glucopyranosyl) oxy]jujubogenin	*Trans*-ebelin lactone *Cis*-ebelin lactone	D-Glucose, L-Arabinose	Rastogi et al., 1994
Bacoside B m.p. 203°C (decomp.)	Same as bacoside A—dextro-rotatory	Mixture of bacogenins A_1 to A_4	Glucose, arabinose	Basu et al., 1967; Chatterji et al., 1963
Bacopasaponin A m.p. 256°C	3-*O*-α-L-arabinopyranosyl-20-*O*-α-L-arabinopyranosyl-jujubogenin	Ebelin lactone	L-Arabinose	Garai et al., 1996a
Bacopasaponin B m.p. 283°C	3-*O*-[α-L-arabinofuranosyl (1→2)α-L-arabinopyranosyl] pseudojujubogenin	Bacogenin A_1	L-Arabinose	Garai et al., 1996a
Bacopasaponin C m.p. 222°C	3-*O*-[β-D-glucopyranosyl (1→3){α-L-arabinofuranosyl (1→2)}α-L-arabinopyranosyl] pseudojujubogenin	Bacogenin A_1	L-Arabinose, D-Glucose	Garai et al., 1996a
Bacopasaponin D m.p. 250°C	3-*O*-[α-L-arabinofuranosyl (I→2)β-D-glucopyranosyl] pseudojujubogenin	Bacogenin A_1	D-Glucose, L-Arabinose	Garai et al., 1996b

Table 5.1. *(Contd.)*

Saponin	Description	Aglycone	Sugar	Reference
Bacopasaponin E	3-*O*-[α-L-Arabinofuranosyl (I→2)β-D-glucopyranosyl]-20-*O*-(α-L-arabinofuranosyl) jujubogenin	Major ebelin lactone	D-Glucose, L-Arabinose	Mahato et al., 2000
Bacopasaponin F	3-*O*-[α-L-Arabinofuranosyl (I→2)β-D-glucopyranosyl]-20-*O*-(α-L-arabinopyranosyl) jujubogenin	Major ebelin lactone	D-Glucose, L-Arabinose	Mahato et al., 2000
Bacopasaponin G	3-O-[α-L-Arabinofuranosyl-(1→2)]-α-L-arabinopyranosyl jujubogenin	Ebelin lactone	Arabinose	Hou et al., 2002
Bacopaside I	3-O-[α-L-Arabinofuranosyl-(1→2)]-[6-O-sulfonyl-β-D-glucopyranosyl-(1→3)]-α-L-arabinopyranosyl pseudojujubogenin	Bacogenin A_1	Arabinose, glucose	Chakravarty et al., 2001
Bacopaside II	3-O-α-L-Arabinofuranosyl-(1→2)-[β-D-glucopyranosyl-(1?3)]-β-D-glucopyranosyl pseudojujubogenin	Bacogenin A_1	Arabinose, glucose	Chakravarty et al., 2001
Bacopaside III	3-O-[6-O-sulfonyl-β-D-glucopyranosyl-(1→3)]-α-L-arabinopyranosyl pseudojujubogenin	Bacogenin A_1	Arabinose, glucose	Hou et al., 2002
*Bacopaside III	3-O-α-L-Arabinofuranosyl-(1→2)-β-D-glucopyranosyl jujubogenin	Ebelin lactone	Arabinose, glucose	Chakravarty et al, 2003
Bacopaside IV	3-O-β-D-Glucopyranosyl-(1→3)-α-L-arabinopyranosyl jujubogenin	Ebelin lactone	Arabinose, glucose	Chakravarty et al, 2003
Bacopaside V	3-O-β-D-Glucopyranosyl-(1?3)-α-L-arabinopyranosyl pseudojujubogenin	Bacogenin A_1	Arabinose, glucose	Chakravarty et al, 2003

Table 5.1. *(Contd.)*

Saponin	Description	Aglycone	Sugar	Reference
Bacopaside A	(3R)-1-octan-3-yl-(6-O-sulfonyl)-β-D-glucopyranoside	(3R)-3-Hydroxy-1-octene	Glucose	Hou et al., 2002
Bacopaside B	3,4-Dihydroxyphenylethyl alcohol (2-O-feruloyl)-β-D-glucopyranoside	3,4-Dihydroxyphenyl ethyl alcohol	Glucose	Hou et al., 2002
Bacopaside C	Phenylethyl alcohol[5-O-p-hydroxy benzoyl-β-D-glucopyranoside	Phenylethyl alcohol	Glucopyranose, apiofuranose	Hou et al., 2002
Monnierin m.p. 263-5°C	A triterpene saponin	A triterpenoid sapogenin	Glucose, arabinose	Dutta and Basu, 1963

*Bacogenins A_1, A_2 are acid catalysed transformation products of the genuine sapogenin pseudojujubogenin, during hydrolysis. Ebelin lactone is an artifact formed from the genuine sapogenin jujubogenin, during the acid hydrolysis.

in certain parts of the brain and increase RNA and protein turnover, suggesting a probable induction of membrane dephosphorylation by the bacosides (Singh et al., 1990; Dhawan and Singh, 1996). *B. monnieri* was shown to protect from phenytoin-induced cognitive deficit in mice (Vohora et al., 2000). Different extracts of the plant had anticholinesterase and cognitive enhancing activities and antidementic property (Das et al., 2002).

Aqueous and ethanolic extracts of the plant did not protect from the chemical or electric shock-induced seizures but did reduce their severity. They did not produce any sedation or antidepressant effect, but potentiated the barbiturate-induced hypnosis. They also increased the pentylenetetrazole-induced seizure latency, suggesting that a longer duration of treatment with the extracts of the plant may be beneficial in petit mal epilepsy (Rao and Karanth, 1992). It was earlier shown that hersaponin, when isolated from the plant, showed sedative and hypnotic potentiating activity in mice (Malhotra et al., 1960). It also reduced the noradrenaline and 5-HT content in rat brain (Malhotra et al., 1961).

The ethanolic extract and, to a lesser extent, the aqueous extract, showed a tranquillizing effect in albino rats and dogs. It was found to be a smooth muscle relaxant and an antispasmodic. However, it did not interfere with responses to adrenaline, acetylcholine and carotid occlusion (Aithal and Sirsi, 1961). An active fraction of the extract has tranquillizing properties, though they are much weaker than chlorpromazine (Ganguly and Malhotra, 1967), while bacosides A and B have a certain anti-depressant property (Singh et al., 1990).

B. monniera showed anti-ulcerogenic activity through its effect on mucosal defensive factors (Sairam et al., 2001), anti-*Helicobacter pylori* activity and through increase of prostanoids (PgE, PG12) levels as shown in *in vitro* studies (Goel et al., 2003). Bacosides exhibited antistress effect and the experiments indicate that they have the potential to modulate the activities of Hsp70, cytochrome P450 and superoxide dismutase in the brain, there by preparing the brain to act under adverse conditions such as stress (Chowdhuri et al., 2002). Standardized extract of the plant exhibited a potent adaptogenic activity as indicated by the response in acute and chronic stress models (Rai et al., 2003). Methanolic extract showed significant antidepressant activity in forced swim and learned helplessness models of depression and the activity was comparable to that of imipramine (Sairam et al., 2002).

The alcoholic extract has cardiotonic, vasoconstrictor, sedative and neuromuscular blocking action (LD_{50} 0.33 g/kg, i.p. in albino rats), while glycosidic saponin fraction has cardiotonic action in normal and hypodynamic frog heart, sedative action in rats and guinea pigs and spasmodic action in smooth muscle of ileum of rabbit and guinea pig and the uterus of rat (LD_{50} 25 mg/kg, ip in albino rats). In these experiments, the death of the animal in higher doses was attributed to a possible respiratory failure due to neuromuscular blocking (Malhotra and Das, 1959). The alcoholic extract showed a positive inotropic effect in frog heart. In rabbit heart, it increased the coronary outflow, heart rate and amplitude of contraction. In rabbit jejunum, it decreased the motility. The mechanism of action is explained to be through β-adrenergic receptors (Khanna and Ahmad, 1992). In another experiment the crude extract of the plant and different fractions showed broncho-vasodilatory activity, which was attributed to the inhibition of calcium ion movement (Channa et al., 2003).

The total extract of the plant and alkaloid fraction inhibited respiration in rat brain homogenate in a dose-dependent manner and it was shown that 5-HT and LSD-25 had no influence on this effect (Dhalla et al., 1961). On the other hand, the total alkaloid fraction produced spasm of the skeletal muscle, initial stimulation and an eventual depression of respiration, initial stimulation of autonomic ganglia followed by blockade, rigidity and convulsions in mice (Das et al., 1961). Aqueous extract of the leaf exhibited dose-dependant depressive activity, as measured by actophotometer (Indurwade and Biyani, 2000). *B. monniera* has a protective role on morphine-induced brain mitochondrial enzyme activity where the enzyme levels were maintained at normal levels (Sumathi et al., 2002a). It also reduced the effects of morphine withdrawal in guinea-pig ileum *in vitro* (Sumathi et al., 2002b). *B. monniera* has thyroid stimulating activity along with lowering of lipid peroxidation, and increasing the activity of superoxide dismutase and catalase (Dar et al., 2002).

The ethanolic extract of the plant exhibited smooth muscle relaxant action when tested on ring segments of pulmonary arteries of guinea pig and rabbit, aorta of rabbit and tracheal preparation of guinea pig, with a possible mediation of prostacyclin compounds in all the tissues and *β*-adrenoceptors. The action was shown to be independent of endothelium and muscarinic receptors (Dar and Channa, 1997a). It also caused bronchodilation in experimental animals (Dar and Channa, 1997b) and this spasmolytic activity was shown to be due to inhibition of Ca^{+2} influx via both voltage- and receptor-operated Ca-channels of the cell membrane (Dar and Channa, 1999).

The ethanolic extract of the plant has also been found to have anticancer activity against Walker carcinosarcoma 256 in rat (Bhakuni et al., 1969) and Sarcoma-180 cell culture (Elangovan et al., 1995). Different extracts of the plant and bacoside A showed potent activity in brine shrimp lethality assay (bacoside A LD_{50} 38.3 μg/ml) (D'Souza et al, 2002). The methanolic extract showed potent mast cell stabilization activity which was comparable to that of the mast cell stabilizer disodium cromoglycate (Samiulla et al., 2001). The ethanolic extract of the plant has potent antioxidant properties (Tripathi et al., 1996). The extract inhibited $FeSO_4$ and cumene hydroperoxide induced lipid peroxidation in a dose-dependent manner (significant protection obtained at the dose of 500 μg/ml). A standardized extract of the plant (containing 50% bacoside A, CDRI, Lucknow) increased the levels of superoxide dismutase, catalase and glutathione peroxidase activities in the frontal cortical, striatal and hippocampus of rat brain (5-10 mg/kg, oral) and the activity was found to be comparable to deprenyl (Bhattacharya et al., 2000).

Methanolic extract inhibited the formation of reactive species and prevented DNA damage in astrocytes *in vitro* (Russo et al., 2003a). Seventy per cent methanolic extract and its butanolic fraction inhibited superoxide radical in polymorphonuclear cells *in vitro* (IC_{50} 69.33 μg/ml, 82.66 μg/ml respectively). Bioassay guided fractionation of butanolic fraction led to the isolation of bacoside A_3 with an IC_{50} of 10.22 μg/ml, with a maximum of 91.66 % inhibition at 50 μg/ml. Bacopasaponin C, a structural analogue of bacoside A_3 also inhibited the superoxide in PMN cells, though it was found to be less potent with an IC_{50} of 54.16 μg/ml (Pawar et al., 2001). Methanolic extract of the plant had free radical scavenging activity and provided protection against DNA damage in human non-immortalized fibroblasts (Russo et al, 2003b).

Since saponins are known to have antiprotozoal property, bacopasaponin C was studied for its anti-leishmanial property both in free form and in different delivery modes *viz.*, niosomes, microspheres and nanoparticles, and it was found to be very active in all forms. However, free form showed nephrotoxicity while in vesicular form it was found to be safe. Further, the activity was found to be parasite specific (Sinha et al, 2002).

Alcoholic extract of the plant exhibited hepatoprotective effect (Sumathi et al., 2001). Bacosine exhibited a moderate analgesic activity and was found to be opioidergic in nature (Vohora et al., 1997).

Clinical trials

In a study conducted on 35 cases of anxiety neurosis, brahmi was shown to provide a significant relief in symptoms and a quantitative reduction in the anxiety level and also improved mental functions. The treatment also reduced the levels of urinary vinyl mandelic acid (V.M.A.) and corticoids. The anti-anxiety effect was attributed to a possible adaptogenic property of brahmi (Singh and Singh, 1980). The plant is also effective in revitalizing intellectual functions in children (Sharma et al., 1987). In a double blind placebo-controlled clinical trial, *B. monnieri* significantly improved speed of visual information processing, learning rate, memory consolidation and anxiety state, which indicate that it may improve higher order cognitive processes that are critically dependent on learning and memory (Stough et al., 2001). In another double-blind randomized placebo controlled clinical study (76 adults – 40 to 65 years old) the plant showed significant effect on retention of new information while attention, verbal and visual short-term memory and the retrieval of pre-experimental knowledge, everyday memory function and anxiety levels were unaffected (Roodenrys et al., 2002). Bacosides isolated from the plant were well tolerated by normal healthy male human volunteers in a single dose and multiple doses administered for 4 weeks in double blind placebo controlled and phase-1 clinical trial (Singh and Dhawan, 1997).

BIOTECHNOLOGY

With the worldwide increase in the demand for herbal medicine and the concommitant increase in demand for herbal raw materials, the availability of good quality raw material has become a major bottleneck. In this context, biotechnology has plant tissue culture systems to offer as an alternative source of plant material as well as phytopharmaceuticals. In the last three decades, tissue cultures of a large number of medicinal plants have been developed and the production of secondary metabolites of interest in the tissue cultures has been demonstrated.

However, very little work has been reported on tissue culture aspects of *B. monnieri*. Initially, abnormal plantlets were regenerated from *B. monnieri* (Thakur et al., 1976). Later, the plant has been used to study the mechanisms underlying the process of morphogenesis in plant cells and tissues, leading to adventitious organogenesis and an eventual regeneration of whole plant (Thakur and Ganapathy, 1978). A correlation between the explant size/ orientation of the explant and the extent of adventitious shoot and root induction was studied. Internodal segments (>4 mm long) differentiated shoot buds mostly at the distal end and roots at the proximal end in Nitsch and Nitsch medium.

A rapid protocol for the micropropagation of the plant from nodal explants was developed by Ali et al. (1996) from the nodal segments. The regenerants were found to be superior to normal plants with the former producing higher amounts of both primary and secondary metabolites (Tejavathi and Rao, 1998). Whole plants were regenerated from stem, leaf and flower bud explants in the presence of combinations of different plant growth regulators, with an intervening stage of callus. The callus formation offers a possible somaclonal variation in the population of the regenerated plants. The best response was obtained with IAA(0.1 mg/ml) + kinetin (1.0 mg/l) in stem and leaf explants and IAA (0.1 mg/l) + 6-iso-pentyladenine (0.1 mg/l) in flower bud explants (Tejavathi and Shailaja, 1999) with multiple shoots and roots being produced in the same medium.

Bacopa monnieri has a high morphogenic potential as the explants readily regenerated shoot buds in medium containing small amounts of cytokinins like benzyladinine or kinetin (Shrivastava and Rajani, 1999). Of the different auxins and cytokinins used, either singly or in combination, the best response—from both stem and leaf explants—was obtained with 2 μM benzyladenine in Murashige and Skoog medium gelled with 0.2% gelrite, where a profuse induction of multiple shoot buds was obtained within 9 days of incubation. The same medium supported the eventual growth of the shoot cultures as well. Further, after 3 weeks of growth of shoot cultures, when the shoots were harvested and the explant base was transferred into a fresh medium, it continued to expand exponentially and regenerate new shoot buds. Further, the leaf explants from shoot cultures proved to be of superior explant material than the leaf explants from field-grown plants. The shoot cultures performed better when grown on a medium gelled with gelrite (0.2%) rather than with agar (0.7%). The phytochemical profile of the regenerated shoots was found to be similar to that of field-grown plants, as revealed by TLC analysis. With the high regeneration potential, the stem and leaf explants showed a tendency to regenerate shoots and/or roots even in media which are known to support callus initiation and growth (e.g. media supplemented with 2,4-dichlorophenoxyacetic acid), while the callusing response of the explants was very poor (Shrivastava and Rajani, 1999).

From the above discussion, it is clear that only certain aspects of morphogenesis and micropropagation of *B. monnieri* have so far been worked out. Extensive research in the lines of detailed phytochemical characterization of the different types of cultures, especially with respect to the active principles and the influence of culture conditions and other factors on the chemical constituents, etc., is needed.

CONCLUDING REMARKS

Brahmi has been extensively investigated for its chemical constituents, especially the saponins, and to a certain extent, its pharmacological activities. However, not much attention has been paid on the tissue culture aspects of the plant. In Ayurveda texts, the plant is prescribed to be collected in the months of October-November, which limits the availability of good quality of raw material to a specific time of the year. In this scenario, more extensive work on the tissue culture systems of the plant can be expected to offer a commercially-viable alternative, with an assurance of a continuous supply of good quality raw material throughout the year.

ACKNOWLEDGEMENTS

The authors are thankful to Prof. Harish Padh, Director, B.V. Patel, PERD Centre, Ahmedabad, for providing facilities. The authors also thank Mr Ajay D. Pillai for his help, especially in preparing sketches of the structures.

REFERENCES

Aithal, H.N. and Sirsi, M., 1961, Pharmacological investigation on *Herpestis monniera*. Indian J. Pharm. 23(1):2-5.

Ali, G., Purohit, M., Mughal, M.H., Iqbal, M. and Srivastava, P.S., 1996, A rapid protocol for micropropagation of *Bacopa monnieri* (L.) Wettst., a medicinal plant. Plant Tiss. Cult. Biotech. 2:208-211.

Anonymous 1, 1999, The Ayurvedic Pharmacopoeia of India, Part-I, vol. II, pp. 25-26, Govt. of India, Ministry of Health & Family Welfare, Dept. of Indian Systems of Medicine, New Delhi, India.

Anonymous 2, 1994, BRAHMI (Mal. BRAHMI). In:Ayurvedic drugs and their plant sources, 97-99. Sivarajan, V.V. and Balachandran, I. (eds). Oxford & IBH Publishing Co. Pvt. Ltd., New Delhi.

Anonymous 3, 1978, In:Ayurvedic Formulary of India, Part I, I[st] edn, Government of India Press, Faridabad.

Anonymous 4, 1997, In: Indian Medicinal Plants:A sector study. Occasional paper No. 54, 2. Export-Import Bank of India, Quest Publications, Bombay, India.

Anonymous 5, 1998, In: Indian Herbal pharmacopoeia. Vol. I., 30-37. S.S. Handa, Mundkinajeddu and A.K. Mangal. (eds.) Publ. Regional Research Laboratory, Jammu Tawi & Indian Drug Manufacturers' Association, Mumbai.

Anonymous 6, 1976, BACOPA (Scrophulariaceae) In: Indian Medicinal Plants, Vol. I, 118-122. Satyavati G.V., Raina M.K. and Sharma M. (eds). Indian Council of Medical Research, New Delhi.

Basu, N., Rastogi, R.P. and Dhar, M.L., 1967, Chemical examination of *Bacopa monniera* Wettst. Part III. Bacoside B. Indian J. Chem. 5:84 –86.

Basu, N.K. and Walia, J.S., 1944, The chemical investigation of the leaves of *Herpestis monniera*. Indian J. Pharm. 6:84-85, 91.

Basu, N.K. and Pabrai, P.R., 1947, Investigations on Indian medicinal plants. *Monniera cuneifolia*. Quart. J. Pharm. Pharmacol. 20:137-139.

Bhakuni, D.S., Dhar, M.L., Dhar, M.M., Dhawan, B.N. and Mehrotra, B.N., 1969, Screening of Indian plants for biological activity: Part II. Indian J. Exp. Biol. 7:250-262.

Bhattacharya, S.K., Bhattacharya, A., Kumar, A. and Ghosal, S., 2000, Antioxidant activity of *Bacopa monniera* in rat frontal cortex, striatum and hippocampus. Phytother. Res. 14:174-179.

Bose, K.C. and Bose, N.K., 1931, J. Indian Med. Assoc. 1: 60 — cited in Sastry et al. (1959) and Rastogi and Dhar (1960).

Chakravarty, A.K., Garai, S., Masuda, K., Nakane, T. and Kawahara, N., 2003, Bacopasides III-V: Three new triterpenoid glycosides from *Bacopa monniera*. Chem. Pharm. Bull. 51(2): 215-217.

Chakravarty, A.K., Sarkar, T., Masuda, K., Shiojima, K., Nakane, T. and Kawahara, N., 2001, Bacopaside I and II: Two pseudojujubogenin glycosides from *Bacopa monniera*. Phytochemistry 58(4): 553-556.

Chakravarty, A.K., Sarkar, T., Nakane, T., Kawahara, N. and Masuda, K., 2002, New phenylethanoid glycosides from *Bacopa monniera*. Chem. Pharm. Bull. 50(12): 1616-1618.

Chandel, R.S., Kulshreshtha, D.K. and Rastogi, R.P., 1977, Bacogenin-A_3: A new sapogenin from *Bacopa monniera*. Phytochemistry 16: 141-143.

Channa, S., Dar, A., Yaqoob, M., Anjum, S., Sultani, Z. and Atta-ur-Rahman., 2003, Broncho-vasodilatory activity of fractions and pure constituents isolated from *Bacopa monniera*. J. Ethnopharmacol. 86(1): 27-35.

Chatterji, N., Rastogi, R.P. and Dhar, M.L., 1963, Chemical examination of *Bacopa monniera* Wettst.: Part I — Isolation of chemical constituents. Indian J. Chem. 1:212-215.

Chatterji, N., Rastogi, R.P. and Dhar, M.L., 1965, Chemical examination of *Bacopa monniera* Wettst.: Part II — The constitution of bacoside A. Indian J. Chem. 3:24-29.

Chopra, I.C., Handa, K.L. and Sobti, S.N., 1956a, Need for the cultivation of vegetable drugs used in Ayurvedic and Unani medicines. Indian J. Pharm. 18: 364-377.

Chopra, R.N., Nayar, S.L. and Chopra, I.C., 1956b, In: Glossary of Indian Medicinal Plants, 32. Council of Scientific and Industrial Research, New Delhi.

Chowdhuri, D.K., Parmar, D., Kakkar, P., Shukla, R., Seth, P.K. and Srimal, R.C., 2002, Antistress effects of bacosides of *Bacopa monnieri*: Modulation of Hsp70 expression, superoxide dismutase and cytochrome P450 activity in rat brain. Phytother. Res. 16(7): 639-645.

Dar, A. and Channa, S., 1997a, Relaxant effect of ethanol extract of *Bacopa monniera* on trachea, pulmonary artery and aorta from rabbit and guinea pig. Phytother. Res. 11:323-325.

Dar, A. and Channa, S., 1997b, Bronchodilatory and cardiovascular effects of an ethanol extract of *Bacopa monniera* in anaesthetized rats. Phytomedicine 4(4):319-323.

Dar, A. and Channa, S., 1999, Calcium antagonistic activity of *Bacopa monniera* on vascular and intestinal smooth muscles of rabbit and guinea-pig. J. Ethnopharmacol. 66:167-174.

Das, P.K., Malhotra, C.L. and Dhalla, N.S., 1961, Studies on alkaloids of *Herpestis monniera* Linn. Indian J. Physiol. Pharmacol. 5:136-143.

Das, A., Shanker, G., Nath, C., Pal, R., Singh, S and Singh, H., 2002, A comparative study in rodents of standardized extracts of *Bacopa monniera* and *Ginkgo biloba*: Anticholinesterase and cognitive enhancing activities. Pharmacol. Biochem. Behav. 73(4): 893-900.

Dhalla, N.S., Sastry, M.S. and Malhotra, C.L., 1961, Some *in vitro* effects of *Herpestis monniera* Linn. On the respiration of rat brain. Indian J. Med. Res. 49:781-787.

Dhawan, B.N. and Singh, H.K., 1996, Pharmacology of ayurvedic nootropic *Bacopa monniera*. International Convention of Biological Psychiatry, 8-10, January, 1996, Bombay, India: 21 pp. Abstr. No. NR 59.

D'Souza, P., Deepak, M., Rani, P., Kadamboor, S., Mathew, A., Chandrashekar, A.P. and Agarwal, A., 2002, Brine shrimp lethality assay of *Bacopa monnieri*. Phytother Res. 16(2): 197-198.

Dutta, T. and Basu, U.P., 1963, Terpenoids: Part II — Isolation of a new triterpene saponin, monnierin, from *Bacopa monniera* Wettst. Indian J. Chem. 1:408-409.

Elangovan, V., Govindasamy, S., Ramamoorthy, N., Balasubramanian, K., 1995, In vitro studies on the anticancer activity of *Bacopa monnieri*. Fitoterapia 66(3):211-215.

Ganguly, D.K. and Malhotra, C.L., 1967, Some behavioral effects of an active fraction from *Herpestis monniera* Linn. (Brahmi). Indian J. Med. Res. 55:473-482.

Garai, S., Mahato, S.B., Ohtani, K. and Yamasaki, K., 1996a, Dammarane-type triterpenoid saponins from *Bacopa monniera*. Phytochemistry 42(3):815-820.

Garai, S., Mahato, S.B., Ohtani, K. and Yamasaki, K., 1996b, Bacopasaponin D — A pseudojujubogenin glycoside from *Bacopa monniera*. Phytochemistry 43(2):447-449.

Gupta, A.P., Mathur, S., Gupta, M.M. and Sushil Kumar., 1998, Effect of the method of drying on the bacoside-A content of the harvested *Bacopa monniera* shoots revealed using a high performance thin layer chromatography method. J. Med. Arom. Plant Sci. 20:1052-1055.

Handa, S.S., 1996, Rasaayana drugs. In: Supplement To Cultivation and Utilization of Medicinal Plants, 510-524. S.S. Handa and M.K. Kaul (eds). Publ. RRL, Jammu-Tawi.

Hou, C.C., Lin, S.J. and Hsu, F.L., 2002, Bacopaside III, bacopasaponin G, and bacopasides A, B, and C from *Bacopa monniera*. J. Nat. Prod. 65(12): 1759-1763.

Indurwade, N.H. and Biyani, K.R., 2000, Evaluation of comparative and combined depressive effect of brahmi, shankhpushpi and jatamansi in mice. Indian J. Med. Sci. 54:339-341.

Jain, P. and Kulshreshtha, D.K., 1993, Bacoside A_1, a minor saponin from *Bacopa monniera*. Phytochemistry 33(2):449-451.

Kar, A., Panda, S. and Bharti, S., 2002, Relative efficacy of three medicinal plant extracts in the alteration of thyroid hormone concentrations in male mice. J. Ethnopharmacol. 81(2): 281-285.

Kawai, K., Iitaka, Y., Shibata, S., Kulshreshtha, D.K. and Rastogi, R.P., 1973, Crystal and molecular structure of bacogenin-A_1 dibromoacetate. Acta Crystallogr. Sect. B.(Pt. 12) 29:2947-2953.

Khanna, T. and Ahmad, B., 1992, Some β-adrenergic activity of saponins isolated from ethanolic extract of *Bacopa monniera*. Proc. of the Conference on Trends in Molecular & Cellular Cardiology, Lucknow, 4-5 May 1992. 20 pp.

Kulshreshtha, D.K. and Rastogi, R.P., 1973a, Chemical examination of *Bacopa monniera*. IV. Bacogenin—A_1: A novel dammarane triterpene sapogenin from *Bacopa monniera*. Phytochemistry 12:887-892.

Kulshreshtha, D.K. and Rastogi, R.P., 1973b, Identification of ebelin lactone from bacoside A and the nature of its genuine sapogenin. Phytochemistry 12:2074-2076.

Kulshreshtha, D.K. and Rastogi, R.P., 1974, Bacogenin A_2: A new sapogenin from bacosides. Phytochemistry 13:1205-1206.

Mahato, S.B., Garai, S. and Chakravarty, A.K., 2000, Bacopasaponins E and F: two jujubogenin bisdesmosides from *Bacopa monniera*. Phytochemistry 53:711-714.

Malhotra, C.L. and Das, P.K., 1959, Pharmacological studies of *Herpestis monniera* Linn. (Brahmi). Indian J. Med. Res. 47:294-305.

Malhotra, C.L., Das, P.K. and Dhalla, N.S., 1960, Some neuropharmacological actions of hersaponin — an active principle from *Herpestis monniera* Linn. Arch. Int. Pharmacodyn. 129:290-302.

Malhotra, C.L., Prasad, K., Dhalla, N.S. and Das, P.K., 1961, Effect of hersaponin and Acorus oil on noradrenaline and 5-hydroxytryptamine content of rat brain. Indian J. Pharm. Pharmacol. 13:447-448.

Nadkarni, KM., 1954, The Indian Materia Medica, 3rd ed. Vol. I. Popular Book Depot., Bombay., 1315. pp.

Pal, R. and Sarin J.P.S., 1992, Quantitative determination of bacosides by UV-spectrophotometry. Indian J. Pharm. Sci. 54:17-18.

Pawar, R., Gopalakrishnan, C. and Bhutani, K.K., 2001, Dammarane triterpene saponin from *Bacopa monniera* as the superoxide inhibitor in polymorphonuclear cells. Plant Med. 67: 752-754.

Prakash, J.C. and Sirsi, M., 1962, Comparative study of the effects of Brahmi (*Bacopa monnieri*) and chlorpromazine in motor learning in rats. J. Sci. Ind. Res. 21C:93-96.

Proliac, A., Chabaud, A. and Raynaud, J., 1991, Pharm. Acta Helv. 66:153 — cited in Anonymous 1998, Indian Herbal pharmacopoeia. Vol. I.

Rai, D., Bhatia, G., Palit, G., Pal, R., Singh, S. and Singh, H.K., 2003, Adaptogenic effect of *Bacopa monniera* (Brahmi). Pharmacol. Biochem. Behav. 75(4): 823-830.

Rao, G.M.A. and Karanth, K.S., 1992, Neuropharmacological activity of *Herpestis monniera*. Fitoterapia 63:399-404.

Rastogi, R.P. and Dhar, M.L., 1960, Chemical examination of *Bacopa monniera* Wettst. J. Sci. Ind. Res. 19B:455-456

Rastogi, S., Pal, R. and Kulshreshtha, D.K., 1994, Bacoside A_3 — A triterpenoid saponin from *Bacopa monniera*. Phytochemistry 36(1):133-137

Rastogi, S., and Kulshreshtha, D.K., 1998, Bacoside A_2 — A triterpenoid saponin from *Bacopa monniera*. Indian J. Chem. 38B:353-356

Renukappa, T., Roos, G., Klaiber, B. and Kraus, W.V., 1999, Application of high-performance liquid chromatography coupled to magnetic resonance spectrometry, mass spectrometry and bioassay for the determination of active saponins from *Bacopa monniera* Wettst. J. Chromatogr. A 847:109-116.

Roodenrys, S., Booth, D., Bulzomi, S., Phipps, A., Micallef, C. Smoker, J., 2002, Chronic effects of Brahmi (*Bacopa monnieri*) on human memory. Neuropsychopharmacology 27(2): 279-281.

Russo, A., Borrelli, F., Campisi, A., Acquaviva, R., Raciti, G. and Vanella, A., 2003a, Nitric oxide-related toxicity in cultured astrocytes : Effect of *Bacopa monniera*. Life Sci. 73(12): 1517-1526.

Russo, A., Izzo, A.A., Borrelli, F., Renis, M. and Vanell, A., 2003b, Free radical scavenging capacity and protective effect of *Bacopa monniera* L. on DNA damage. Phytother. Res. 17(8): 870-875.

Sairam, K., Dorababu, M., Goel, R.K. and Bhattacharya, S.K., 2002, Antidepressant activity of standardized extract of *Bacopa monniera* in experimental models of depression in rats. Phytomedicine, 9(3): 207-211.

Sairam, K., Rao, C.V., Babu, M.D. and Goel, R.K., 2001, Prophylactic and curative effects of *Bacopa monniera* in gastric ulcer models. Phytomedicine 8(6): 423-430.

Samiulla, D.S., Prashanth, D. and Amit, A., 2001, Mast cell stabilizing activity of *Bacopa monnieri*. Fitoterapia 72:284-285.

Sastry, M.S., Dhalla N. S. and Malhotra, C.L., 1959, Chemical investigation of *Herpestis monniera* Linn. (BRAHMI). Indian J. Pharm. 21:303-304.

Schulte, K.E., Ruecker, G. and El-Kersch, M., 1972, Components of medicinal plants. XXXVI. Nicotine and 3-formyl-4-hydroxy-2H-pyran from *Herpestis monniera*. Phytochemistry 11(8):2649-51.

Sharma, R., Chaturvedi, C. and Tewari, P.V., 1987, Efficacy of *Bacopa monnieri* in revitalising intellectual function in children. J. Res. Edu. Indian Med. 6(1-2):1-10.

Shrivastava, N. and Rajani, M., 1999, Multiple shoot regeneration and tissue culture studies on *Bacopa monnieri* (L.) Pennell. Plant Cell Rep. 18:919-923.

Shukla, B., Khanna, N.K. and Godhwani, J.L., 1987, Effect of brahmi rasayan on central nervous system. J. Ethnopharmacol. 21:65-74.

Singh, H.K. and Dhawan, B.N., 1978, The effect of *Bacopa monnieri* on the learning ability of rats. Indian J. Pharmacol. 10:72.

Singh, H.K. and Dhawan, B.N., 1982, Effect of *Bacopa monniera* Linn. (Brahmi) extract in avoidance responses in rats. J. Ethnopharmacol. 5:205-214.

Singh, H.K. and Dhawan, B.N., 1994, Preclinical neuro-psychopharmacological investigations on bacosides: A nootropic memory enhancer. Update Ayurveda-94, Mumbai, India, 74 pp. 24-26th February 1994 .

Singh, H.K. and Dhawan, B.N., 1997, Neuropsychopharmacological effects of the ayurvedic nootropic *Bacopa monniera* Linn. (*BRAHMI*). Indian J. Pharmacol. 29:S359-S365.

Singh, H.K., Rastogi, R.P., Srimal, R.C. and Dhawan, B.N., 1988, Effect of bacosides A and B on avoidance responses in rats. Phytother. Res. 2:70-75.

Singh, H.K., Srimal, R.C., Srivastava, A.K., Garg, N.K. and Dhawan, B.N., 1990, Neuro-psychopharmacological effects of bacosides A & B. Proceedings of the Fourth Conference on the Neurobiology of Learning and Memory, California, 17-20 October, 1990. 79 pp.

Singh, R.H. and Singh, L., 1980, Studies on the anti-anxiety effect of the medhya rasayana drug, brahmi (*Bacopa monniera* Wettst.) — Part I. J. Res. Ayur. Siddha (Quarterly) 1(1):133-148.

Sinha, J., Raay, B., Das, N., Medda, S., Garai, S., Mahato, S.B. and Basu, M. K., 2002, Bacopasaponin C: Critical evaluation of anti-leishmanial properties in various delivery modes. Drug Delivery 9: 55-62.

Stough, C., Lloyd, J., Clarke, J., Downey, L.A., Hutchison, C.W., Rodgers, T. and Nathan, P.J., 2001, The chronic effects of an extract of *Bacopa monniera* (Brahmi) on cognitive function in healthy human subjects. Psychopharmacology (Berl). 156(4): 481-484.

Sumathi, T., Govindasamy, S., Balakrishna, K. and Veluchamy, G., 2002a, Protective role of *Bacopa monniera* on morphine-induced brain mitochondrial enzyme activity in rats. Fitoterapia73(5): 381-385.

Sumathi, T., Nayeem, M., Balakrishna, K., Veluchamy, G. and Devaraj, S.N., 2002b, Alcoholic extract of '*Bacopa monniera*' reduces the *in vitro* effects of morphine withdrawal in guinea-pig ileum. J. Ethnopharmacol. 82(2-3): 75-81.

Sumathi, T, Subramanian, S., Govindasamy, S., Balakrishna, K. and Veluchamy, G., 2001, Protective role of *Bacopa monniera* on morphine induced hepatotoxicity in rats. Phytother Res. 15(7): 643-645.

Tejavathi, D.H. and Rao, N.N., 1998, Comparative studies between normal and tissue cultured plants of *Bacopa monnieri* (L.) Pennell. In: Role of Biotechnology in Medicinal and Aromatic Plants. Vol. I. 239-247. Khan I.A. and Khanum A. (eds). Ukaaz Publications, Hyderabad, India.

Tejavathi, D.H. and Shailaja, K.S., 1999, Regeneration of plants from the cultures of *Bacopa monnieri* (L.) Pennell. Phytomorphology 49:447-452.

Thakur, S. and Ganapathy, P.S., 1978, Morphogenesis of organ differentiation in *Bacopa monnieri* stem cultures. Indian J. Exp. Biol. 16:514-516.

Thakur, S., Ganapathy, P.S. and Johri, B.M., 1976, Differentiation of abnormal plantlets of *Bacopa monnieri*. Phytomorphology 26:422-425.

Tripathi, Y.B., Chaurasia, S., Tripathi, E., Upadhyay, A. and Dubey, G.P., 1996, *Bacopa monniera* Linn. as an antioxidant: mechanism of action. Indian J. Exp. Biol. 34(6):523-526.

Vohora, S.B., Khanna, T., Athar, M. and Ahmad, B., 1997, Analgesic activity of bacosine, a new triterpene isolated from *Bacopa monnieri*. Fitoterapia 68(4):361-365.

Vohora, D., Pal, S.N. and Pillai, K.K., 2000, Protection from phenytoin-induced cognitive deficit by *Bacopa monniera*, a reputed Indian nootropic plant. J. Ethnopharmacol. 71(3): 383-390.

CHAPTER 6

Chlorophytum borivilianum: An Endangered Aphrodisiac Herb

Dilip K. Arora, Ashok K. Jain, K.G. Ramawat and *J.M. Merillon*

INTRODUCTION

Chlorophytum comosum (commonly known as spider plant) along with several other species comprise indoor plants. A few *Chlorophytum* species possess strong pharmacological activity and are little known. These include *C. arundinaceum, C. attenuatum, C. borivilianum, C. laxum, C. tuberosum,* and *C. breviscapum,* all commonly known as 'Safed musli' because of their white tuberous roots (Fig. 6.1). Out of these six species, *C. arundinaceum* and *C. borivilianum* are the most important species of commerce. The economic part of the herb is roots, which are used to cure general debility. The roots are usually consumed in a powdered form with milk or mixed with sweet preparations. The price of dried roots has increased in recent years because tuberous roots of *C. borivilianum* are widely used as an important ingredient in several aphrodisiac formulations in India (Kirtikar and Basu, 1975; Ramawat et al., 1998).

Chlorophytum borivilianum is of economic importance to India, since its dried roots are exported in bulk quantity to the Middle East. The present demand is met from the forests, but the plant population is dwindling at an alarming rate. Tribals in Rajasthan, Gujarat and Madhya Pradesh (provinces of India) collect the roots from forest before the onset of autumn, dry them in the sun and sell the produce to either co-operatives or directly to local traders, who in turn, despatch these to larger markets to the end-user industries. The size and colour of the roots determine their cost; white roots fetch higher prices than the pale-yellow or brown roots. The current annual demand is estimated around 500 tonnes. Increasing demand, however, has raised serious questions about over-exploitation of this species. A few of these plants have become 'endangered species' because of over-exploitation (Nayar and Sastri, 1988). The increasing demand can be met by domestication of the plant, which requires a study of several aspects of

Fig. 6.1 Tuberous roots produced by a mature plant of *C. borivilianum*

the plant's biology, agronomy, reproduction and vegetative multiplication. Much work has been carried out in recent years on cultivation trials (Bordia et al., 1995), biology (Ramawat et al., 1996), phytochemistry (Tandon and Shukla, 1995) and the pharmacology (our unpublished results) of *Chlorophytum* species. Due to the scarcity of the expensive produce,

C. borivilianum, the cheap material, *Asparagus ascendense* and *C. tuberosum* are mixed to it (Tandon and Shukla, 1995). This chapter attempts to present a synoptic coverage of the available technology about the scientifically little-described medicinal herb.

GEOGRAPHIC DISTRIBUTION

The family Liliaceae contains about 254 genera comprising about 4075 species, scattered in both the hemispheres of the earth. In India, this family is represented by 35 genera and about 189 species occurring chiefly in the Himalayas. Important plants of this family include *Allium, Aloe, Asparagus, Chlorophytum, Cholchicum, Gloriosa, Yucca*, etc. Several species of *Chlorophytum* are cultivated for their ornamental value. There are about 215 species in the genus *Chlorophytum* (Liliaceae). Most species have originated from Africa and are distributed throughout the warmer regions of the world (Li et al., 1990; Kativu and Nordal, 1993). Thirteen species are known to occur in India (Sheriff and Chennaveeraiah, 1972), while eight are considered endemic to the subcontinent (Nair, 1974). Four species are described in local ethnobotanical uses, these are: *C. arundinaceum* Bak., *C. borivilianum* Sant. and Fern., *C. laxum* and *C. tuberosum* Bak. (Jain, 1991). In India, *C. borivilianum* is restricted to Gujarat, Maharashtra, Madhya Pardesh and Rajasthan provinces with sporadic occurrence. In Rajasthan, it occurs in moist forests or reserve sanctuaries in the Aravalli Hills of Udaipur, Banswara and Chittorgarh districts. *C. arundianaceum* is mainly distributed in the foothills of North-Eastern Himalayas (Assam, Bihar and West Bengal), Orissa, parts of the Vindhya, the Satpura and Himalayas upto 5300 feet. *C. attenuatum* in the West Peninsula, *C. breviscapum* in Sikkim, Western Ghats, *C. laxum* in the Deccan Peninsula and *C. tuberosum* is most widely distributed throughout India along with other species. *C. tuberosum* is not commonly used, but certainly administered as an adulterant with the commercial material. Perhaps this is the reason that this species is still present in abundance throughout the country. Other species have restricted distribution as they vanished from common sites in major parts of the country. Commercial cultivation of 'Safed musli' is just beginning in the country.

TAXONOMY

Santapau and Fernandes (1955) described *C. borivilianum* for the first time from India (Fig. 6.2). According to Nayar and Sastry (1988), important taxonomic features of *C. borivilianum* are: rhizomatous herb with 1-9 sessile roots; tubers upto 5 cm long, 1 cm in diameter; leaves radially, cariaceous, spirally imbricated at the base, sessile, linear, acute, flat, 8-22 × 0.7-1.8 cm; the flower is a raceme, white, 2.5-3.5 cm across, bracteate, pedicillate, in

Fig.6.2 Flowering in *C. borivilianum*

clusters of 3, bracts linear, papery, purplish; capsules greenish-yellow, obcordate, trimerous, loculicidal, seeds orbicular, discoid and black. More recently, the taxonomic position of genus *Chlorophytum* has been revised (Marais and Reilly, 1978). It has been placed close to the genus *Anthericum* and is distinguished by the enlarged middle part of stamen filament and capsule with three wings.

CYTOLOGY

Cytological studies of the genus *Chlorophytum* suggest 7 and 8 basic chromosome number. *Chlorophytum* species are reported to be diploid, triploid, tetraploid and polyploid. *C. laxum*, a Liberian species has been described as diploid with 2n=14 (Baldwin and Speese, 1951). Comparison of mitotic and meiotic chromosomes of a number of collections showed similar karyomorphology within a species. These authors considered *C. laxum* as diploid and other species as tetraploids or derived diploids. As compared to this, the Indian material of *C. laxum* was found to be diploid with 2n=16 (Boraiah, 1966). The diploid chromosome number of some other Indian species recorded was 2n=16 for *C. tuberosum* and *C. laxum* (Kumar and Rao, 1958; Datta and Mitra, 1968) and *C. borivilianum* (Kumar and Subramanium, 1986). Pahuja and Kumar compared the chromosome number of several collections and recorded the chromosome number for *C. arundinaceum* (2n=42) and *C. tuberosum* (2n=16). They observed six pairs of satellite chromosomes in *C. arundinaceum* and one pair in *C. tuberosum*. Sherif and Chennaveeraiah (1972) made detailed studies of karyotype and idiograms in four diploid species of *Chlorophytum* and reported n=7 (*C. heyneanum*) and n=8 (*C. glaucum* and *C. tuberosum*) as basic chromosome numbers in this genus.

The various features of chromosomes of different species indicated the primitiveness of the karyotype under 8 basic number. However, Baldwin and Speese (1951) considered 8 as the basic number on the basis of Liberian material for the genus through duplication of one of the chromosomes. In our laboratory, *Chlorophytum borivilianum* plants raised through organogenesis contained 2n=24 chromosomes, while those derived from somatic embryogenesis contained 2n= 21 to 27 chromosomes. The parent plants are supposed to be triploid with 8 as the basic number (Arora et al., 1999).

BIOLOGY OF THE PLANT

Flowers emerge after 4-6 days of first rainfall during June in mature plants (having a few tuberous roots). Generally, one plant produces only one inflorescence but sometimes 3-4 inflorescences may be produced, depending upon the age of the mother plant. Older and thick root-bearing plants produce more inflorescences. The number of flowers per inflorescence increases with the increase in size of the underground tuberous roots. Ripened seeds are ready in two weeks after flowering. This is an insect-pollinated crop (Dalal et al., 1987) and the capsule matures in succession in a raceme inflorescence. Therefore, seed collection from the forest plants is difficult. Poor seed set and germination is common in *C. borivilianum* (Bordia, 1993; Ramawat et al., 1996). This may be due to chromosomal abnormalities at meiosis in the plant as various levels of ploidy, which is reported in the natural populations.

Chlorophytum borivilianum produced about 50-110 seeds per plant with 7-10 seeds per fruit; depending on the agro-climatic conditions and age of the plant, and 15% seeds were abortive. Existing embryos were of two different sizes, i.e. 31% were 1 mm and 69% were 2 mm long. Most of the small-sized embryos were non-viable (triphenyl tetrazolium chloride test). Viability of all the combined embryos was 79%, but the seed germination was only 15%. Moreover, viability of embryos was reduced to 27% on storage of seeds for one year at 5°C. This may be due to a lack of dormancy factor in the seeds (Ramawat et al., 1996). Application of 25 mg/l GA_3 increased the germination of seeds to 36%. Poor viability and germination of the seeds has made this plant an endangered species. Therefore, alternative methods of propagation are the necessary. Bio-technological approaches are best tool available for rapid vegetative multiplication.

PHYTOCHEMISTRY

Tuberous roots of *C. arundinaceum* contains steroidal saponins (neohecogenin, neotigogenin, stigmasterol, tokorogenin), a bibenzyl xyloside (2,4,4'-trihydroxy-2-xylopyranosyl bibenzyl), a disubstituted tetrahydrofuran (4-hydroxy-8,11-oxidoheneicosanol), nonacosone, tetracosanoic and triacontanoic acid, pentacosyl docosanoate, galactoglucan and sugars (arabinose, glucose and xylose) (Fig. 6.3, Tandon and Shukla, 1992, 1993; Tandon et al., 1992; Gupta et al., 1979). Saponins are common in most of the plants belonging to tribe anthericinae. They have been reported from *C. elatum* (Villar, 1948) and *C. malayense* (Li et al., 1990). Neohecogenin and neotigogenin are the usual genins for the genus *Chlorophytum*. Steroidal saponins, chloromaloside A-D were isolated from the tubers of *C. malayense* (Li et al., 1990). A steroidal glycoside, chloromaloside A isolated from *C. malayense*, has been shown to be cytotoxic against several human cancer cell lines (Qiu et al., 2000). Three new spirostanol pentaglycosides were isolated from the tuberous roots of *C. comosum* together with four known saponins (Mimaki et al., 1996). Comparison of aglycone structures of the saponins from *C. comosum* with those of *C. malayense*, showed that the former were based on the (25 R)-spirostan series, such as in neotigogenin and neohecogenin (Mimaki et al., 1996). The economic interest raised in these plants is from the use of steroidal saponins as a starting material for the industrial production of sex hormones, corticosteroids and steroid derivatives in general. The main source of saponins has been from higher plants, especially those belonging to the family Liliaceae, Solanaceae and Scrophulariaceae. Bibenzyl are commonly present in Orchidaceae, Cannabinaceae and Bryophytes (Sharma and Ramawat, 2000).

PHARMACOLOGY

The tuberous roots (mostly powdered) are widely used in the Indian system of medicine for the following properties: (1) as a non-hormonal

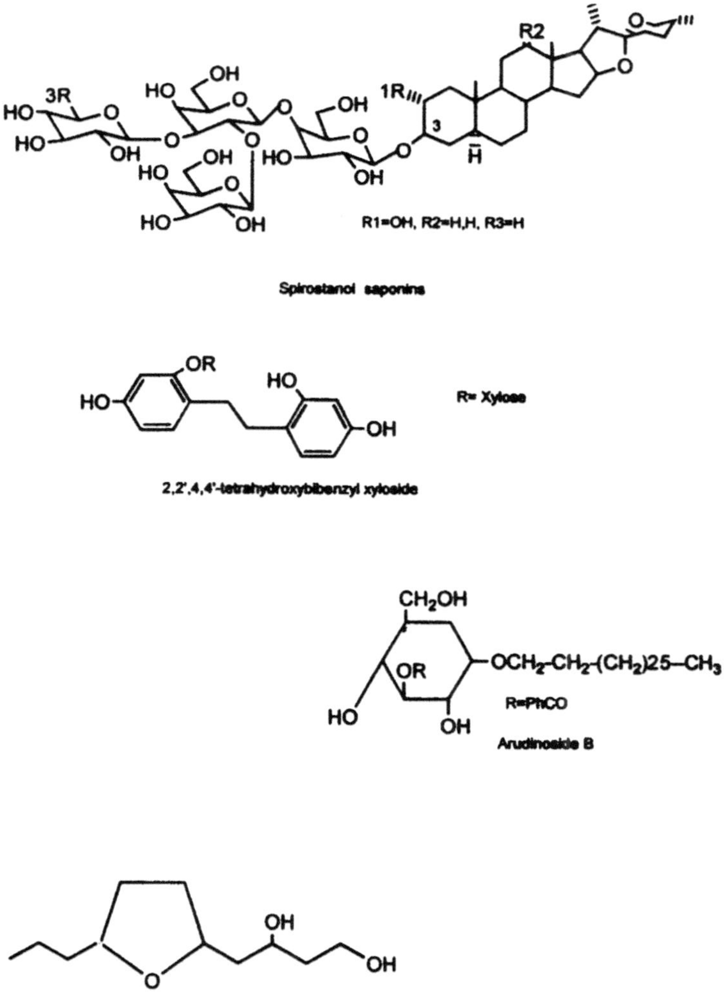

Fig. 6.3 Structures of metabolites isolated from *Chlorophytum* species

restorative tonic; (2) in fatigue, general debility, weakness and as a general purpose tonic; (3) used in impotency and sterility and to enhance male potency; (4) as cardiac and brain tonic; (5) as a curative agent in various diseases like piles, diabetes, albuminorrhoea-leucorrhoea, menorrhoea; and (6) as anti-pyretic, sialogogue, galactogue, diuretic, hemostatic properties.

These properties have been inferred on the basis of this root's use in folk medicine (Jain, 1991), traditional medicine and experiments on man volunteers (Ojha, 1987). Generally, *C. borivilianum* is used along with *Withania somnifera* and *Asparagus racemosus* in various commercial formulations (Ramawat et al., 1998). Three new saponins isolated from *C. comosum* (Mimaki et al., 1996) showed an inhibitory activity against ^{32}P incorporation into phospholipids of HeLa cells as the primary screening test to identify new anti tumour promoter compounds.

The extract prepared from the dried root tubers of *C. borivilianum* inhibited ^{3}H-dopamine — uptake in striatal synaptosomes, and thereby, could lead to enhanced dopamonergic tone in the central nervous system as also act as psychostimulants. This has a beneficial effect on the brain and resultantly human body by increasing alertness, mental ability and intelligence, sexual and maternal characters. The active principle, saponins, are known as stimulants and metabolic enhancers. Our experiments are in progress to specify more clearly the effect of these active principles (our unpublished results). Wang et al. (1996) demonstrated strong spermicidal activity of chloromalocides-A and -B obtained from *C. malayense.*

PHARMACOGNOSY

Due to the high price of roots of *C. borivilianum,* relatively cheaper root material of *Asparagus ascendense* is used to adulterate the 'Safed musli'. Therefore, pharmacological characters were investigated to distinguish these materials, leaving aside the aerial plant parts, which are quite different. All these species have tuberous roots, which are either sessile (*C. borivilianum*) or attached to fibrous roots (*C. tuberosum*). Roots of *C. borivilianum* are soft, without any clear distinction of vascular (absence of well-developed xylem, thick-walled cells in medullary ray and pith) and cortical zones, poorly-developed vascular tissue as these are non-functional roots, and without xerophytic anatomy. These roots are produced on the maturity of the plant as a storage and perennation organ (50-80 × 5-8 mm). Whereas, in *A. ascendense,* the tuberous roots are present from the very beginning and grow with the age of plant as functional roots (25-30 × 1-1.5 cm), having clear-cut distinction of central vascular and cortical regions and exhibiting xerophytic characters.

CONVENTIONAL PROPAGATION AND CULTIVATION

Attempts were made to develop certain methods for cultivation of *C. borivilianum* by seeds and root tubers (Bordia, 1993). Degeneration of tuberous roots during storage reduces the effective vegetative multiplication of the plant for the following season. Factors affecting storage and sprouting of underground tuberous roots (containing modified and reduced stem) of *C. borivilianum* have been studied (Bordia, 1993). Separation of tuberous roots along with the stem results in the formation of new plants. Such plants produce a new crop of tuberous roots towards the end of the growing season. The roots are collected from forests before the end of monsoon season (September-October); the skin is removed by peeling before the roots become completely dry, and are finally sun-dried and bleached. Seed-raised plants produce small tuberous roots in their first season and large roots of commercial value are produced at the end of the second season. Roots without stem-disc do not sprout (as they lack apical bud). Sprouting of the material is governed by the presence of atmospheric moisture and season. Details of crop production and management, viz., propagation by seeds and tuberous roots (having a part of the stem), use of fertilizers, storage of tuberous roots and pests involved, were given by Bordia et al. (1995). The plants require well-drained loamy to sandy soils rich in organic matter. Warm and humid climatic conditions with high soil moisture during the growing period favour luxuriant growth and facilitate the development of fleshy roots. The plants mature in about 90 days followed by withering of the leaves. An experimental field at Rajasthan College of Agriculture, Udaipur produced 1 tonne/hectare of fresh roots, which yielded 200 kg dried roots of commerce (Bordia et al., 1995). Natural variability was also observed in the plants collected from different localities in India. *C. borivilianum* (56 localities) and *C. tuberosum* (29 localities) collected from different parts of Rajasthan, Madhya Pardesh and Gujarat showed a variation in the root tuber yield per plant, days of maturity, number of roots per plant, carbohydrate, protein and saponin contents. More diversity was observed in *C. tuberosum* samples (Bordia et al., 1995). The authors concluded that these variations may be due to different genotypes and also due to edaphic and environmental factors. Roots, after drying for 10-20% water content, can be stored in dry soil (15 cm deep) in shade for planting in the next season (Bordia, 1993).

The actual problem is the non-availability of saplings of *C. borivilianum*. The government forests departments are transplanting saplings of the plant to nursery sites for small-scale cultivation, although with a high rate of mortality. The younger plants are difficult to locate in the dense forest grasses in which these plants grow. Also, plants with tuberous roots show a high rate of mortality if transplanted during the growing season. The plants are difficult to locate in the forests before or after the growing season as the aerial parts wither away. Therefore, plants are raised through seeds or tuberous roots of the previous year.

IN VITRO APPROACHES

Organogenesis

In vitro multiplication of *C. comosum* var. *folis medio-variegatis* (Apollionio et al., 1990) was achieved by apical meristem culture. The stem-disc posed a serious problem of contamination whenever used as an explant to raise cultures (Apollonio et al., 1990; Purohit et al., 1994; Suri et al., 1998). The use of carbendazim, a broad-spectrum systemic fungicide or a mixture of a systemic fungicide and an antibiotic (amoxicillin, 100 mg/l) before sterilization for 12 hours reduces the percentage of contamination effectively. B_5 medium (Gamborg et al., 1968) supplemented with benzyl adenine (BA, 2.5-5.0 mg/l) supported shoot proliferation from the stem disc explants. Cultures maintained by regular subculture on B_5 medium with 5.0 mg/l BA produced 60 shoots per flask (100 ml capacity with 40 ml medium, Fig. 6.4) after 3 subcultures. Higher concentrations (> 7.5 mg/l) were inhibitory to the shoots, growth and multiplication. B_5 medium with indole butyric acid (IBA, 0.25 mg/l) was optimal for rooting of the isolated shoots (Suri et al., 1998). Earlier, Purohit et al. (1994) reported a few shoots formation from the stem disc explants. Cuttings of *C. elatum* were grown in Knop's nutrient solution for raising plants (Bienfait et al., 1985).

In an effort to develop technology for large-scale multiplication of *C. borivilianum*, conditions for shoot multiplication were optimised using 22.2 μM benzyladenine with phytagel (0.2%) incorporated in MS medium and 15000 plantlets were produced. These plantlets showed about 90% survival in field conditions and produced tubers. The tubers of previous year were used as propagule for field plantations next season. Multi-location trials of *in vitro* raised plants showed similar performance in different field conditions. The micropropagated plants were cost effective for commercial production (Dave et al., 2003a, Purohit et al., 2003). *In vitro* produced shoot buds were encapsulated in 3% sodium alginate to save time and labour required for regular subculture. Storage conditions, gel matrix medium and duration of storage influenced the *in vitro* regrowth potential of shoot buds. Freshly encapsulated buds showed sprouting readily while decline in sprouting was observed with increase in storage period irrespective of the light or dark condition used (Dave et al., 2003b).

In vitro tuberous root formation

Plantlets produced 3-4 tuberous roots when kept in vitro for 2-3 more months after induction of fibrous roots on in vitro produced shoots. These plantlets were maintained on a filter paper bridge in liquid B_5 medium containing IBA (0.1 mg/l) and sucrose (60 g/l). Fresh liquid medium was added at a regular interval of 15 days. Such plantlets, on transfer to the soil, showed 100% survival, and after over-wintering, sprouted in the following

Fig. 6.4 Multiple shoots through organogenesis

season. These experiments showed that tuberous roots are produced *de novo* from the stem on attaining maturity and not formed by tuberization of fibrous roots (Suri et al., 1998). Organogenetic cultures have been maintained in our laboratory for the past 6 years without loss of any regenerative potential.

Somatic embryogenesis

An improved method has been developed for large-scale rapid multiplication of *C. borivilianum* through the somatic embryogenesis in callus cultures. Somatic embryos were obtained on Murashige and Skoog (MS, Murashige and Skoog, 1962) medium containing 0.5 mg/l 2,4-D (2,4-dichlorophenoxyacetic acid) and 0.25 mg/l kinetin (Jain et al., 1997). The concentration and combination of KNO_3 and $(NH_4)_2SO_4$ in the medium had a marked effect on the somatic embryogenesis. Ammonium to nitrate nitrogen in the ratio of 1:4 at a low total nitrogen level of 250 and 500 mgN/l in the medium was favourable for the growth and somatic embryogenesis, respectively (Fig. 6.5). Among the various cytokinins tested, 2-isopentenyl adenine (2iP) had a maximum stimulatory effect on somatic embryogenesis. The decreasing order of the effectiveness of various cytokinins incorporated in the MS-9 medium [i.e. MS medium with 500 mgN/l (100 mgN/l

$(NH_4)_2SO_4$ and 400 mgN/l KNO_3)] supplemented with 0.5 mg/l 2,4-D exhibited 2-iP > TDZ > kinetin > BAP (TDZ, thidiazuron). Incorporation of amino acids, viz., glutamine and proline promoted embryogenesis. Maximum number of somatic embryos (63 per inoculum) were observed on MS-9 medium supplemented with 100 mg/l proline, 0.25 mg/l 2,4-D and 1.5 mg/l 2-iP. Germination of somatic embryos, isolated from young cultures (1-5 months old), was maximum (48%) on MS-9 medium supplemented with 4.0 mg/l BA (Fig. 6.6). On the other hand, somatic embryos developed in long-term cultures were abnormal and showed a wide range of variation in the number of chromosomes, ranging from 23 to 56 (i.e. 3X-3 to 7X+2) along with several structural chromosomal abnormalities. This may be the reason for the decline in the germination potential of somatic embryos in long-term cultures. Thus, the callus produced from shoot cultures obtained from stem-disc explants on B_5 medium containing 5.0 mg/l TDZ and 25 mg/l adenine, provides an excellent source for obtaining fresh cytologically-stable embryogenic cultures. Plantlets developed in vitro through embryo germination were successfully transferred to the soil with 90% survivability (Fig. 6.7). The technique developed is highly efficient to get miniature plantlets for field transfer in 2 months, starting from the callus (Arora et al., 1999).

Cyclic somatic embryogenesis

Leaf explants obtained from in vitro maintained shoots/plantlets serve as a continuous source of aseptic material and proved useful for cyclic

Fig. 6.5 Callus with germinating somatic embryos

Fig. 6.6 Isolated and germinated somatic embryos

Fig. 6.7 Plantlets transferred in field conditions

embryogenesis. However, in spite of use of a large number of combinations and concentrations of different cytokinins and auxins, along with explants from different positions of leaf, nothing concrete was achieved in the sense that per cent explant response was erratic and low when associated with slow callus growth. Little callus produced on certain treatments did not survive on the subculture. Callus produced on high concentration of TDZ from a few explants grew profusely and produced a large number of embryos on the transfer to embryo induction medium. These embryos were germinated on MS-9 medium containing BA and later compared. However, these attempts could not form a system for a continuous production of somatic embryo, neither in the callus nor in a liquid medium (Arora, 1999).

Root culture on static and in liquid medium

Chlorophytum borivilianum is a root crop. Attempts were also made to grow isolated roots in culture as an alternative to natural roots. In-vitro roots produced higher amounts of saponins as compared to natural roots on the dry weight basis. Therefore, it was attempted to produce root biomass. Roots were initiated from callus cultures and germinating embryos.

These experiments provided the basis to select root culture medium. Large quantities of roots (200 g fresh weight/l) were obtained by inoculating these roots to 1 L flasks (Fig. 6.8) and 2 L bioreactor grown in MS-9

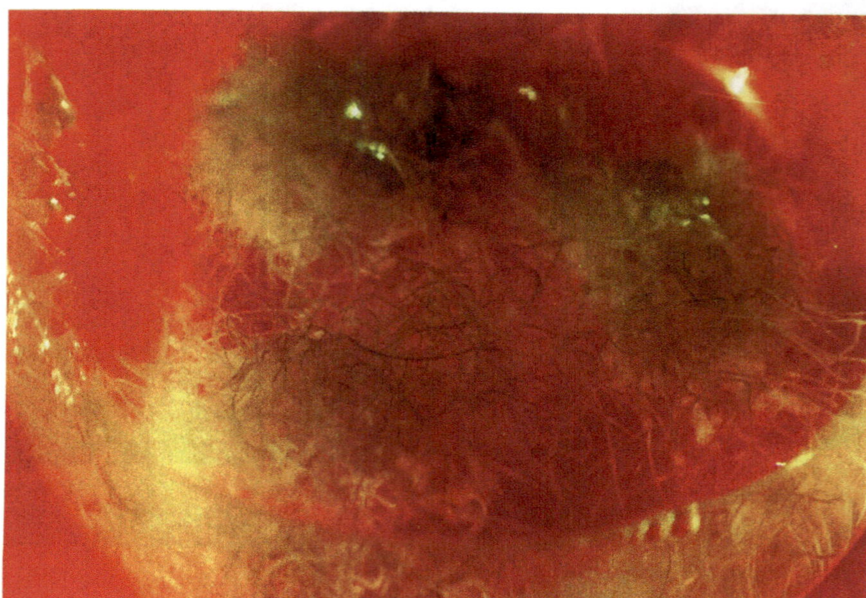

Fig. 6.8 Root culture in liquid medium

medium containing 0.1 mg/l each of NAA and kinetin. In the stirred tank, bioreactor roots were broken into pieces by the stirrer, thereby increasing the inoculum. The roots grew as thin fibrous roots and no thickening of roots was observed on any of the treatments used (Arora, 1999).

Somaclonal variation

The results obtained with DNA polymorphism using different probes showed that plants obtained from stem disc derived callus through somatic embryogenesis were not genetically uniform, but the plants obtained from a single leaf explant generated callus were genetically uniform. This may be due to the fact that the callus of leaf origin was regenerated in the second subculture. Variation set in only during a prolonged culture period. There were conclusive evidences of morphological characteristics, cytological observations and DNA polymorphism analysis that a certain degree of genetic variability exists in the plants of *C. borivilianum* regenerated through long-term somatic embryogenesis. Uniform plants of the selected material can be produced through organogenesis or somatic embryogenesis from the leaf explant. Variability sets in with a prolonged culture period; therefore, culture period can be manipulated either way to obtain variants or no variability (Arora, 1999).

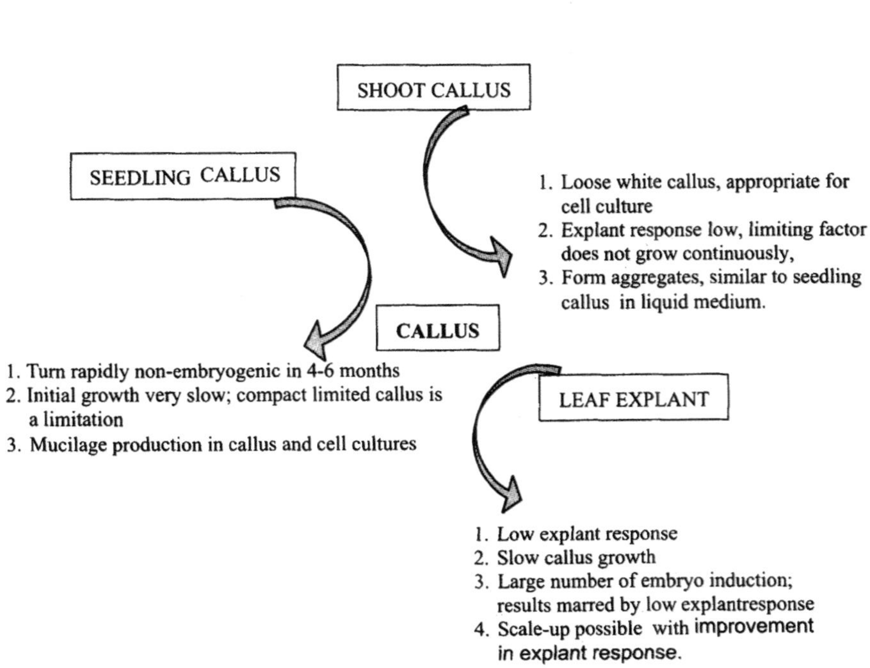

Fig. 6.9 Summary of results obtained with three sources of callus of *C. borivilianum*

RESEARCH NEED

Presently, tuberous roots of 'Safed musli' are fetching a high price and there is a great demand in the market. Biotechnological methods are being developed to meet the high demand of planting material for the domestication of the plant. A summary of the results obtained with three sources of callus is presented in Fig. 6.9. There are still some bottlenecks before these methods may be successfully employed for large-scale multiplication. Research in the following areas is required for better cultivation and uses:

1. Propagation of selected genotypes and improvement of agronomic practices for the tuberous root's yield.
2. Production of active principle in cell and root cultures.
3. Phytochemical investigations of in vitro and in vivo materials.
4. Improvement in biotechnological methods for scale-up multiplication.
5. Pharmacological investigations of the plant extracts and isolated active principles.

REFERENCES

Apollonio, G., Scaramuzzi and D'Emerico, S., 1990, Observations on the stability of ploidy and on the behaviour of variegation transmission in the cultures and related subcultures of *Chlorophytum comosum* Bak. var. *foliis medio variegatis*. Hort. C.R. Seances Soc. Biol. Filiales 184 (3-4):240-250.

Arora, D.K., 1999, Micropropagation and selection of variants in certain medicinal plants. Ph.D. Thesis, M.L. Sukhadia University, Udaipur, pp. 1-96.

Arora, D.K., Suri, S.S., Ramawat, K.G. and Merillon, J.M., 1999, Factors affecting somatic embryogenesis in long term callus cultures of 'safed musli' (*Chlorophytum borivilianum*), an endangered wonder herb. Indian J. Exp. Biol. 37:75-82.

Baldwin, J.T. Jr., and Speese, B.M., 1951, Cytogeography of *Chlorophytum* in Liberia. Amer. J. Bot. 38:153.

Bienfait, H.F., Van den Brie, W. and Mesland-Mul, N.T., 1985, Free space iron pool in roots, generation and mobilization. Plant Physiol. 78(3): 596-600.

Boraiah, G., 1966, A note on cytotaxonomy of *Chlorophytum laxum* Roxbur. Curr Sci. 35:314.

Bordia, P.C., 1993, Development of suitable culture of safed musli (*Chlorophytum* species) for root yield and acceptable medicinal quality, 1988-93. Final Technical Report, ICAR Project, RCA, Udaipur.

Bordia, P.C., Joshi, A. and Simlot, M.M., 1995, Safed musli, In: Advances in Horticulture. Vol. 11—Medicinal and Aromatic Plants. Chadha, K.L. and Gupta, R. (eds). Malhotra Pub. House, New Delhi. pp. 429-451.

Dalal, K.C., Patel, D.H. and Hirpara, B.V., 1987, Information on floral biological aspects and improvement, propagation and collection availability from Gujarat. All India Co-ordinated Project on Medicinal and Aromatic Plants. ICAR VII Workshop Report, Rajasthan Agriculture University, Udaipur., pp. 45-48.

Datta, N. and Mitra, K., 1968, Karyotype analysis in *Chlorophytum tuberosum* Baker and *C. laxum* Roxbur. Bull Bot. Survey India 10:228.

Dave, A., Bilochi, G. and Purohit, S.D., 2003a, Scaling-up production and field performance of micropropagated medicinal herb 'Safed musli' (*Chlorophytum borivilianum*). *In vitro* Cell Dev. Biol., 39: 419-425.

Dave, A., Joshi, N. and Purohit, S.D., 2003b, *In vitro* propagation of *Chlorophytum borivilianum* using encapsulated shoot buds. Europ. J. Hort. Sci., 68 (6) (in press).

Gamborg, O.L., Miller, R.A. and Ojima, K., 1968, Nutrient requirements of suspension cultures of soybean root cells. Exp. Cell Res. 50:151-158.

Gupta, R., Gupta, O.C.D., Gupta, P.C., and Panda, C.S., 1979, A new galactoglucan from the fruits of *Chlorophytum arundinaceum*. Planta Med. 37(1):94-95.

Jain S. K., 1991, The Dictionary of Indian Folk Medicines and Ethnobotany. Deep Pub., New Delhi.

Jain, S., Ramawat, K.G. and Sonie, K.C., 1997, Somatic embryogenesis in *Chlorophytum borivilianum*—A medicinal plant of Aravalli Hills. In: Biotechnological Applications of Plant Tissue and Cell Culture. Ravishankar, G.A. and Venkataraman, L.V. (eds). Oxford & IBH Pub. Co., New Delhi, pp.199-203.

Kativu, S. and Nordal, I., 1993, New combinations of African species in the genus *Chlorophytum* (Anthericaceae). Nordic J. Bot. 13(1):59-65.

Kirtikar, K.R. and Basu, B.D., 1975, Liliaceae: *Chlorophytum*. In: Indian Medicinal Plants. K.R. Kirtikar and B.D. Basu (eds). Published by L.M. Basu, Allahabad, pp. 2508-2509.

Kumar, L.S.S. and Rao, H.K.S., 1958, Chromosome number in *Chlorophytum tuberosum*. Curr. Sci. 27:406.

Kumar, V. and Subramanium, B. (eds), 1986, Chromosome Atlas of Flowering Plants of Indian Subcontinent. Vol. II. Monocot. Botanical Survey of India Publications, Calcutta.

Li, X.C., Wang, D.Z. and Yang, C.R., 1990, Steroidal saponins from *Chlorophytum malayense*. Phytochem., 29(12): 3893-3898.

Marais, W. and Reilly, J., 1978, *Chlorophytum* and its related genera (Liliaceae). Kew Bull. 32(3):653-664.

Mimaki, Y., Kanmoto, T., Sashida, Y., Nishino, A., Satomi, Y. and Nishino, H., 1996, Steroidal saponins from the underground parts of *Chlorophytum comosum* and their inhibitory activity on tumor promoter-induced phospholipids metabolism of HeLa cells. Phytochem. 41:1405-1410.

Murashige, T. and Skoog, F., 1962, A revised medium for rapid growth and bioassay with tobacco tissue cultures. Physiol. Plant. 15:473-497.

Nair, V. N., 1974, Cytological studies in *Chlorophytum bruchae* Ansari. Curr. Sci. (India) 43:161.

Nayar, M.P. and Shastri, A.R.K., 1988, *Chlorophytum borivilianum*. In: Red Data Book of Indian Plants, Vol. 2, M.P. Nayar and A.R.K. Shastri (eds), Botanical Survey of India, pp. 142.

Ojha, A., 1987, Botanical Study of safed musli (Aphrodisiac effects). MD Thesis, MMR Ayurvedic College, Udaipur (India).

Pahuja, A.N. and Kumar, V., 1969, Cytogeography of three species of *Chlorophytum*. Curr. Sci. 38:469.

Purohit, S.D., Dave, A. and Kukda, G., 1994, Micropropagation of safed musli (*Chlorophytum borivilianum*), a rare Indian medicinal herb. Plant Cell Tiss. Org. Cult. 39:93-96.

Purohit, S.D., Dave, A., Bilochi, G. and Joshi, N., 2003, Opportunities for scaling–up production and commercial cultivation of micropropagated 'Safed musli' (*Chlorophytum borivilianum*). Proc. First National Interactive Meet Med. Arom. Plants. A.K. Mathur, S. Dwivedi, D.D. Patra, G.D. Bagechi, N.S. Sangwan, A Sharma and S.P.S. Khanuja (eds), CIMAP, Lucknow, pp387-398.

Qiu, S.X., Li, X.C., Xiong, Y., Dong, Y., Chai, H., Farnsworth, N.R., Pezzuto, J.M. and Fong, H.H.S., 2000, Isolation and characterization of cytotoxic saponin chloromaloside A from *Chlorophytum malayense*. Planta Med. 66:587-590.

Ramawat, K.G., Jain, S., Suri, S.S. and Arora, D.K., 1998, Aphrodisiac plants of Aravalli Hills with special reference to safed musli. In: Role of Biotechnology in Medicinal and Aromatic Plants. I. Khan and A. Khanum (eds),. Ukaz Pub., Hyderabad., pp. 210-223.

Ramawat, K.G., Sharma, R., Jain, S. and Suri, S.S., 1996, Problem of propagation of desert plants : Seed Biology and Biotechnological Approaches. In: Seed Technol., G. S. Paliwal (ed.), Srinagar. pp. 98-100.

Santapau, H. and Fernandes, R.R., 1955, A new species of *Chlorophytum* from Salsette island. J. Bombay Nat. Hist. Soc. 52:897.

Sharma, P.V., 1978, Dravyaguna Vigyan, Vol. 2., pp. 559–560.

Sharma, R. and Ramawat, K.G., 2000, Stilbenes: bioactive plant phenolics. In: Role of Biotechnology in Medicinal and Aromatic Plants. III. Khan and A. Khanum (eds), Ukaz Pub., Hyderabad. pp.106-115.

Sheriff, A. and Chennaveeraiah, M.S., 1972, Karyomorphology of four diploid species of *Chlorophytum*. Nucleus 15(1):39-45.

Suri, S. S., Jain, S., Arora, D. K. and Ramawat, K.G., 1998, In vitro high frequency regeneration of plantlets and tuberous root formation in *Chlorophytum borivilianum*. Gartenbau, 64(3):106-110.

Tandon, M. and Shukla, Y.N., 1992, Sapogenins from *Asparagus adscendens* and *Chlorophytum arundinaceum*. J. Indian Chem. Soc. 69:893.

Tandon, M. and Shukla, Y.N., 1993, A bibenzyl xyloside from *Chlorophytum arundinaceum*. Phytochem. 32:1624-1625.

Tandon, M. and Shukla, Y.N., 1995, Phytoconstituents of *Asparagus adscendens, Chlorophytum arundinaceum* and *Curculigo orchioides*: A review. Curr. Res. Med. Arom. Plants 17:42-50.

Tandon, M., Shukla, Y.N. and Thakur, R.S., 1992, 4-hydroxy-8,11-oxidohemicosanol and other constituents from *Chlorophytum arundinaceum* roots. Phytochem. 31(7):2525-2526.

Villar, P., 1948, Saponins in the plant kingdom. II. Experimental part. Farmacognosia (Madrid) 8:305-340.

Wang, Y., Yang, H.Y., Wang, J.J. and Yang, C.R., 1996, Inhibitory effect of some steroidal saponins on human spermatozoa *in vitro*. Plant Med.62:130-132.

CHAPTER 7

Fagaronine: An Antileukemic Alkaloid with Cell Differentiation Activity

E. COUILLEROT[1], C. DUPONT[2], C. CARON [3], J.C. CHENIEUX[4] and C. TRENTESAUX[2]

INTRODUCTION

Fagara zanthoxyloides is a good example to illustrate the interest of a typical medicinal plant of the African pharmacopeae. Searching for a new active molecule on blood cell differentiation, Pr Léopold Comoë (deceased) focussed on this plant due to the isolation of an antileukemic alkaloid from its roots as also its properties to intercalate with DNA. He pointed out that fagaronine was in effect an efficient molecule for the differentiation of human erythroleukemic cell lines with a high inhibition of cell growth (Comoë et al., 1987). Due to the difficult obtention of fagaronine using plant root extraction or chemical synthesis, plant and tissue cultures of *Fagara zanthoxyloides* have been investigated in order to study the biosynthesis, the production of the potent antileukemic molecule in vitro and to explain its original activity. This chapter complements our precedent chapter on the in vitro studies (Couillerot et al., 1999) and presents our recent work on the explanation of the mechanisms by which fagaronine induces the differentiation of human erythroleukemic cells.

[1] Laboratoire de Stress, Défenses et Reproduction des Plantes, URVVC EA2069, UFR Sciences, Université de Reims Champagne-Ardenne, BP 1039, F-51687 Reims cedex 2, France

[2] JE Oncopharmacologie, UFR Pharmacie, Université de Reims Champagne-Ardenne, 51 rue Cognacq-Jay, F-51096 Reims cedex, France

[3] Laboratoire de Pharmacognosie, UPRES-A 6013, UFR Pharmacie, Université de Reims Champagne-Ardenne, CRPB, BP 1039, F-51687 Reims cedex 2, France

[4] EA 2106, Biomolécules, Biotechnologies Végétales, UFR Pharmacie, Université de Tours, 31 avenue Monge, F-37200 Tours, France

BOTANY

The genus of *Fagara* is represented essentially in Africa and South America and is composed of at least 206 different species (Molinatti, 1994). *Fagara zanthoxyloides* Lam. (synonyms: *Zanthoxylum zanthoxyloides* Waterm., *Fagara senegalensis* A. Chev., *Zanthoxylum senegalense* A. Chev., *Zanthoxylum polyganum* Schum.; common names: ouo, kaingué, busan, goro-gna, for example in African languages) is a West African thorny shrub or tree, 3 to 8 m high, and its area of distribution is the littoral from Senegal to Nigeria (Fig. 7.1). Its cylindric stems exhibit a very hard and pale yellow wood with a reddish-brown coloured bark. Branchlets and leaf-rachis are armed with sharp recurved prickles.

The leaves are imparipinnate alternate (5 to 9 pairs of leaflets). The leaflets are not accuminate (a characteristic of *F. zanthoxyloides*), oblong or oblong-oblanceolate, sometimes emarginate at the apex, subacute at the base, 5-10 cm long, 2-4 cm broad and glabrous. The flowers, clustered panicles, are white and unisex. The fruits, the size of a pea, are encapsuled with two valve-liked openings containing a single shiny black seed. The germination rate is low and germination relatively slow (Hutchinson et al., 1958; d'Aubreville, 1959; Bouquet and Debray, 1974; Ake-Assi and Adjanohoun, 1979).

CHEMISTRY

Various groups of secondary metabolites have been isolated from the genus *Fagara*, i.e. alkaloids, coumarins, amides, flavonoids, lignans, alcohols, ketons, organic acids (Comoë, 1987; Molinatti, 1994). The group of alkaloids appears to be the most interesting from the point of view of biological activity. Major alkaloids of *F. zanthoxyloides* are the benzophenanthridines chelerytrine and fagaronine, the furoquinolines skimmiannine and γ-fagarine. The different families of alkaloids are derivatives of the two aromatic aminoacids tyrosine and tryptophan and from anthranilate (Couillerot et al., 1999). Skimmianine has been described as a potential antileishmanial drug (Fournet et al., 1993) and as an antagonist of 5-hydroxytryptamine receptors in animals (Cheng et al., 1994). Chelerythrine occurs in several species of Rutaceae and Papaveraceae (Krane et al., 1984) and has been described as a specific protein kinase C inhibitor (Herbert et al., 1990).

So far, fagaronine was only isolated from the roots of *F. zanthoxyloides* (Messmer et al., 1972; Comoë, 1987) and it is the only benzophenanthridine to possess a phenolic group at C-3. The molecule could be considered as the most active compound from the plant and has been the subject of numerous studies.

The biosynthetic pathway of benzophenanthridine has been elucitated at the enzymatic level with the biosynthesis of sanguinarine and macarpine

a)

b)

Fig. 7.1 *Fagara zanthoxyloides* Lam. (Rutaceae); a) Branch, ♂ and ♀ flowers, fruits; b) Tree.

(review in Kutchan et al., 1991; Kutchan and Zenk, 1993). (S)-norcoclaurine, and not norlaudanosoline, is implicated in the biosynthesis leading to S-reticuline (Stadler et al., 1989; Müller and Zenk, 1992). Transformation of reticuline by the berberine bridge enzyme (BBE, EC 1.5.3.9) leads to the specific benzophenanthridine pathway. Molecular characterization of BBE genes has been realized (Facchini et al., 1996). The high stereoselectivity of the N-methyltransferase (methylation of the tetrahydroprotoberberine with the S-configuration) is, moreover, dependent of the substitution patterns of rings A and D: the N-methylation should only occur when the methylendioxy groups have been formed (Kutchan et al., 1991). The formation of benzophenanthridines which possess no or only one methylendioxy group (as, respectively, fagaronine and chelerythrine) may involve different or modified steps of biosynthesis. The 6-hydroxylation of the protropine (Kutchan et al., 1991) or of the (S)-cis-N-methylstylopine (Battersby et al., 1975) appears to be a key of benzophenanthridine pathway.

IN VITRO CULTURES

We have recently described our work on the in vitro production of alkaloids from *F. zanthoxyloides* (Couillerot et al., 1999). Plant tissue culture studies have shown the ability of both cell suspension and hairy root cultures to accumulate the potent antileukemic alkaloid fagaronine, as well as the other benzophenanthridine chelerythrine and the furoquinoline skimmianine and γ-fagarine. Transformed roots showed a better stability of alkaloid production but the suspension yielded a higher biomass. This work confirmed the ability of plant cell cultures to synthesize valuable products independently from the plant and its environment and avoiding cutting the tree and its resulting consequences. We are screening new cell suspension lines, working on the new establisment of hairy root cultures and optimization of culture and alkaloid production through the use of a 14 L bioreactor.

BIOLOGY

In West Africa, the roots of *Fagara zanthoxyloides* are commonly used for tooth cleaning as a chewing stick (Lewis, 1983): the root extract was shown to exhibit a high antimicrobial activity against oral flora (El Said et al., 1971). In traditional medicine, the plant has various uses, some examples of which have been summarized recently (Couillerot et al., 1999). The antisickling activity of root extracts on sickle cell anaemia has been associated with the presence of benzoic acid derivatives (Edwards et al., 1979). Clinical experiments have shown the positive antisickling effect obtained by oral treatment of children with hard capsules of an association of *Fagara zanthoxyloides* and *Calotropis procera* (Ouattara et al., 1992).

Nevertheless, the major interest of this plant appears to be the presence in the roots of the specific alkaloid fagaronine (Messmer et al., 1972; Barret and Sauvaire, 1992) (Fig. 7.2) which exhibits antileukemic activities.

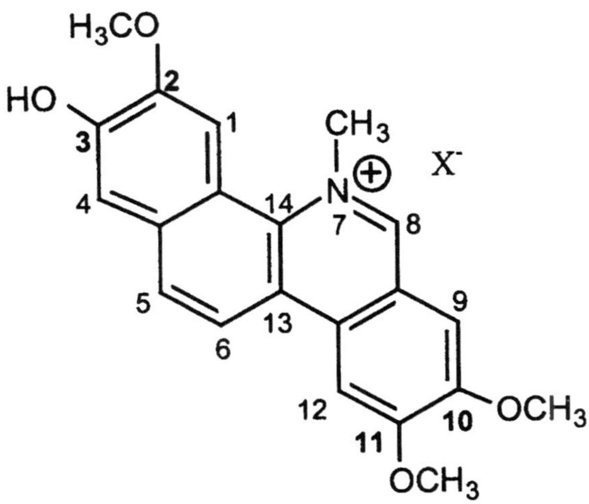

Fig. 7.2 Structure of fagaronine

Fagaronine and leukemia

Tumour cells, particulary leukemic cells, can be considered to be maturation-arrested cells which continue to proliferate and rapidly accumulate, escaping the normal controls regulating cell proliferation and differentiation. Numerous physiological as well as non-physiological agents, and especially antitumour drugs, have been described as inducers of leukemic cell differentiation (Reiss et al., 1986). Complete remission in patients with acute promyelocytic leukemia has been succesfully obtained by all trans retinoic acid (ATRA). Low dose cytotoxic chemotherapy is now commonly used in combination with ATRA to consolidate the remission period. This protocol represents the earliest model of differentiation therapy (Degos et al., 1995).

In vitro, our group has previously demonstrated that anthracycline antitumour drugs such as aclacinomycin (ACLA) and doxorubicin (DOX), at subtoxic concentrations, were effective in controlling erythroid differentiation of human leukemic K562 cells, leading to the appearance of haemoglobinized cells. Indeed, ACLA stimulates this process by increasing the transcription of genes involved in the haemoglobin synthesis. This transcriptional mechanism occurs by the recruitment of erythroid-specific transcription factors, notably GATA-1 and NF-E2 (Aries et al., 1996; Morceau et al., 1996a).

By contrast, DOX stimulates the haemoglobinization of this cell line by a post-transcriptional mechanism, leading to an increased stability of the erythroid transcripts (Morceau et al., 1996b).

Among the other compounds reported to induce the hemoglobinization of K562 cells, fagaronine (FAG) displays an antileukemic activity against murine leukemia P388 in vivo (Messmer et al., 1972). Other works have established that FAG inhibits DNA polymerase in murine embryos (Sethi, 1976), nucleic acid and protein synthesis in KB cells, respectively, by intercalation in DNA and interaction with the ribosomal system (Casiano Torres and Baez, 1986). FAG also inhibits the activity of the DNA topoisomerases I and II (Larsen et al., 1993; Wang et al., 1993; Fleury et al., 2000), human DNA ligase I (Tan et al., 1996) and reverse transcriptases from RNA virus (Sethi and Sethi, 1975; Sethi, 1977, 1979), especially the human HIV-1 reverse transcriptase in vitro (Tan et al., 1991). As DNA intercalation and inhibition of topoisomerase I activity seem to be common molecular targets of FAG and ACLA (Sorensen et al., 1994), it appeared interesting to determine whether the molecular events implicated in the activation of haemoglobin synthesis induced by FAG were dependent of a transcriptional activation. So, erythroid gene expression induced by FAG was studied by RT-PCR and reporter gene analysis.

Effects of FAG on erythroid differentiation and growth of K562 cells

Haemoglobinized cell content and growth inhibition of human leukemic K562 cells were evaluated after 3 days of treatment by various FAG concentrations (Table 7.1). It appeared that FAG induced a significant haemoglobinization of K562 cells which was dose-related in nature. The optimal differentiating effect was obtained at 6 µmol/L FAG and maintained at 10 µM. Indeed, at 6 µM, the haemoglobinization reached 65% of benzidine-positive cells versus 1% in untreated cells. This induction of haemoglobin synthesis was accompanied by a marked growth

Table 7.1 Effects of FAG on growth, differentiation and lethality of the K562 human erythroleukemic cells after 3 days of treatment. Results are means ± S.D. (standard deviation) of three independent experiments.

	Growth inhibiton (%)	Cell hemoglobinization (%)	Cell lethality (%)
Control	0	2 ± 1	2 ± 1
FAG, 1 µM	63 ± 5	36 ± 3	3 ± 1
FAG, 3 µM	72 ± 4	49 ± 6	7 ± 2
FAG, 6 µM	82 ± 2	65 ± 2	6 ± 1
FAG, 10 µM	95 ± 1	62 ± 4	13 ± 3

inhibition varying from 61% for the lowest concentration to 95% for the highest. The percentage of cell death, evaluated by trypan blue staining, did not exceed 5% at the optimal differentiating concentration, 6 μM versus 1% in control cells. Taken together, these data demonstrated that FAG requires a strong inhibition of cell growth (about 80%) to induce an efficient haemoglobinization of the K562 cell line without subsequent toxicity.

Influence of fagaronine on erythroid gene expression

The expression of erythroid mRNAs was studied after 3 days of treatment by 6 μM FAG. Total RNAs were extracted and submitted to RT-PCR analysis using specific primers. The analysis of PCR products by electrophoresis showed that FAG induced an overexpression of γ- and α-globin transcripts, of porphobilinogene deaminase (PBGD)mRNAs, a key enzyme of the heme synthesis, and of Epo-R mRNAs, binding site for erythropoietin an important hormone which regulates the erythroid differentiation process (Fig. 7.3). This increase in mRNA expression is specific to erythroid genes since the expression of GAPDH ubiquitous transcripts has remained constant after FAG treatment.

Beside the role GATA-1 and NF-E2 transcription factors in the regulation of erythroid gene expression, we also examined their expression following FAG treatment (Fig. 7.4). Similarly, FAG induced an overexpression of GATA-1 and NF-E2 transcripts after 3 days of treatment. These results suggest that FAG may exert its differentiating effects by a transcriptional regulation of erythroid gene expression.

PROTOCOLS

Cell line and induction of erythroid differentiation

The human erythroleukemic K562 cells were grown in RPMI 1640 Glutamax medium (Life technologies) supplemented with 10% heat-inactivated foetal calf serum (Life technologies). FAG was reconstituted in 70% ethanol and diluted in culture medium immediately prior to use. Various FAG concentrations were added to the K562 cell suspensions at the beginning of the exponential growth phase. Cell cultures were incubated in a 5% CO_2 humidified atmosphere at 37°C during 72 hours. Erythroid differentiation was scored by benzidine staining after 3 days of culture. Growth Inhibition and cell viability were evaluated as previously described (Trentesaux et al., 1993).

RNA extraction and RT-PCR analysis

Total RNAs were extracted using TriZOL reagent (Life Technologies) in

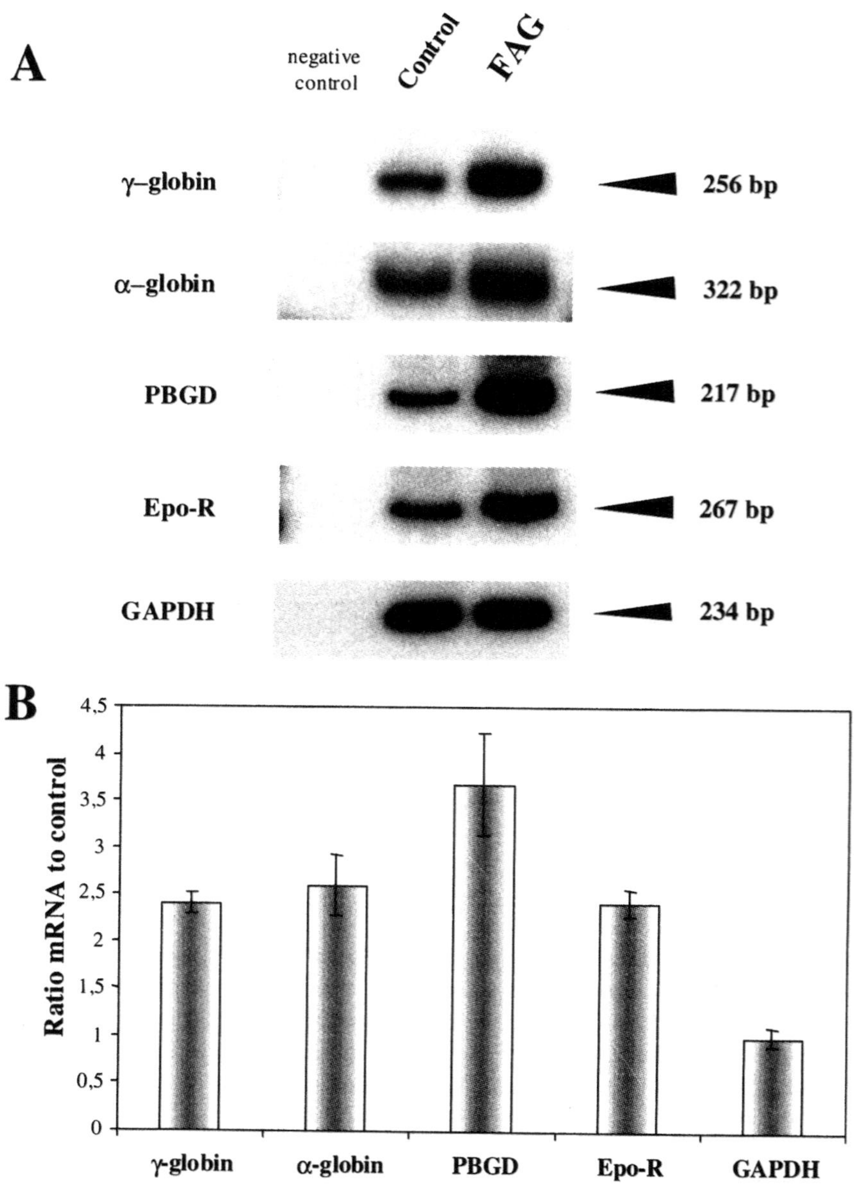

Fig. 7.3. **(A)**: Expression of erythroid gene mRNAs in K562 cells treated with 6 µM FAG during 3 days. Total RNA was purified from cells and subjected to RT-PCR analysis with primers specific for γ-globin, α-globin, PBGD, Epo-R. The PCR product of GAPDH was used as a control.
(B): Quantification of the experiment of panel A using a GS-363 Molecular Imager. Similar results were obtained in 3 independent experiments.

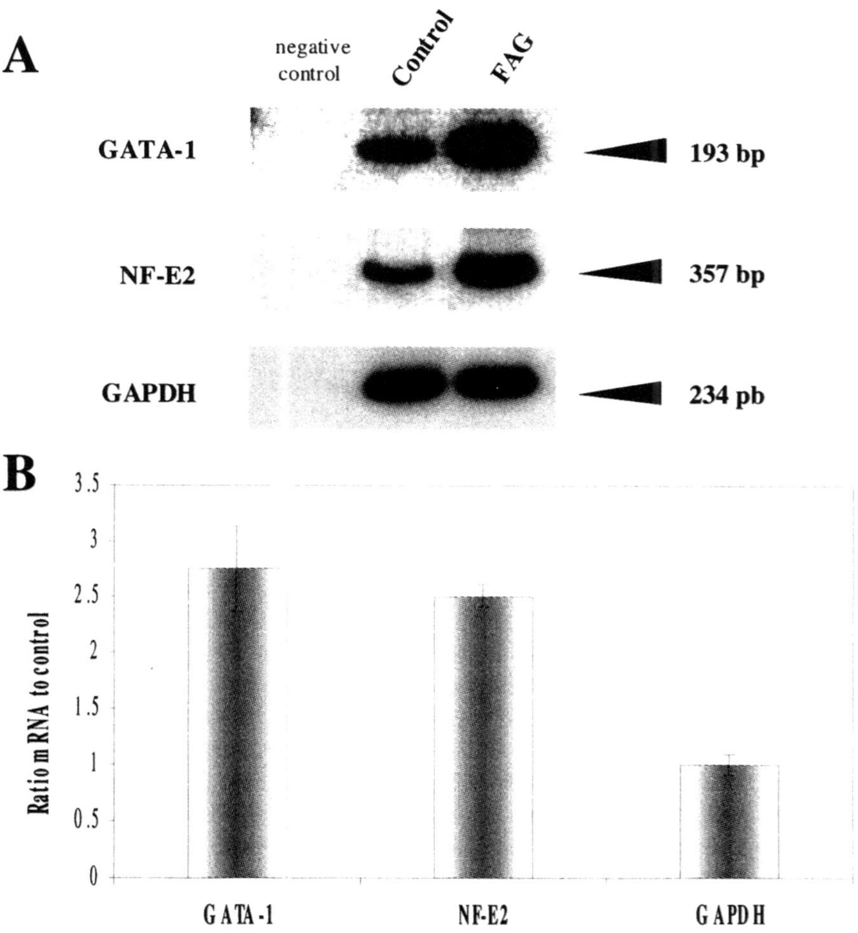

Fig. 7.4 **(A)**: Expression of GATA-1 and NF-E2 mRNAs in K562 cells treated with 6 µM FAG during 3 days. Total RNA was purified from cells and subjected to RT-PCR analysis with primers specific for these erythroid transcription factors. The PCR product of GAPDH was used as a control. **(B)**: Quantification of the experiment of panel A using a GS-363 Molecular Imager. Similar results were obtained in 3 independent experiments.

accordance with Chomczinsky et al. (1987). cDNA was synthesized from 1 µg of total RNAs using oligo-dT primers and *Superscript II* reverse transcriptase (Life technologies) according to the manufacture protocol. The RT products were used as templates for PCR amplification using $\alpha\,^{32}$P-dCTP (NEN) and gene specific primers of: *γ-globin* (sense: 5'-GGCAACCT

GTCCTCTGCCTC-3'; antisense: 5'-GCCAGGAAGCTTGCACCTCA-3') (Leonard et al., 1993); α-*globin* (sense: 5'-TGGGGTAAGGTCGGCGCGCA-3' ; antisense: 5'-TGCACCGCAGGGGTG AACTC-3') (Privitera et al., 1995); *PBGD* (sense: 5'-GGTCCTACTATCGCCTCCCTC-3'; antisense: 5'-AGAATCTTGTCCCCTGTGGTG-3') (Dessen et al., 1990); *Epo-R* (sense: 5'-AGCCTGTGTCGCTGCTG ACGC-3'; antisense: 5' GGTCCTCCGTGA-AGGGG GTGC-3') (Dessen et al., 1990); *GATA-1* (sense: 5'-GATCCTGC-TCTGGTGTCCTCC-3'; antisense: 5'-ACAGTTGAGCAATGGGTACAC-3') (Leonard et al., 1993); *NF-E2* (sense: 5'-ATTTGAGCCCCAAGCCCCAGC-3' ; antisense: 5'-CCAGCCT CTGTCCCCTCCAGC-3) (Dessen et al., 1990); *GAPDH* (sense: 5'-CTCTGCCC CCTCTGCTGATGC-3' ; antisense: 5'-CCATCACGCCACAGTTTCCCG-3' (234 bp)). PCR was performed using *Taq DNA polymerase* (Life technologies). PCR reaction mixture was subjected to 15-25 successive cycles of amplification consisting of heat denaturation (94°C for 2 min.), annealing (60 °C for 30 sec.), and extension (72°C for 1 min.). The initial heating step was performed at 60°C for 30 sec. and 72°C for 1 min. PCR products were analysed on a polyacrylamide gel. After electrophoresis, the gels were exposed on a GS-363 Molecular Imager (Bio Rad).

SUMMARY AND CONCLUSIONS

In this study, we showed that fagaronine induces a significant hemoglobinization of the human erythroleukemic cell line K562. This hemoglobin synthesis was accompanied by a strong increase of erythroid m RNA expression such as γ-and α-globin, PBGD, an enzyme of heme synthesis. In addition, the Epo-R transcripts were also stimulated, demonstrating that cells are engaged in a maturation process. Both transcription factors, GATA-1 and NF-E2 which play an important role in the regulation of genes implied in the erythroid differentiation, were also activated at the transcriptional level.

All these results suggest that fagaronine exerts its differentiating activity by a specific activation of regulatory regions which control the differenciation programme in erythroleukemic cells. This mechanism seems to involve the participation of erythroid transcription factors (studies are in progress). Fagaronine may represent a new family of natural products which would act by modulating the activity of genes important for proliferation, differentiation and apoptosis control.

REFERENCES

Ake-Assi, L. and Adjanohoun, E., 1979, I. Contribution au recensement des plantes médicinales de Côte d'Ivoire. *CRES* Université d'Abidjan.

Aries, A., Trentesaux, C., Ottolenghi, S., Jardillier, J.C., Jeannesson, P., and Doubeikovski, A.,

1996, Activation of erythroid-specific promoters during anthracycline-induced differentiation of K562 cells. Blood, 87:2885-2890.

Aubreville d', A., 1959, La flore forestiére de la Côte d'Ivoire. Centre Tech. Forest. Natl, 2:107-112.

Barret, Y. and Sauvaire, Y., 1992, Fagaronine, a novel antileukaemic alkaloid. Phytother. Res. 6:59-63.

Battersby, A.R., Staunton, J., Wiltshire, H.R., Francis, R.J., and Southgate, R., 1975, Biosynthesis part XXIII: the origin of chelidonine and of other alkaloid derived from the tetra hydroprotoberberine skeleton. J. Chem. Soc. Perkin Trans., I:1147-1156.

Bouquet, A. and Debray, M., 1974, Plantes médicinales de la Côte d'Ivoire. Travaux et documents de l'ORSTOM, 32:156-158.

Casiano Torres, C.A., and Baez, A., 1986, Effects of the antitumor drugs 3-nitrobenzothialozo {3,2-a}quinolium and fagaronine on nucleic acid and protein synthesis. Biochem. Pharmacol., 35:679-685.

Cheng, J.T., Chang, T.K., and Chen, I.S., 1994, Skimmianine and related furoquinolines function as antagonist of 5-hydroxytryptamine receptors in animals. J. Auton. Pharmacol., 14:365-374.

Chomczynski, P. and Sacchi, N., 1987, Single-step-method of RNA isolation by acid guanidinium-thiocyanate-phenol chloroform extraction. Anal. Biochem., 162:156-159.

Comoë, L., 1987, Apports de la médecine traditionnelle africaine à la chimiothérapie anticancéreuse. Activité antitumorale et différenciante de la fagaronine (*Fagara xanthoxyloïdes* Lam.). Thèse Sci Pharma, Reims: 402.

Comoë, L., Jeannesson, P., Trentesaux, C., Desoize, B., and Jardillier, J.C., 1987, The antileukemic alkaloid fagaronine and the human K 562 leukemic cells:effects on growth and induction of erythroid differentiation. Leuk. Res., 11:445-451.

Couillerot, E., Caron, C., Trentesaux, C., Chénieux, J.C., and Audran J.C., 1999, *Fagara zanthoxyloides* Lam. (Rutaceae): In vitro culture and the production of benzophenanthridine and furoquinoline alkaloids. In: Biotechnology in Agriculture and Forestry 43—Medicinal and Aromatic Plants XI, 136-156. Y.P.S. Bajaj (ed.). Springer-Verlag, Berlin-Heidelberg.

Degos, L. Dombret, H., Chomienne, C., Daniel, M.T., Micléa, J.M., Chastang, C., Castaigne, S., and Fenaux, P., 1995, All-trans-retinoic acid as a differentiating agent in the treatment of acute promyelocytic leukemia. Blood, 85:2643-2653.

Dessen, P., Frondrat, C., Valencien, C., and Mugnier, C., 1990, BISANCE: A French service for access to biomolecular sequence databases. Comput. Appl. Biosci., 6:355-356.

Edwards, A.R., Heading, V.E., and Wheeler, J.W., 1979, Isolation and identification of new-antisickling agents from *Fagara zanthoxyloides*. Clin. Res., 27:239A.

El Said, F., Fadulu, S.O., Kuye, J.O., and Sofowora, E.A., 1971, Native cures in Nigeria II. The antimicrobial properties of buffered extracts of chewing-sticks. Lloydia, 34:172-174.

Facchini, P.J., Penzes, C., Johnson, A.G., and Bull, D., 1996 Molecular characterization of berberine bridge enzyme genes from opium poppy. Plant Physiol., 112:1669-1677.

Fleury, F. Sukhanova, A., Ianoul, A., Devy, J., Kudelina, I., Duval, O., Alix, A.J., Jardillier, J.C., and Nabiev, I., 2000, Molecular determinants of site-specific inhibition of human DNA topoisomerase I by fagaronine and ethoxidine. Relation to DNA binding. J. Biol. Chem., 275:3501-3509.

Fournet, A., Barrios, A.A., Munoz, V., Hocquemiller, R., Cave, A., and Bruneton, J., 1993, 2-substituted quinoline alkaloids as potential antileishmanial drugs. Antimicrob. Agents Chemother., 37:859-863.

Herbert, J.M., Augereau, J.M., Gleye, J., and Maffrand, J.P., 1990, Chelerythrine is a potent and specific inhibitor of protein kinase C. Biochem. Biophysic. Res. Commun., 172: 993-999.

Hutchinson, J. and Dalziel, J.M., 1958, Flora of West Tropical Africa, 1 (2): 683-685. R.W.J.

Keay (ed.). Whitefriars Press, London–Tonbridge.

Krane, B.D., Fagbule, M.O., Shamma, M., and Gozler, B., 1984, The benzophenanthridine alkaloids. J. Nat. Prod., 47:1-43.

Kutchan, T.M., Dittrich, H., Bracher, D., and Zenk, M.H., 1991, Enzymology and molecular biology of alkaloid biosynthesis. Tetrahedron, 47:5945-5954.

Kutchan, T.M. and Zenk, M.H., 1993, Enzymology and molecular biology of benzophenanthridine alkaloid biosynthesis. J. Plant. Res., **3** (special Issue): pp.165-173.

Larsen, A.K., Grondard, L., Couprie, J., Desoize, B., Comoë, L., Jardillier, J.C., and Riou, J.F., 1993, The antileukemic alkaloid fagaronine is an inhibitor of DNA topoisomerases I and II. Biochem. Pharmacol., 46:1403-1412.

Leonard, M., Brice, M., Engel, J.D., and Papayannopoulou, T., 1993, Dynamics of GATA transcription factor expression during erythroid differentiation. Blood, **82:** 1071-1079.

Lewis, J.R., 1983, Biological activity of some Rutaceous compounds. In: Chemistry and Chemical Taxonomy of the Rutales, pp.301-318. Waterman, P.G. and Grundon MF (eds). Academic Press, London.

Messmer, W.M., Tin-WA, M., Fong, H.H.S., Bevelle, C., Farnsworth, N.R., Abraham, D.J., and Trojanek, J., 1972, Fagaronine, a new tumor inhibitor isolated from *Fagara zanthoxyloïdes* Lam. (Rutaceae). J. Pharm. Sci., 61:1858-1859.

Molinatti, P., 1994, Le genre *Fagara* (Rutaceae): Etude bibliographique; étude chimique et structurale des métabolites secondaires de *Fagara zanthoxyloides* obtenus in vitro, et de *Fagara rhoifolia*. Thèse de Doctorat Pharm, Reims, France.

Morceau, F., Aries, A., Lahlil, R., Devy, L., Jardillier, J.C. ,Jeannesson, P. and Trentesaux, C., 1996a, Evidence for distinct processes in the aclacinomycin- and doxorubicin-mediated differenciation of human erythroleukemic cells. Biochem. Pharmacol., 51:839-845.

Morceau, F., Chenais, B., Gillet, R., Jardillier, J.C., Jeannesson, P., and Trentesaux, C., 1996b, Transcriptional and posttranscriptional regulation of erythroid gene expression in anthracycline-induced differentiation of human erythroleukemic cells. Cell Growth Diff., 7:1023-1029.

Müller, M.J. and Zenk, M.H., 1992, The norcoclaurine pathway is operative in berberine biosynthesis in *Coptis japonica*. Planta Med., 58:524-527.

Ouattara, A., Guissou, I.P., Sawadogo, A., and Sawadogo, M., 1992, Etude comparée du traitement de la crise drépanocytaire par une présentation galénique de deux plantes médicinales (*Fagara xanthoxyloides* Lam. et *Calotropis procera* Ait) et un médicament usuel de référence (dihydroergotoxine=hydergine) chez les enfants de 5 à 15 ans à l'hôpital de Ouagadougou. Le Pharmacien d'Afrique, 68:19-27.

Privitera, E., Schiro, R., Longoni, D., Ronchi, A., Rambaldi, A., Bernasconi, S., Ottolenghi, S., Masera, G., and Biondi, A., 1995, Constitutive expression of GATA-1, EPOR, α-globin, and γ-globin genes in myeloid clonogenic cells from juvenile chronic myelocytic leukemia. Blood, 86:323-328.

Reiss, M., Gamba-Vitalo, C., and Sartorelli, A.C., 1986, Induction of tumor cell differentiation as a therapeutic approach: preclinical models for hematopoietic ans solid neoplasms. Cancer Treat. Rep., 70:201-218.

Sethi, V.S. and Sethi, M.L., 1975, Inhibition of reverse transcriptase activity of RNA-tumor viruses by fagaronine. Biochem. Biophys. Res. Comm., 63:1070-1076.

Sethi, V.S., 1976, Inhibition of mammalian and oncornavirus nucleic acid polymerase activities by alkoxybenzophenantridine alkaloids. Cancer Res., 36:2390-2395.

Sethi, V.S., 1977, Base specificity in in the inhibition of oncornavirus reverse transcriptase and cellular nucleic acid polymerases by antitumor drugs. Ann. N.Y Acad. Sci., 284:508-524.

Sethi, M.L., 1979, Inhibition of reverse transcriptase activity by benzophenantridine alkaloids. J. Nat. Prod., 42:187-196.

Sorensen, B.S., Jensen, P.B., Sehested, M., Jensen, P.S., Kjeldsen, E., Nielsen, O.F., and Alsner,

J., 1994, Antagonistic effect of aclarubicin on camptothecin induced cytotoxicity: role of topoisomerase I. Biochem. Pharmacol., 47:2105-2110.

Stadler, R., Kutchan, T.M., and Zenk, M.H., 1989, (S)-norcoclaurine is the central intermediate in benzylisoquinoline alkaloid biosynthesis. Phytochemistry, 28:1083-1086.

Tan, G.T., Pezzuto, J.M., Kinghorn, A.D., and Hughes, S.H., 1991, Evaluation of natural products as inhibitors of human immunodeficiency virus type 1 (HIV-1) reverse transcriptase. J. Nat. Prod., 54:143-154.

Tan, G.T., Lee, S., Lee, I.S., Chen, J., Leitner, P., Bersterman, J.M., Kinghorn, A.D., and Pezzuto, J.M., 1996, Natural-product inhibitors of human DNA ligase I. Biochem. J., 314:993-1000.

Trentesaux, C., Ngo Nyoung, M., Aries, A., Morceau, F., Ronchi, A., Ottolenghi, S., Jardillier, J.C., and Jeannesson, P., 1993, Increased expression of GATA-1 and NFE-2 erythroid-specific transcription factors during aclacinomycin-mediated differentiation of human erythroleukemic cells. Leukemia, 7:452-457.

Wang, L.K., Johnson, R.K., and Hecht, S.M., 1993, Inhibition of Topoisomerase I Function by Nitidine and Fagaronine. Chem. Res. Toxicol., 6:813-818.

CHAPTER 8

Ginkgo biloba: A Medicinal and Ornamental Tree of Great Interest

DOMINIQUE LAURAIN-MATTAR

INTRODUCTION

The *Ginkgo biloba* tree is the only surviving genus and species of the Ginkgoaceae family. Its origin can be traced back to the Jurassic period. It is a dioecious tree, native to China. Resistant to city smoke, industrial fumes, insects and fungi, *G. biloba* has become a favourite ornamental tree in parks and streets of numerous towns allover the world.

Ginkgo biloba is also of great interest for therapeutics because its leaves possess pharmaceutical properties such as radical scavenging, improved blood flow, vasoprotection and anti-PAF activity (DeFeudis, 1991; O'Reilly, 1993). The Ginkgo leaf extract for pharmaceutical purposes was registered for human use in 1974 in France under the code-name 'Egb 761' (DeFeudis, 1998). This extract is used for the treatment of cerebral and peripheral vascular diseases.

Plantations of *G. biloba* trees with an industrial vocation have been established in France, the United States and China in order to meet the growing demand for *G. biloba* leaves. Two types of compounds of pharmaceutical importance accumulated in the leaves: terpenes (ginkgolides and bilobalide) and flavonoids. These compounds have also been detected in the roots (Nakanishi, 1988). Seasonal variation of their concentration in leaves has also been reported (van Beek and Lelyveld, 1992; Flesh et al., 1992).

Works on plant cell biotechnology of Ginkgo have been achieved in order to study the biosynthesis of active substances: mainly terpenes like ginkgolides and bilobalide in undifferentiated cell cultures; in embryogenic cell cultures (Laurain et al., 1993a, 1993b, 1996); and in transgenic cell cultures (Laurain et al., 1997; Dupré et al., 2000). In addition, metabolic studies, via in vitro culture, have shown that these unique secondary metabolites are not produced from mevalonate but via an alternative 'triose/pyruvate' pathway (Schwarz, 1994).

In this chapter, the biology, utilization and chemical constituents of the leaves, pharmacological activities and clinical uses of leaf extract and a compilation of the most recent results in plant cell biotechnology of *G. biloba* have been presented.

EVOLUTIONARY HISTORY

The first definitive description of the genus and species of *Ginkgo biloba* L. was given by Linnaeus (1771) (cited by Melzheimer and Lichius, 2000). However, this tree is also called 'Maidenhair tree' in relation with the similarity to the foliage of the 'Maidenhair fern' (*Adiantum*). It is called 'ginkyo' in Japan and 'l'arbre aux quarante écus' in France. This French name is based on the price of five plants of Ginkgo sold by a Frenchman, Mr Pétigny, to an English horticulturist, in 1780. Almost all the ginkgos growing in France have descended from these five plants.

The Ginkgo is one of the best known trees with a long geological history of fossil antecedents (Seward, 1938). This tree is the sole living survivor of a genus that has changed little for more than 80 million years. The genus *Ginkgo*, one of the major groups in ginkgophytes, appeared in the Early Jurassic (Tralau, 1968). During the lower Cretaceous, this genus appeared to have reached the peak of its diversity. This was not only true of the number of different species and genera occurring, but also of their nearly worldwide distribution and high number of habitats (Melzheimer and Lichius, 2000). During the Upper Cretaceous, this group showed a decline in the diversity and distribution when worldwide temperatures decreased dramatically. Today, they have disappeared from all the continents except one, wherein only one single species has survived, the Asia (Dorf, 1958).

Ginkgo biloba is generally classified in its own division, the Ginkgophyta (Del Tredici, 2000). This taxon is distinguished from the Coniferophyta on the basis of its reproductive structures, most notably, its multiflagellated sperm cells, and from the Cycadophyta on the basis of its vegetative anatomy. Recent molecular analysis of the Ginkgo genome suggests much closer relationship to the cycads rather than to the conifers (Hasebe, 1997).

BIOLOGY

Ginkgo is a dioecious species, with separate male and female individuals.

This tree possesses deciduous leaves and can exceed 30 m in height. Trees raised from ovules are composed of an orthotropic monopodial trunk with plagiotropic branches per tier. The form of these vigorous young trees is distinctly pyramidal with a dominant central leader and widely-spaced whorls of lateral branches that grow out at a diagonal orientation to the trunk. As the tree ages, the pyramidal shape is lost and a broad, spreading form develops (Peter del Tredici, 1991). Ginkgo trees produce two types of shoots: long shoots with widely-spaced leaves that subtend axillary buds (Figure 8.1) and short shoots (Figure 8.2) with clustered leaves that lack

Fig. 8.1 Long shoots of *Ginkgo biloba* with spaced bilobed leaves.

Fig. 8.2 Short shoots of *Ginkgo biloba* with clustered leaves bearing ovules that are female reproductive structures.

both internodes and axillary buds (Critchfield, 1970). Long shoots make up the woody framework of the tree and are responsible for generating new growing points (Del Tredici, 2000), while the short shoots produce the majority of leaves and, when the Ginkgo is 25-30 years old, they bear the reproductive structures. In response to a variety of environmental and physiological factors, some short shoots are transformed into long shoots and the terminal growth of a long shoot can fail to elongate and become a short shoot (Gunckel et al., 1949). This ability is frequent in the conifers when a reduction or a deletion of the carrier axis dominance appears after a traumatism (Nozeran, 1986).

Ginkgo trees have characteristic leaves with one or several lobes, a distinctive dichotomous venation and a long petiole (Figure 8.2). On dwarf shoots and at the basal regions of long shoots, the leaf blades are either entire or are divided into two lobes by a distal notch. By contrast, the blades of leaves on the upper parts of long shoots are always divided by a very deep sinus into two major lobes (Gifford and Foster, 1989). It is a well-known fact that the Ginkgo leaf has open dichotomous venation, which is devoid of anastomoses. However, Arnott (1959) described finding leaves with anastomoses. Long shoots had a significantly higher average percentage of leaves with anastomoses (13.4%) rather than dwarf shoots (8.2%). Arnott wrote that the higher anastomosis index of these leaves was in harmony with the data on the diffusibility of auxin in long shoots reported by Gunckel and Thimann (1949). However, there is no experimental evidence as yet to directly link the concentration of auxin and the frequency of anastomosis in Ginkgo (Hara, 1997).

The male catkins (Figure 8.3) emerge before the leaves and fall off immediately after shedding their pollen. Wind pollination occurs anywhere from early April in areas with mild winters to late May in regions beset with severe winters. Ginkgo ovules are produced in pairs at the end of stalks, which are 1 to 1.5 centimetres long (Figure 8.4). When the ovule is receptive, it secretes a small droplet of mucilaginous fluid from its micropyle which functions to capture airborne pollen. Once inside the ovule, the male gametophyte commences a four-month long development period that culminates with the production of a pair of multiflagellated spermatozoids,

Fig. 8.3 Male catkins of *Ginkgo biloba*.

Fig. 8.4 Ovules of *Ginkgo biloba*.

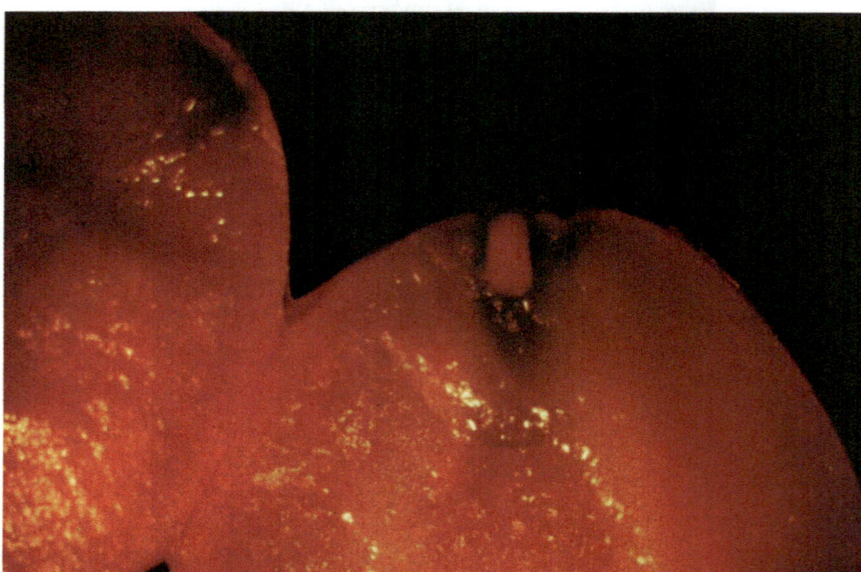

Fig. 8.5 Zygotic embryo of *Ginkgo biloba* in the tissue of a mature ovule.

one of which fertilizes a waiting egg cell (Friedman, 1987), while the ovules are still on the tree (Holt and Rothwell, 1997). The fertilization can occur anytime between late August to late September. The mature ovule of Ginkgo consists of an embryo (Figure 8.5) embedded in the tissue of the female gametophyte, surrounded by a thick seed coat. The developing ovules are green until they mature in the autumn by which time they turn the same yellow colour of the leaves. Typically, they fall from the tree about a month after fertilization. The foul odour associated with Ginkgo ovules develops only when they are fully mature, and is the result of the presence of two volatile compounds, butanoic and hexanoic acids, localized in the sarcotesta (Parliment, 1995). These ovules also contain phenolic compounds known to cause contact dermatitis in humans (Kochibe, 1997).

ECOLOGY

As a wild species, Ginkgo is native to China. One of Ginkgo's last wild refugia was thought to be in Zhejiang Province, on the west peak of Tianmu Mountain. There are also reports of wild Ginkgo populations in other parts of China (for review, see: Del Tredici, 2000). Many Tianmu Ginkgos possessed more than one trunk. Most of these secondary trunks originated from lignotubers or 'basal chichi' located at, or just below, ground level. Anatomically, lignotubers develop in all Ginkgo seedlings from buds located in the axils of the two cotyledons. When stimulated by some traumatic event, one of these cotyledonary buds often grows out from the trunk to form a woody, positively-geotropic lignotuber which has the capability of producing both aerial shoots and adventitious roots (Del Tredici, 1992, 1997). The sprouting ability of Ginkgo may well have played a role in the survival and morphological stability of the genus since the Tertiary (Del Tredici, 2000).

CULTIVATION

Ginkgo has been cultivated in China for several thousand years (Del Tredici, 2000). The tree was introduced into western Japan from eastern China about eight hundred years ago. Both in China and Japan, the largest trees are growing in the vicinity of Buddhist or Daoist temples and shrines (Tsumura et al., 1992). From Japan, Ginkgo was introduced into Europe at the Botanical Gardens in Utrecht (Netherlands) around 1730, and into Kew Gardens near London (England) around 1754. From England, the tree was imported into North America in 1784 to Philadelphia (Del Tredici, 1991).

The Ginkgo thrives in a great variety of soil such as sandy humus soils or heavy loamy (Melzheimer and Lichius, 2000). The tree tolerates a soil of pH = 5.5-7.7. Under favourable conditions, young plants quickly develop a vertical axis with less robust horizontal branches. After 5 or 6 years, the plants can reach a height of 2-3 m. Then the growth slows down, and it takes several decades before the tree achieves a certain size which can exceed 30 m in height. Flesh et al. (1991) reported that higher light intensity and higher temperatures activate the growth of long shoots and the sprouting of lateral buds. Ginkgo also shows an extraordinary ability to adjust to a wide variety of climates. It grows in southern Canada and other regions where winter temperature may drop to –25°C (Rensselaer, 1969). Ginkgo does not perform particularly well in subtropical climates or on soils that are overly wet or dry during the growing season (Santamour et al., 1983; Del Tredici, 1991). Ginkgo grows upto an altitude of 1000 metres (Michel, 1985).

UTILIZATION

Ornamental uses

Ginkgo is cultivated throughout the temperate zones of the world for ornamental purposes. Its interest for horticulture is linked to its wonderful foliage which takes on a bright yellow colour in autumn (Figure 8.6). This tree became a symbol of indominability and hope by reason of the regeneration of a shoot of Ginkgo the year following its destruction by the Hiroshima bomb (Michel, 1985). In addition, Ginkgo has been said to be the species of tree the most resistant to air pollution. However, little is known about the biochemical mechanisms for this tolerance (Kim et al., 1997). Many investigations have shown that Ginkgo exhibits a considerable tolerance for gaseous air polluants and, more precisely, for sulphur dioxide, nitrogen oxides and ozone. The leaves of Ginkgo were collected for morphological and chemical analysis (Kim and Lee, 1990). The leaf analysis showed that Ginkgo growing in heavy traffic areas contained five times more sulphur than that growing in traffic-free areas (1.81% vs 0.36%). However, no visible injuries were detected in Ginkgo leaves collected from areas with heavy road traffic. In addition, no significant degradation of cuticular waxes was shown with scanning electron microscopy while a degradation of stomatal wax crystalloids was found in various conifer needles. A continuous wax layer could be observed on both sides of the Ginkgo leaf surface (Gülz, 1994). The epicuticular wax contains many of the usual wax lipids of angiosperm trees and the secondary alcohol nonacosan-10-ol, which is a characteristic component of conifer needle waxes (Gülz et al., 1992).

Although Ginkgo is resistant to gaseous dry deposition, this tree is susceptible to damage from acid rain (for review, see: Kim et al., 1997). In contrast, no injuries appeared in most gymnospermous trees like *Abies holophylla* and *Pinus thungergii* (Oh et al., 1987, 1986). The reason of the

Fig. 8.6 *Ginkgo biloba* tree in autumn in the Botanical Garden of Epernay (France).

sensibility of Ginkgo to simulated acid rain has not yet been clarified. Ginkgo is a broad-leaved tree, unlike most other gymnosperms. Broad-leaved trees are much more susceptible to acid precipitation than gymnospermous trees (Kim et al., 1997). The epicuticular wax has many protective functions on plant surfaces. In Ginkgo, the epidermis has relatively thin walls and hypodermal sclerenchyma is absent. The stomata occur almost exclusively on the lower epidermis of the leaves while most of the conifers, which are not affected by acid rain, bear numerous stomata on both sides of their needles. In addition, the cuticle is not well developed in Ginkgo.

Fig. 8.7 Plantation of *Ginkgo biloba* in Champaubert la Bataille (France).

Ginkgo is also one of the least vulnerable trees as far as insects are concerned. In Japan, for example, of 344 pest species found on 40 species of the landscaped tree, only 6 species were found on the Ginkgo tree (for review: Honda, 1997). No species-specific lepidopterous defoliators for this tree have been recorded. However, the most common lepidopterous pests that damaged the Ginkgo tree are: *Phassus excrescens* Butler, *Holcocerus vicarius* Walker, *Setora postornata, Ocneria disper* L., *Cryptothelea variegata* Snellen, *Cnidocampa flavecens* Walker *and Vapocheima robiniae.* In addition, coleopterous pests living on trees are extremely few compared with those on other species of forest and garden trees. Some authors reported the Ginkgo tree to be a host plant for *Acalolepta sejuncta.* The last group of insect pests on Ginkgo trees in Japan is the homopterous insect group, of which the most common species is *Pseudococcus comstoki,* which infests the leaves and twigs of Ginkgo. Severe infestations by these scale insects on the Ginkgo tree were clearly associated with environmental pollution in areas where they were planted. A new species of tortorix, *Pammene ginkgoicola,* has been identified on buds, twigs and seeds of Ginkgo tree damaged by this pest (Zhang and Li, 1981). *Criconemoides morgensis* Taylor is the only nematode that attacks Ginkgo tree (for review, see Honda, 1997).

Extracts of Ginkgo leaves have been used as agents to control agricultural pest insects in China and Japan (Konishi, 1996; Honda, 1996; Zheng, 1983; Yagi, 1996, cited by Honda, 1997). These facts suggest the existence of some chemical barriers against pest insects in Ginkgo. Investigations have been carried out on the insecticidal properties of the tree. Alcohol extracts of ground whole roots of the Ginkgo tree strongly inhibited the growth of the European corn borer, *Pyrausta nubilalis* (Beck, 1960). Major (1967) reported that this inhibitory effect might be due to acidity from oxalic acid and malic acid. Gingko leaves are very acidic in nature and 2-hexenal is produced when Ginkgo leaves are damaged in the presence of oxygen (Major et al., 1963). Bevan et al. (1961) have reported that this aldehyde is an insect repellent. Ginkgolide A has been found to be the most active of several compounds which included bilobalide and ginkgolic acid, possessing antifeedant activity against larvae of the cabbage butterfly (*Pieris rapae crucivora*) (Matsumoto and Sei, 1987). From the studies reported above, it is clear that Ginkgo contains certain chemicals that are toxic to at least some insects. In addition, some investigations have shown that Ginkgo leaves lack feeding stimulants or attractants for inducing beetle feeding (Major, 1967; Major and Tietz, 1962 ; Hirano, 1971).

Some fungal species have been recorded as parasites, i.e. *Ceuthospora melaleuca, Fomes meliae, Glomerella cingulata, Irpex lacteus, Trametes hirsuta, Trametes versicolor* (Seymour, 1929) and *Pestalotia sinensis* (Guba, 1961).

Leaf production

Ginkgo leaves are of great interest for therapeutics because they possess pharmaceutical properties such as radical scavenging, improved blood flow, vasoprotection and anti-PAF activity (DeFeudis, 1991; O'Reilly, 1993). Therefore, the pharmaceutical industry needs huge quantities of leaves; two thousand tonnes of leaves are used each year in the world (Masood, 1997). So, to meet the growing demand for Ginkgo leaves, it is necessary to set up plantations. Some plantations of *Ginkgo biloba* with an industrial vocation have been established in the world. China is the major producer worldwide of Ginkgo leaves (Fusheng et al., 1997). In this country, leaves are harvested both from trees growing for an ornamental purpose and also from trees growing in plantations with an industrial vocation. In Europe, one plantation has been established in the Bordeaux region (France) since 1982. A smaller plantation is established in Champaubert la Bataille (France) since 1992 with organic farming techniques (Figure 8.7). When grown for this purpose, seedlings are spaced 40 centimetres apart in rows 1 metres apart, producing a density of approximately 25,000 plants per hectare (Del Tredici, 2000). Special pruning techniques are used to keep the trees below 3 metres tall, thereby allowing the use of mechanical hervesting equipment. Green leaves are harvested in mid- to late-summer. Leaves are used to prepare extracts for pharmaceutical purposes. The 'EGb 761' extract is made from dried Ginkgo leaves and has a standardized content of 22-27% flavonol glycosides and 5-7% terpene trilactones (DeFeudis, 1998). This extract was registered for human use in 1974 in France.

CHEMICAL CONSTITUENTS

Many phytochemical studies have been carried out on *Ginkgo biloba*, more especially, on root bark, wood, sarcotesta and mainly on leaves which are used for phytomedicines and also on ovules which, when roasted, are consumed as food. Terpenes, flavonoids, organic acids, polyacetate-derived compounds, carbohydrates, miscellaneous organic compounds and inorganic compounds were found in *G. biloba* (Boralle et al., 1988; Hasler, 2000). In this chapter, the two groups of compounds, at least partially responsible for the beneficial activity of Ginkgo leaf extracts, as the standardized extract Egb 761 marketed in Europe, are described: terpenes and flavonoids.

Terpenes

Different groups of terpenes, namely terpene trilactones (ginkgolides, bilobalide), triterpenes (steroids, phytosterols), carotenoids, polyprenols and volatile terpenes (mono- and sesquiterpenes) can be found in Ginkgo (for review, see Hasler, 2000). Till date, ginkgolides (diterpenoids) and bilobalide (sesquiterpene) were exclusively found in *G. biloba*. The concentration of these terpenes in leaves fluctuates during the vegetative season. Van Beek et al. (1991) found the concentration of these terpene trilactones to be the lowest in spring, gradually increasing to reach its maximum in mid-summer. When the leaves turned yellow in late autumn, the concentration started to decline. In contrast, Hasler and Meier (1993) have found a high content of the terpenes in the yellow and fallen leaves, comparable to that of green leaves. Furthermore, the content of these secondary metabolites in Ginkgo leaves showed differences from tree to tree, upto a factor of 160 (van Beek et al., 1991) or upto a factor of 300 (Hasler and Meier, 1993). Flesh et al. (1992) have examined the relative importance of the growth and light level on the terpene trilactones content. Cartayrade et al. (1997) have studied the site of synthesis of these metabolites which appeared to be biosynthesized in roots and accumulated in leaves (see Plant Cell Biotechnology in this chapter). Ginkgolides are a small group of five compounds. They were first isolated from *G. biloba* root bark (Furukawa, 1932). The structure of four of these diterpenoids (Ginkgolides A, B, C and M) were determined by Maruyama et al., (1967) and Okabe et al. (1967). In 1987, Weinges et al. 1987 identified the ginkgolide J in *G. biloba* leaves. The ginkgolides A (Okabe et al., 1967; Sakabe et al., 1967), B and C (Okabe et al., 1967) were also found in the leaves. The only differences between these compounds are the number and the position of hydroxyl groups which may be present on C1, C3 or C7 of the spirononane framework (Figure 8.8). In addition to the diterpenoids (ginkgolides), a sesquiterpene trilactone, named bilobalide, was described in leaves (Major, 1967; Weinges and Bähr, 1969) (Figure 8.9).

Problems in terpene trilactone analysis are met in the initial extraction and even more so in the sample clean up (van Beek, 2000). Extraction solvents usually used are a mixture of water with methanol, alcohol or acetone. As long as the water content in such mixtures is superior or equal

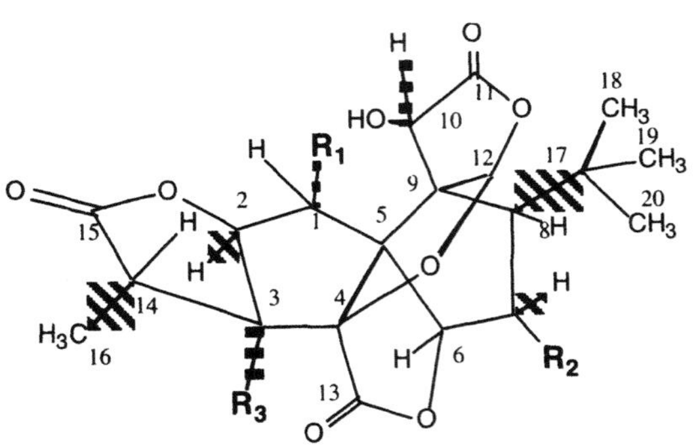

Ginkgolides	R_1	R_2	R_3
A	H	H	OH
B	OH	H	OH
C	OH	OH	OH
J	H	OH	OH
M	OH	OH	H

Fig. 8.8 Ginkgolide structures of *Ginkgo biloba* leaves.

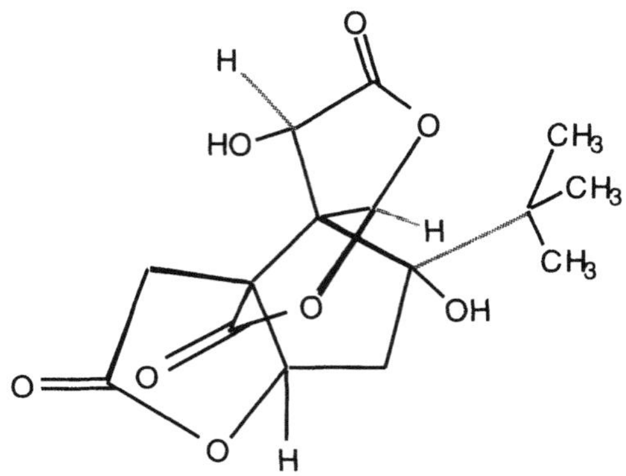

Bilobalide

Fig. 8.9 Bilobalide structure of *Ginkgo biloba* leaves.

to 50%, the solubility of terpene trilactones is sufficient to warrant a good extraction (Stumpf, 1997). However, metabolites (e.g. flavonol glycosides, biflavonoids, alkylphenols and chlorophyll) occurring in Ginkgo leaves are also extracted. Generally speaking, the amount of such compounds increases with an increasing percentage of organic solvent with the exception of the polar proanthocyanidins. Some authors (van Beek, 2000) increase the selectivity of the initial extraction medium further by decreasing the organic content to 10% or even 0%. To obtain a good extraction of ginkgolide B, refluxing is necessary. Camponovo (1996) has shown that with 50% aqueous methanol, the terpene trilactones were extracted and this solvent gave the most clean extract. The addition of a small percentage of acid could be considered to prevent possible decomposition of the less stable bilobalide and to diminish co-extraction of chlorophyll. Many different approaches have been followed in order to remove unwanted impurities from the primary extracts prior to the separation and analysis. Some groups use a defatting step, others one or more partitioning steps and still others use one or two solid phase extractions columns (for review, see van Beek, 2000).

The chemical analysis of terpenes in *G. biloba* is also difficult due to the lack of a good chromophore, the absence of any reactive groups, their relatively low concentrations in leaves (about 0.05%), the difficult separation of some ginkgolides and the occurrence of many other UV-active substances in high concentrations in Ginkgo leaves (van Beek, 2000).

TLC is still used for qualitative purposes. Silica gel is used as stationary phase in combination with relatively apolar solvents like $CHCL_3$-MeOH (8:2) (Okabe et al., 1967) and benzene-Me_2CO (8:2) (Weinges and Bähr, 1969). However with these solvents, the pairs ginkgolides A and B and ginkgolides J and C have almost identical Rf values on silica gel. Another problem is that the ginkgolides are very stable, unreactive compounds without any olefinic double bonds. This makes them difficult to detect on TLC plates (van Beek, 2000). By impregnating silica gel plates with sodium acetate, an excellent resolution of all five terpene trilactones occurring in leaves is nowadays possible (van Beek and Lelyveld, 1993). In addition, the impregnation increases the sensitivity of the acetic anhydride spray reagent, which is the best reagent found so far.

For quantitative analysis of terpenes in Ginkgo leaves, extracts or phytopharmaceuticals, high performance liquid chromatography (HPLC) and gas chromatography (GC) can be used. The biggest advantage of HPLC over GC is that no time-consuming derivatization step with an aggressive reagent is necessary. However, GC gives a better separation in less time than HPLC. Detection of terpene trilactones in HPLC used UV detector at first (Lobstein-Guth et al., 1983; van Beek et al., 1991; Camponova, 1996). However, these compounds do not exhibit a maximum with a significant extinction coefficient above 200 nm. For this reason, the use of refractive index (RI) detection, although less sensitive, was reported (Komoda et al., 1988; Teng, 1988; van Beek et al., 1991; Laurain et al., 1997; Chen et al., 1998). It is still used by several industries for controlling the quality of their Ginkgo products. Two more LC detectors have been experimentally used for Ginkgo terpenes: the mass spectrometer (LC-MS) (Maillard et al., 1993) and the evaporative light scattering detector (ELSD) (Camponovo et al., 1995), but these last techniques are not used for routine analysis. Although Ginkgo terpene trilactones are totally involatile, they can be separated by GC once they have been silylated. As detector the flame ionization detector (FID) was used (Carrier et al., 1990; Carrier et al., 1991; Chauret et al., 1991; Hasler and Meier, 1992; Huh and Staba, 1993; Jeon et al., 1995). Most of the other quantitative methods which have been proposed like: biological method based on the antagonism of PAF by ginkgolides (Steinke et al., 1993), thin layer chromatography method (Xie et al., 1997), capillary electrophoresis (Oerhle, 1995) and NMR (van Beek et al., 1993), are rather specialized and unsuitable for routine analysis (van Beek, 2000). The only alternative separation/detection method which could compete with either LC or GC for routine analysis is supercritical fluid chromatography (SFC) in combination with ELSD (Strode et al., 1996).

Flavonoids

Flavonoids are polyphenolic, low molecular weight compounds, probably found in all green plants (Hasler, 2000). The flavonoids of Ginkgo leaves represent a great variety of flavonol glycosides based on kaempferol and quercetin and are found as mono-, di- and triglycosides. Only glucose and rhamnose can be found as sugar molecules. Minor flavonoid compounds are derived from isorhamnetin, myricetin, 3′-methylmyricetin (with flavonol structures) as well as from apigenin and luteolin (with flavone structures) (Sticher et al., 2000). In leaf extracts of *G. biloba*, Hasler et al. (1992) identified 22 flavonoid glycosides. In addition, non-glycosidic biflavonoids, like ginkgetin, bilobetin, isoginkgetin and sciadopitysin (Figure 8.10) and proanthocyanidins have been isolated. Chang et al. (1993) have demonstrated that the total amount of biflavones increased during the growing season to reach its maximum in yellow autumn leaves. Hasler et al. (1990) investigated the influence of the season on the flavonoid concentration. Between June and November, no significant differences in the flavonoid content were found except for the month of May. It was shown that dried Ginkgo leaves had a calculated flavonol glycoside content of 0.7 to 0.8% from June to November and this content reached 1.76% during May (Sticher et al., 2000). The flavonol glycosides and the proanthocyanidins contribute to the free radical scavenging activity of Ginkgo extracts (DeFeudis, 1991).

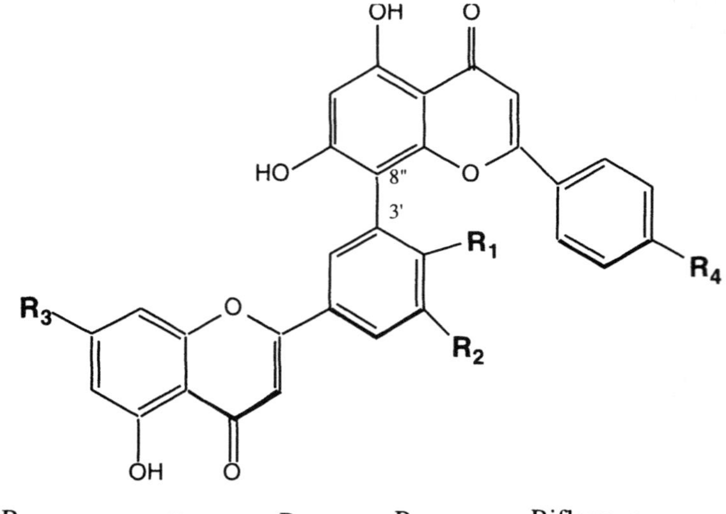

R_1	R_2	R_3	R_4	Biflavone
OH	OH	OH	OH	Amentoflavone
OCH_3	OH	OH	OH	Bilobétine
OCH_3	OH	OCH_3	OH	Ginkgétine
OCH_3	OH	OH	OCH_3	Isoginkgétine
OH	OH	OCH_3	OH	7-Methoxyamentoflavone
OCH_3	OCH_3	OH	OH	5'-Methoxybilobétine
OCH_3	OH	OCH_3	OCH_3	Sciadopitysine

Fig. 8.10 Biflavonoid structures of *Ginkgo biloba* leaves.

Supercritical fluid extraction can be used to remove the flavonoids from leaves (Yao et al., 1995). The method for a selective analysis of flavonol glycosides as their aglycones after hydrolysis was reversed-phase high-performance liquid chromatography (RP-HPLC) combined with diode-array detection (Sticher et al., 2000). The flavonoid profile of Ginkgo leaf extract is very complex. The great variety of genine flavonol glycosides is reduced by hydrolysis to the corresponding aglycones with methanolic HCL. The UV spectra of the aglycones allow a selective detection at wavelengths above 300 nm. The main aglycones quantified were kaempferol, quercetin, isorhamnetin and the minor aglycones were myricetin and luteolin. The aglycone mixtures could be separated after pre-purification over Sephadex LH-20, using a LiChroCart RP-18 column with RP-8 guard column and methanol-water-formic acid (55:41:4) (Schennen, 1988) or a Nucleosil C_{18} column and a methanol-water gradient (Wagner et al., 1989), respectively, as mobile phase. Another HPLC method has been developed for the analysis of flavonol glycosides after their hydrolysis in Ginkgo leaf extracts and phytomedicines (Hasler et al., 1990, 1992). Extraction and hydrolysis were performed by refluxing the pulverized plant material or extract with hydrochloric acid (25%) in methanol for 60 min. Sample clean-up was carried out using C_{18} solid-phase extraction cartridges. Finally, the purified extract was subjected to HPLC analysis. The reduction of the genine glycosides by hydrolysis is an accepted method in the quality control of phytomedicines (Sticher et al., 2000). The aglycone content is converted into a 'Ginkgo flavonol' glycoside content. The calculation of this content square to = sum of aglycones × 2.51 (average conversion factor). Hasler et al. (1992) have developed a second method of analysis using a methanol-tetrahydrofuran gradient with 0.5% (v/v) *ortho*-phosphoric acid in water. Identical quantifications of flavonols resulted for both methods.

RP-HPLC, with diode array detection, was used to study the metabolic fate of the flavonol glycosides present in standardized Ginkgo extracts after administration to rats (Pietta et al., 1995) and humans (Pietta et al., 1997).

In contrast to the ubiquitous flavonol aglycones, biflavonoids are characteristic leaf compounds for the identification of Ginkgo leaves and leaf extracts (Hasler et al., 1992). Biflavonoids were thought not to possess any clear pharmacological activities. However, in 1995, Joyeux et al. reported free radical scavenging properties as well as protective effects against peroxidative reactions. In addition, complexes of biflavonoids and phospholipids have been shown to increase the skin microcirculation after topical application (Bombardelli et al., 1996). HPLC separation and quantitative determination of Ginkgo biflavonoids were reported by Briançon-Scheid et al. (1982, 1983), Song (1986) and Pietta et al. (1988).

Briançon-Scheid et al. (1982) used a LiChrosorb diol column (10 μm) in combination with a tetrahydrofuran-chloroform gradient. However, the proposed system was neither suitable for the separation nor for the quantification of the biflavones. Pietta et al. (1988) developed an isocratic RP-HPLC separation system using a Novapak C_{18} column with a Guard-Pak μBondapak C_{18} pre-column and tetrahydrofuran-propanol-water (21:10:69) as a mobile phase. With this system, both the sample clean-up and the separation of four biflavonoids were selective enough. Bilobetin, ginkgetin/isoginkgetin, the most difficult pair to separate, and sciadopitysin could be separated and determined with the system using a LiChroCART 125-4, Superspher 100 RP-18 column with a Lichroprep RP-18 guard column and gradient elution with acetonitrile and water containing 10% phosphoric acid (Lobstein et al., 1991). Gobbato et al. (1996) used a SupelcoSil LC 18 column equiped with an RP-18 Newguard guard column and gradient elution with acetonitrile-water containing 0.3% phosphoric acid. This system led to a good separation of six biflavones with good reproducibility. They identified also the main biflavones of *G. biloba* by HPLC-MS using a thermospray interface. Thin layer chromatographic methods have been described by Vanhaelen and Vanhaelen-Fastre (1988).

PHARMACOLOGY

The medicinal properties of *Ginkgo biloba* leaves have been known in China since ancient times. The first publication concerning the therapeutic use of the leaves of this tree dates back to the sixteenth century in a text written by a Chinese author, Liu Wen Tai (Di Renzo, 2000). However, in Europe, extracts of *G. biloba* leaves were introduced into medical practice in 1965 (in Germany). The standardized extract, EGb 761, was launched in the French market in 1974 and in the German market in 1978 (Christen, 2000). Since, some thirty years ago, a large number of pharmacological and clinical studies have been reported. In 1991, DeFeudis described the polyvalent action of the *G. biloba* leaf extract which results from the combined activities of several active principles among these flavonoids (ginkgo flavonol glycosides) and terpene trilactones (ginkgolides and bilobalide). This polyvalent action may be classified on the one hand as a vaso- and tissue-protective action, which explains the peripheral and central effects of the extract; and on the other hand as a cognition-enhancing action which contributes to the beneficial effect on subjects with senile cognitive decline (van Beek et al., 1998). Different mechanisms of action seem to underlie the pharmacological activity of the extract such as stimulation of the release of EDRF/NO and PGI_2, antioxidant activity, PAF-antagonism, increased release of neurotransmitters and inhibition of biogenic amine uptake (DeFeudis, 1991).

Activity on EDRF and PGI_2

EDRF, one of the endogenous mediators for vessel relaxation, has now been identified as nitric oxide (NO), which induces vasodilatation by the activation of a soluble guanylate cyclase in the vascular smooth muscle (van Beek et al., 1998). The vasorelaxing effect of *G. biloba* extract on isolated rabbit aorta with an intact endothelium seems to be related to the release of EDRF (Auguet et al., 1986; Delaflotte et al., 1984; Auguet et al., 1994). The relaxing effect on isolated blood vessels could also involve an increased release of prostacyclin (PGI_2) from intact endothelium (DeFeudis, 1991). Prostacyclin is the major relaxing and antithrombotic prostaglandin. Since EDRF/NO is inactivated by superoxide anion, the *G. biloba* extract may protect the EDRF/NO mechanism by scavenging the free radicals of oxygen (DeFeudis, 1994).

Antioxidant activity

The free radical-scavenging activity of *G. biloba* extract against hydroxyl radical and superoxide anion has been reported by Pincemail et al. (1989) and is probably due to the flavonoid moiety of the extract (DeFeudis, 1994). However, the presence of the terpenes could contribute indirectly to this effect (Marcocci et al., 1994). The free-radical scavenging properties of the extract that protect the membranes from peroxidative degradation and damage have been demonstrated in various in vitro models (Pincemail et al., 1989; Barth et al., 1991; Dumont et al., 1995; Marcocci et al., 1994). Reactive oxygen species released by actived neutrophils are involved in various pathological processes such as inflammation, autoimmune diseases, ischemia and reperfusion and adult respiratory distress syndrome (van Beek et al., 1998). Recent evidence has focussed on the role played by oxidative stress in various acute and chronic neurodegenerative diseases. However, the precise mechanism by which oxidative stress can trigger the degenerative process is still a matter of debate (Di Renzo, 2000). Ni et al. (1996) investigated the effect of EGb on the oxidative stress-induced apoptosis in cerebellar neuronal cells. Cells were exposed to hydroxyl radicals that produced neuronal apoptotic death. Pre-treatment of neurons with EGb reduced the percentage of dead cells. It has been suggested that the protective effects exerted by EGb on oxidative stress-induced cell death may occur via the direct scavenging of hydroxyl radicals (Di Renzo, 2000).

Platelet-activating factor antagonist

Platelet-activating factor (PAF), a potent phospholipid inflammatory mediator, has been implicated in several pathophysiological disorders. Numerous studies have shown the PAF antagonistic actions of the ginkgolides, particularly ginkgolide B (for review, see Braquet, 1997). PAF stimulates the generation of free radicals and may amplify the neurotoxicity of the excitatory neurotransmitter glutamate, produced in large quantities during neuronal injury (Clark et al., 1992). The terpenic fraction of the extract, which contains the PAF-antagonist ginkgolide B, may contribute to the neuroprotective properties of *G. biloba* extract, through the free radical scavenging activity of the flavonoidic component may also be involved (Smith et al., 1996).

Effects on neurotransmitter receptors and neurotransmission

The ageing process is generally associated with a decrease of most neurotransmitters and of their receptors (Clostre, 1999). Several authors have shown that EGb might restore age-induced decrease in neurotransmitter receptor binding (Taylor, 1986; Huguet and Tarrade, 1992; Huguet et al., 1994; Kristofikova et al., 1992). In the pathological cerebral ageing, particularly Alzheimer's disease, the cholinergic system is mainly affected. It has been shown that EGb, at high concentrations, inhibits the uptake of tritiated norepinephrine, dopamine and 5-hydroxytryptamine into synaptosome-enriched fractions of rat brain (Taylor, 1990). The flavonoid glycosides and/or proanthocyanidins are the components of EGb responsible for these effects on amine uptake (Di Renzo, 2000). EGb can prevent the decrease of synaptosomal uptake of tritiated dopamine and 5-hydroxytryptamine uptake, that occurs in conditions of prolonged incubation in media of usual composition (Ramassamy et al., 1992; Ramassamy et al., 1994). This effect has been attributed to a free radical scavenging property. Flavonoids have been shown to be responsible for this effect of EGb, whereas the terpene fraction failed to show any activity (Di Renzo, 2000).

Effect on cerebral ischaemia

In various cerebral ischaemia models, the administration of EGb to rats and Mongolian gerbils has reduced the mortality rate, the extent of neurological deficits and the development of cerebral oedema; the mitochondrial respiration and energy metabolism were improved (Le Poncin-Laffitte et al., 1980; Spinnewyn et al., 1986; Spinnewyn, 1992). In the mechanism underlying the neuroprotective effect of EGb during ischaemia, the free radical scavenging and antioxidant activites of its constituents, including flavonoids, ginkgolides with anti-PAF activitie and bilobalide, could be involved (Prehn and Krieglstein, 1993).

CLINICAL APPLICATIONS

The major therapeutic indications for the *G. biloba* extract are: disturbances of cerebral function and peripheral vascular insufficiency. Many clinical studies concerning the therapeutic application of the extract have been reported (for review, see van Beek et al., 1998; Juretzek, 1997; Bulling et al., 2000; van Dongen et al., 2000). The influence of the *G. biloba* extract on the cerebral insufficiency was investigated in some controlled trials (Kleijnen and Knipschild, 1992). The term 'cerebral insufficiency' does not indicate a precise syndrome but a collection of cerebral disturbances such as impairment of recent memory, confusion, change in social behaviour, lack of initiative, affective and somatic troubles. Sometimes, cerebral insufficiency symptoms are considered to be early signs of dementia, both of the degenerative type (Alzheimer's disease) and vascular origin (multi-infarct dementia) (van Beek et al., 1998). The results obtained in some trials suggested that Ginkgo extract may be effective in relieving the symptoms associated with dementia of the Alzheimer type (Kanowski et al., 1996; Maurer et al., 1998; Wettstein, 1999/2000), with cerebral insufficiency (for review, see van Beek et al., 1998; Rigney et al., 1999; Brautigam et al., 1998) or even with intermittent claudication (for review, see van Beek et al., 1998; Bulling et al., 2000; van Dongen et al., 2000).

PLANT CELL BIOTECHNOLOGY

There are some reports on the biotechnological studies on *Ginkgo biloba* for the production of ginkgolides and bilobalide which represent, together with the flavonoids, the active constituents of the drug. Although ginkgolides have been chemically synthesized (Corey et al., 1988), their commercial supply can only arise from biomass sources.

Micropropagation

Highly productive trees have been selected in view to micropropagate them (Drieu and Jaggy, 2000). Surprisingly, it has been observed that the variability of ginkgolides and bilobalide content in leaves was very low (one to five) compared to what could have been expected for a dioecious species (Balz et al., 1999). Selected genotypes were micropropagated in determined in vitro conditions (stimulation of bud development) (Sohier, 2002). A greater diversity of terpene trilactone content has been observed; however, this method was not economical in comparison to ovule multiplication.

Somatic and gametophytic embryogenesis

Others biotechnogical studies on Ginkgo have been reported in view to obtain a greater variability via somatic and gametophytic embryogenesis and protoplast culture.

Somatic embryogenesis has been observed from the in vitro culture of immature zygotic embryos (Laurain et al., 1996). It was determined that the stage of development of the immature zygotic embryo influenced the rate of induction of embryogenic lines. Young cotyledonary stage embryos were found to be most appropriate for induction of Ginkgo embryogenic tissues. Direct embryogenesis was observed only on hypertrophic cotyledons cultivated on Murashige and Tücker (1969) (MT) or half-strength Murashige and Tücker mineral solution (MT/2) induction medium supplemented with 10 μM of benzyl amino purine (BA) and yielded, on an average, 9.6 somatic embryos per explant. Such embryos developed to the cotyledonary stage (Figure 8.11); however, they were unable to complete their development when isolated from their tissue of origin.

Embryogenesis obtained from haploid tissue (male and female gametophytes) has been also reported by Laurain et al. (1993a, 1993b). Interest in these investigations is justified by the production of homozygous plants and by the isolation of mutants producing large quantities of medicinal substances (Sangwan-Norreel et al., 1986). In Ginkgo, direct embryogenesis has been obtained from both male and female gametophytes. As for the somatic embryogenesis, the degree of maturation of the explant material has limited the rate of induction of haploid embryogenic tissue. For male Ginkgo gametophytic material, only microspores isolated at the uninucleate stage led to embryogenesis (Laurain et al., 1993a). In the presence or absence of exogenous growth regulators, microspores divided and formed microclones which directly evolved into embryos. After ten weeks of culture, embryogenesis was observed with an efficiency factor (defined as the number of embryos divided by the number of microspores originally plated) of 1.8%. After cultivation for a period of four to five months in Bourgin and Nitsch (1967) liquid medium (BN) supplemented with coconut milk, embryo-clusters and embryos were transferred onto BN solid madia devoid of growth regulators or containing various combinations of kinetin (Kn) and indol acetic acid (IAA). Embryo-clusters (Figure 8.12) showed signs of growth, particularly when cultured on BN solid medium supplemented with 11.4 μM Kn. Cell suspensions have been established from these embryogenic clusters and produced embryogenic cells in MT liquid medium devoid of growth regulators or supplemented with 18.8 μM naphthalene acetic acid (NAA) (Trémouillaux-Guiller et al., 1996). Protoplasts were isolated from a six-day-old subculture of these cell suspensions. After four weeks of cultivation, the efficiency of pro-embryo and embryo formation ranged from 0.6% to 4%, depending on the media tested. For the embryogenesis derived from female Ginkgo, only prothallus characterized by an ovoid hollow bag, reaching a length of 0.5 to 3 mm, could produce haploid protoplasts (Laurain et al., 1993b). In France, ovules presenting such prothallus can be harvested from mid May to the end of June. Direct embryogenesis was observed with frequenties between 0.7% to 1.9% when using modified MT liquid medium (without ammonium)

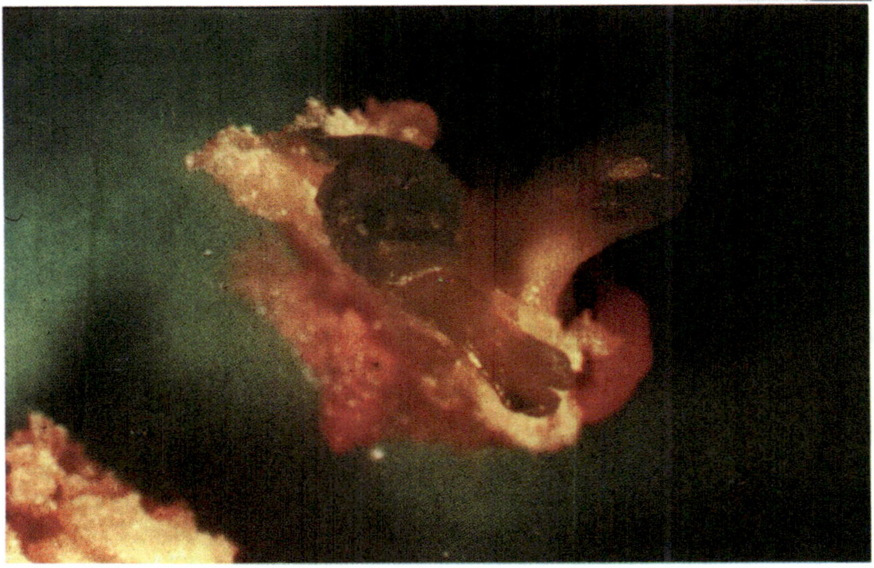

Fig. 8.11 Somatic embryo of *Ginkgo biloba*.

and of 0.2% when using BN liquid medium. As for the embryos derived from the microspore cultures, embryos obtained from protoplasts of male and female gametophytes exhibited slow growth, rendering the transfer to

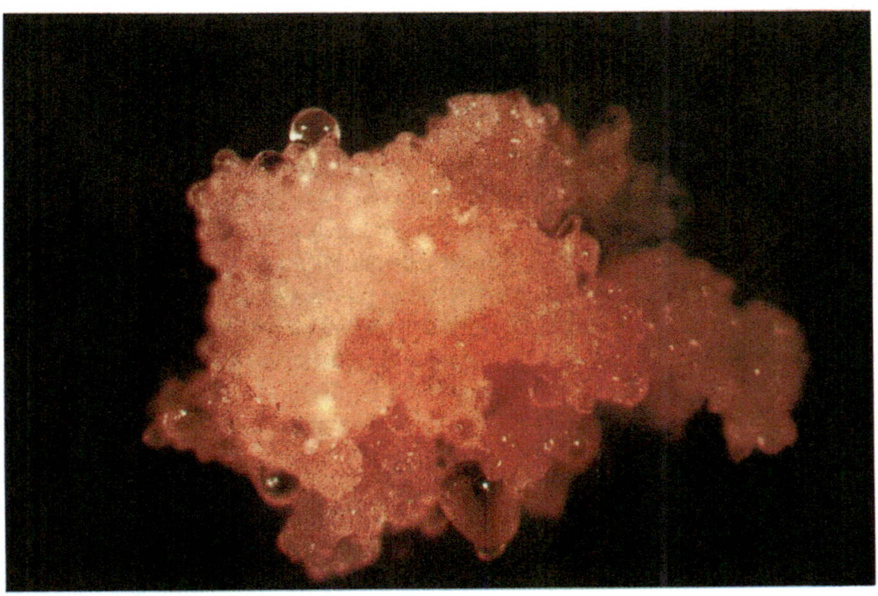

Fig. 8.12 Embryo-clusters from microspores of *Ginkgo biloba*.

solid media impossible. *G. biloba* female protoplasts produced new cell walls and divided in BN medium supplemented solely with coconut milk. It is possible that the coconut milk contained cytokinins which acted as a supply and contributed to the initiation of embryogenesis in *G. biloba* microspore and female prothallus protoplast cell lines.

Heart-shaped embryos have been also obtained from in vitro cultures of the female gametophyte; however, no further development has been observed (Fontanel et al., 1994).

Actually, no propagation of Ginkgo tree can be considered via somatic or gametophytic embryogenesis. However, these cell cultures are a material of choice in the study of the biosynthetic pathway of terpenes in *G. biloba*.

Ginkgolide and bilobalide content of various cell cultures

Many cell cultures, established from various explants, have been screened for their ability to produce ginkgolides and bilobalide. Cell suspensions derived from a prothallus and from microspores and callus derived from a zygotic embryo, respectively, were analyzed (Laurain et al., 1997). Only the first one was found to contain ginkgolides and bilobalide (870 μg gdw^{-1}). These results indicated that a synthesis of terpenes in Ginkgo cell cultures seems to be dependent on the nature of the initial explant.

Cellular differentiation is also a factor influencing secondary metabolites synthesis. Laurain et al. (1999) have reported that morphine and codeine were not found in *Papaver somniferum* undifferentiated cell cultures, but were detected in their somatic embryos and in roots developed on calli. In *G. biloba*, undifferentiated cell cultures have been analysed for their terpene contents. Ginkgolides have also been detected in embryo-derived cells cultivated in shake flasks and 6 L bioreactors using the Murashige and Skoog (1962) medium (MS) (Carrier et al., 1994). The detected quantities were extremely low, ranging in the ng gdw^{-1}. In order to increase ginkgolide production, modifications were made to MS with respect to the ammonium/nitrate ratio of 1:2 ; 1:1 ; 0:1. The highest concentration of ginkgolide was found in cells cultivated in MS where the ammonium/nitrate ratio was 1:2. These results were in accordance with those presented by Jeon et al. (1995) where the maximum ginkgolide B production was obtained in cells derived from Ginkgo leaves cultured in media with an ammonium/nitrogen ratio of 1:2 and 1:3. The biomass and kinetics of ginkgolide production in cells cultured in MS medium in shake flasks and in a 2 L immobilisation bioreactor were also evaluated (Archambault et al., 1990). Ginkgolides were first detected when phosphate was depleted. This possibly indicates that the production of ginkgolides may commence when cells cease their division and initiate their increases in mass. Jeon et al. (1995) also showed that the maximum amount of ginkgolides produced corresponded to the middle of the exponential growth phase. These works showed that when phosphate is depleted from the cultivation medium, cells may halt division and initiate secondary metabolite production. Although the terpene biosynthetic pathway can be expressed in undifferentiated cells, large scale cultures of these cells are not the method of choice for industrial level production of ginkgolides (Carrier and Laurain, 2000).

Genetic transformation of *Ginkgo biloba*

Gene-transfer technology offers new opportunities to improve the synthesis of plant secondary metabolites. Bonhomme et al. (2000a, 2000b) have shown that the *rol ABC* genes and the *rol C* gene alone were sufficient to increase tropane alkaloid production in *Atropa belladonna* hairy roots cultures. Only a few works have been reported on the genetic transformation of *G. biloba*. Transgenic cells have been obtained from the transformation of zygotic embryos inoculated with a wild strain of *Agrobacterium rhizogenes* (A4) (Laurain et al., 1997). The presence of opines such as agropine and mannopine was detected by high voltage paper electrophoresis and showed that transformed cells had most likely integrated T_R-DNA carrying genes responsible for opine synthesis. The opines, compounds never detected in normal cells, are specific biochemical markers for the transformation (Petit et al., 1983). So, *G. biloba* showed a susceptibility to an agropine type *A. rhizogenes* strain; nevertheless, no hairy roots appeared. Terpenoid products have been analysed in the transgenic cells obtained. The combined amounts of bilobalide and ginkgolides were determined to be 650 μg gdw^{-1}. These amounts were higher than the amounts measured in untransformed cell cultures (7-19 ng gdw^{-1}) by Carrier et al. (1994). Shunan et al. (1997) reported the appearance of hairy roots on Ginkgo leaves infected with the strain *A. rhizogenes* A4 that were successfully grown in 2.5 L to 5 L flasks. Although these cultures indicated scale-up possibilities, terpene were not quantified. Moreover, the transformation frequencies were low and their transgenic natures have not been confirmed by molecular assays. Recently, a reproducible protocol has been established for the transformation of *G. biloba* embryos by a modified strain of *Agrobacterium tumefaciens* GV3101 (pGV2260) (Dupré et al., 2000). This strain harboured a binary vector containing, on the one hand, the *gus* reporter gene with an intron in the coding region so that the bacteria cannot express the gene; and, on the other hand, the *nptII* selection gene. The first gene encoded the enzyme β-glucuronidase and the second one encoded the neomycin phosphotransferase. The *gus* gene integration was demonstrated by PCR and Southern blot analysis and its expression was characterized using histochemical and fluorimetric assays. The authors have shown that the age of the explant (embryos) had a marked influence on the response to bacterial infection. The transformation frequency was enhanced with 1.5-month-old embryos compared to 5-month-old embryos. In addition, the composition of the co-cultivation medium was the determinant factor of gene transfer into *G. biloba*. Transfer of T-DNA was only observed when co-cultivation media did not contain mineral elements. It is known that ions are involved in bacterial attachment to plants. For *G. biloba* transformation, a half or quarter reduction of the ionic strength of the co-cultivation medium was not sufficient to obtain a high percentage of T-DNA transfer. All the embryos (100%) were stained after the *Gus* assay with the co-cultivation medium containing distilled water, sucrose, agar and lacking mineral compounds. Under these conditions, 45% of the explants provided transformed calluses. This method could be a valuable tool for introducing various genes of interest into *G. biloba*, in particular those genes that could stimulate the production of secondary metabolites. As cell cultures described previouly, transgenic cells could be a material of choice in the study of the biosynthetic pathways of terpenes in this species.

Biosynthetical pathway of terpenes

Important advances (Schwarz, 1994) were made in the elucidation of the biosynthetical pathway of ginkgolides and bilobalide which are of terpenoid nature. In *G. biloba*, two parallel and mechanistically-distinct biosynthetic pathways govern the formation of isopentenylpyrophosphate (IPP) which is the unit required for the formation of sitosterol and ginkgolides. Sterols were formed following the classical acetate-pathway, namely through the biosynthesis of mevalonic acid and the primary terpenes (geranyl (GPP), farnesyl (FPP) and geranylgeranyl pyrophosphate (GGPP)). These steps were located in the cytosol. Ginkgolides and bilobalide would be synthesized from IPP units arising from a new pathway, namely the 'triose/pyruvate'-pathway, occurring possibly in the plastids. This pathway has also been found in green algae (Schwender et al., 1996) as well as higher plant chloroplasts (Lichtenthaler et al., 1997). Bilobalide, a sesquiterpene, would be derived from the ginkgolide (diterpenes) biosynthetical pathway. These results, combined with those of Carrier et al. (1996) who reported that undifferentiated *G. biloba* cell cultures showed mainly FPP synthase activity, possibly suggest that the 'triose/pyruvate'-pathway is not very active in undifferentiated cultured cells, justifying the low ginkgolide concentrations detected in such cultures (Carrier et al., 1991). Neau et al. (1997), using inhibitors of cytochrome P-450 dependent oxygenases, have shown that ginkgolide and bilobalide biosynthesis proceeds via dehydroabietane.

Site of terpene biosynthesis

Cartayrade et al. (1997) through intricate labelling experiments, shed light on the site of ginkgolide synthesis within the *G. biloba* plant. In distinct $^{14}CO_2$ and (U-^{14}C) glucose labelling experiments, labelled ginkgolides were detected in roots before being traced in the stems and the leaves. Ginkgolides were translocated from the roots to the stems and the leaves. Cytological studies have shown that peripheral cells localized at the tip of roots growing are involved in this synthesis. The concentration of terpene trilactones in leaves increased throughout the summer season (Flesh et al., 1992; van Beek and Lelyveld, 1992; Huh and Staba, 1993). Carrier et al. (1998) have shown that rosette leaves contained high concentrations of terpene trilactones, but yielded crude enzyme preparations incapable of incorporation of the labeled substrate. However, results obtained with capsicum (Kutz et al., 1992) have shown that the activity of GGpp synthase and its mRNA expression varied as a function of plant development.

CONCLUSION

Ginkgo biloba has been extensively studied in different scientific fields: botany, physiology, chemistry and pharmacology. However, many new researches must yet continue. Many more findings are still awaiting to be discovered, particularly in the field of biotechnology. It has been shown that roots are the site of ginkgolide synthesis. This knowledge, combined with expertise in differentiated cell culture and in genetic transformation, could lead to interesting industrial prospects.

REFERENCES

Archambault, J., Volesky, B. and Kurz W., 1990, Development of bioreactors for the culture of surface immobilized plant cells. Biotechnol. Bioeng., 35:702-711.

Arnott, H.J., 1959, Anastomoses in the venation of *Ginkgo biloba*. Amer. J. Bot., 46:405-411.

Auguet, M., Delaflotte, S., Chabrier, P.E. and Braquet P., 1994, *Ginkgo biloba* extract (Egb 761) and the regulation of vascular tone. In:Advances in *Ginkgo biloba* Extract Research, Vol. 3, Clostre F., V. DeFeudis F. (eds), Elsevier, Amsterdam, pp.19-26.

Auguet, M., Delaflotte, S., Hellegouarch A. and Clostre F., 1986, Pharmacological basis for the impact of *Ginkgo biloba* extract on vessels. Presse Méd., 15, 1524-1528.

Balz, J.P., Courtois, D., Drieu, J., Drieu, K., Reynoird, J.P. and Sohier C., 1999, Production of ginkgolides and bilobalide by *Ginkgo biloba* plants and tissue cultures. Planta Medica, 65, 620-626.

Barth, S.A., Inselmann, G., Engemann, R. and Heidemann H.T., 1991, Influences of *Ginkgo biloba* on Cyclosporin A induced lipid peroxidation in human liver microsomes in comparison to Vitamin E, Glutathione and N-Acetylcysteine. Biochem. Pharmacol., 41, 1521-1526.

Beck, S.D., 1960, The European corn borer, *Pyrausta nubilalis* (HUBN), and its principal host plant VII. Larval feeding behavior and host plant resistance. Ann. Entomol. Soc. Amer., 53, 206-212.

Bevan, C.W.L., Birch, A.J. and Caswell, H., 1961, An insect repellant from black cocktail ants. J. Chem. Soc., 96, 488.

Bombardelli, E., Cristoni, A. and Curri, S.B., 1996, Activity of phospholipid complex of *Ginkgo biloba* dimeric flavonoids on the skin microcirculation. Fitoterapia, 67, 265-273.

Bonhomme, V., Laurain-Mattar, D., Lacoux, J., Fliniaux, M.A. and Jacquin-Dubreuil, A., 2000a, Tropane alkaloid production by hairy roots of *Atropa belladonna* obtained after transformation with *Agrobacterium rhizogenes* 15834 and *Agrobacterium tumefaciens* containing *rol A, B, C* genes only. Journal of Biotechnology, 81, 151-158.

Bonhomme, V., Laurain-Mattar, D. and Fliniaux, M.A., 2000b, Effects of the *rol C* gene on hairy root: induction, development and tropane alkaloid production by *Atropa belladonna*. J. Nat. Prod., 63, 1249-1252.

Boralle, N., Braquet, P. and Gottlieb, O.R., 1988, *Ginkgo biloba*: A review of its chemical composition. In: Braquet, P. (ed.) Ginkgolides—Chemistry, Biology, Pharmacology and Clinical Perspectives, J.R. Prous Science Publishers, Barcelona, pp.9-25.

Bourgin, J.P. and Nitsch J.P., 1967, Obtention de *Nicotiana haploïdes* à partir d'étamines cultivées *in vitro*. Ann. Physiol. Vég., 9, 377-382.

Braquet, P.G., 1997, Platelet-activating Factor and Its Antagonists : Scientific Background and Clinical Applications of Ginkgolides. In: Hori, T., Ridge, R.W., Tulecke, W., Del Tredici, P., Trémouillaux-guiller, J., and Tobe, H. (eds), *Ginkgo biloba* — A Global Treasure. Springer-Verlag, Tokyo, pp.359-369.

Brautigam, M.R.H., Blommaert, F.A., Verleye, G., Castermans, J., Jansen Steur, E.N.H. and Kleijnen, J., 1998, Treatment of age-related memory complaints with *Ginkgo biloba* extract: a randomized double blind placebo-controlled study. Phytomedicine, Vol. 5(6), 425-434.

Briançon-Scheid, F., Guth, A. and Anton, R., 1982, High-performance liquid chromatography of biflavones from *Ginkgo biloba* L. J. Chromatogr., 245, 261-267.

Briançon-Scheid, F., Lobstein-Guth, A. and Anton, R., 1983, HPLC separation and quantitative determination of biflavones in leaves from *Ginkgo biloba*. Planta Med., 49, 204-207.

Bulling, B., Clemens, N. and Dankers, V., 2000, Clinical uses of *Ginkgo biloba* extract in the field of peripheral arterial occlusive disease (PAOD). In: van Beek (ed.), *Ginkgo biloba*. Harwood Academic Publishers, Singapore, pp.371-384.

Camponovo, F., 1996, Thesis: Mise au point de procédés pour l'analyse phytochimique et étude comparative de quelques médicaments à base de *Ginkgo biloba*, *Panax ginseng* et *Harpagophytum procumbens*, Université de Lausanne.

Camponovo, F.F., Wolfender, J.-L., Maillard, M.P., Potterat, O. and Hostettmann, F., 1995, Evaporative light scattering and thermospray mass spectrometry: two alternative methods for detection and quantitative liquid chromatographic determination of ginkgolides and bilobalide in *Ginkgo biloba* leaf extracts and phytopharmaceuticals. Phytochem. Anal., 6, 141-148.

Carrier, D.J., Coulombe, P., Mancini, M., Neufeld, R., Weber, M., Archambault, J., 1990, Immobilized *Ginkgo biloba* cell cultures. In: Nijkamp, H.J.J., van de Plas, L.H.W. and Aartrijk, J. (eds), International Congress on Plant Tissue and Cell Culture, Kluwer, Amsterdam, pp.614-618.

Carrier, D.J., Chauret, N., Mancini, M., Coulombe, P., Neufeld, R., Weber, M., Archambault, J., 1991, Detection of ginkgolide A in *Ginkgo biloba* cell cultures. Plant Cell Rep., 10, 256-259.

Carrier, J., Chauret, N., Neufeld, R. and Archambault, J., 1994, *Ginkgo biloba* L. (maiden hair Tree): In vitro culture and the formation of ginkgolides. In: Bajaj, Y., (ed.), Biotechnology in Agriculture and Forestry, Volume 26, Springer-Verlag, Berlin, pp.136-145.

Carrier, D.J., Archambault, J., van der Heijden, R. and Verpoorte, R., 1996, Formation of terpenoid products in *Ginkgo biloba* L. cultivated cells. Plant Cell Rep., 15, 888-891.

Carrier, D.J., van Beek, T.A., van der Heijden, R. and Verpoorte, R., 1998, Formation of terpenoid products in *Ginkgo biloba* L. cultivated cells. Phytochemistry, 48, 89-92.

Carrier, D.J. and Laurain, D., 2000, Plant cell biotechnology of Ginkgo. In: *Ginkgo biloba*, T.A. van Beek, (ed.) Medicinal and Aromatic Plants — Industrial Profiles, Harwood Academic Publishers, Singapore, pp.81-97.

Cartayrade, A., Neau, E., Sohier, C., Balz, J.P., Carde, J.P. and Walter, J., 1997, Ginkgolide and bilobalide biosynthesis in *Ginkgo biloba*. I: Sites of synthesis, translocation and accumulation of ginkgolides and bilobalide. Plant Physiol. Biochem., 35, 859-868.

Chang, S.K., Youm, J.R. and Kang, S.S., 1993, Seasonal variations of biflavone content from *Ginkgo biloba* leaves. Saengyak Hakhoechi, 24, 54-57. Cited from: Chem. Abstr. (1993) 119, 210366, 472.

Chauret, N., Carrier, J., Mancini, M., Neufeld, R., Weber, M. and Archambault, J., 1991, Gas chromatographic-mass spectrometric analysis of ginkgolides produced by *Ginkgo biloba* cell culture. J. Chromatogr., 588, 281-287.

Chen, P., Su, N.-L., Nie, L.-H., Yao, S.-Z., Zeng, J.-G., 1998, Analysis of ginkgolides and bilobalides in *Ginkgo biloba* L. extract for its production process control by high-performance liquid chromatography. J. Chromatogr. Sci., 36, 197-200.

Christen, Y., 2000, The Ginkgo, Past, Present and Future. *In*: *Ginkgo biloba*, van Beek, T.A., (ed.) Medicinal and Aromatic Plants — Industrial Profiles, Harwood Academic Publishers, pp.523-532.

Clark, G.D., Happel, L.T., Zorumski, C.F. and Bazan, N.G., 1992, Enhancement of hippocampal excitatory synaptic transmission by platelet-activating factor. Neuron, 9, 1211-1216.

Clostre, F., 1999, Extrait de *Ginkgo biloba* (EGb 761) état des connaissances à l'aube de l'an 2000. Ann. Pharm. Fr., 57, 1S5-1S88.

Corey, E., Kang, M., Desai, C., Ghosh, A. and Houpis, I., 1988, Total synthesis of (±) ginkgolide B. J. Am. Chem. Soc., 110, 649-651.

Critchfield, W.B., 1970, Shoot growth and heterophylly in *Ginkgo biloba*. Bot. Gaz., 131, 2, 150-162.

DeFeudis, F.V., 1991, *Ginkgo biloba* Extract (EGb 761) : Pharmacological Activities and Clinical Applications, Elsevier, Paris.

DeFeudis, F.V., 1994, Overview of the actions of *Ginkgo biloba* extract (Egb 761) on the cardiovascular system: from basic pharmacology to the clinic. In: Advances in *Ginkgo biloba* Extract Research, Vol. 3, Clostre, F., and DeFeudis F.V. (eds), Elsevier, Paris, pp.105-119.

DeFeudis, F.V., 1998, *Ginkgo biloba* extract (Egb 761). From chemistry to the clinic. Ullstein Medical, Wiesbaden.

Delaflotte S., Auguet, M., DeFeudis, F.V., Baranes, J., Clostre, F., Drieu, K. and Braquet, P., 1984, Endothelium-dependent relaxation of rabbit isolated aorta produced by carbachol and by *Ginkgo biloba* extract. Biomed. Biochim. Acta, 43, S, 212-216.

Del Tredici, P., 1991, The architecture of *Ginkgo biloba* L. In: C. Edelin (ed.), L'Arbre, Biologie et Développement. *Naturalia Monspeliansia* n° h.s., pp.155-168.

Del Tredici, P., 1992, Natural regeneration of *Ginkgo biloba* from downward growing cotyledonary buds (basal chichi). Am. J. Bot., 79, 522-530.

Del Tredici, P., 1997, Lignotuber formation in *Ginkgo biloba*. In: Hori, T., Ridge, R.W., Tulecke, W., Del Tredici, P. Trémouillaux-guiller, J. and Tobe, H. (eds), *Ginkgo biloba* — A Global Treasure. Springer-Verlag, Tokyo, pp.119-126.

Del Tredici, P., 2000, The evolution, ecology, and cultivation of *Ginkgo biloba*. In: van Beek T.A. (ed.), *Ginkgo biloba*. Harwood Academic Publishers, Singapore, pp.7-23.

Di Renzo, G., 2000, *Ginkgo biloba* and the central nervous system. Fitoterapia, 71, S43-S47.

Drieu, K. and Jaggy, H., 2000, History, Development and Constituents of Egb 761. In: van Beek, T.A. (ed.), *Ginkgo biloba*. Harwood Academic Publishers, Singapore, pp.267-277.

Dorf, E., 1958, The geological distribution of the Ginkgo family. Bull. Wagner Free Inst. Sci., 33, 1-10.

Dumont, E., D'Arbigny, P., Nouvelot, A., 1995, Protection of polyunsaturated fatty acids against iron-dependent lipid peroxidation by a *Ginkgo biloba* extract (EGb 761). Meth. Find. Exp. Clin. Pharmacol., 17, 83-88.

Dupré, P., Lacoux, J., Neutelings, G., Mattar-Laurain, D., Fliniaux, M.A., David, A. and Jacquin-Dubreuil, A., 2000, Genetic transformation of *Ginkgo biloba* by *Agrobacterium tumefaciens*. Physiologia Plantarum, 108, 413-419.

Flesh, V., Jacques, M., Cosson, L., Pétiard, V. and Balz, J.P., 1991, Influences de l'éclairement et de la température sur divers paramètres de croissance de jeunes *Ginkgo biloba* L.; cultivés en conditions contrôlées de jours longs. Ann. Sci. For., 48, 133-147.

Flesh, V., Jacques, M., Cosson, L., Teng, B.P., Pétiard, V. and Balz, J.P., 1992, Relative importance of growth and light level on terpene content of *Ginkgo biloba*. Phytochemistry, 31, 1941-1945.

Fontanel, A., Serraf, I. and Pétiard, V., 1994, Essais de régénération chez le *Ginkgo biloba* L.: aptitude à l'embryogenèse du mégagamétophyte selon le stade de développement. C.R. Acad. Sci. Paris, Sciences de la Vie, 317, 149-155.

Friedman, W.E., 1987, Growth and development of the male gametophyte of *Ginkgo biloba* within the ovule (in vivo). Amer. J. Bot. 74, 1797-1815.

Furukawa, S., 1932, Studies on the constituents of the *Ginkgo biloba* L., leaves. Part I and II Scientific Papers Institute Physical Chemical Research, 19, 27-42.

Fusheng, P., Wangou, L., Fugui, H., Shanan, H. and Zijie, P., 1997, Study on the optimizing of the resources of the *Ginkgo biloba* leaves. Proceedings of, 97 International Seminar on Ginkgo, November 10-12, Beijing, China, pp.105-112.

Gifford, E.M., and Foster, A.S., 1989, Morphology and evolution of vascular plants, 3rd edn. WH Freeman, New York.

Gobbato, S., Griffini, A., Lolla, E. and Peterlongo, F., 1996, HPLC quantitative analysis of biflavones in *Ginkgo biloba* leaf extract and their identification by thermospray liquid chromatography-mass spectrometry. Fitoterapia, 67, 152-158.

Guba, E.F., 1961, Monograph of *Monochaetia* and *Pestalotia*. Harvard University, Cambridge, p. 136.

Gülz, P.G., 1994, Epicuticular leaf waxes in the evolution of the plant kingdom. J. Plant Physiol. 143, 453-464.

Gülz, P.G., Müller, E., Schmitz, K., Marner, F.J., Güth, S., 1992, Chemical composition and surface structures of epicuticular leaf waxes of *Ginkgo biloba*, *Magnolia grandiflora* and *Liriodendron tulipifera*. Z. Naturforsch sect. C. Biosci. 47, 516-526.

Gunckel, J.E. and Thimann, K.V., 1949, Studies of development in long shoots and short shoots of *Ginkgo biloba*. III. Auxin production in shoot growth. Am. J. Bot., 36, 145-151.

Gunckel, J.E., Thimann, K.V., Wetmore, R.H., 1949, Studies of development in long shoots and short shoots of *Ginkgo biloba*. IV. Growth habit, shoot expression and the mechanism of its control. Am. J. Bot., 36, 309-316.

Hara, N., 1997, Morphology and anatomy of vegetative organs in *Ginkgo biloba*. In: Hori, T., Ridge, R.W., Tulecke, W., Del Tredici, P., Trémouillaux-guiller, J. and Tobe, H. (eds), *Ginkgo biloba* — A Global Treasure. Springer-Verlag, Tokyo, pp. 3-15.

Hasebe, M., 1997, Molecular phylogeny of *Ginkgo biloba* : close relationship between *Ginkgo biloba* and cycads. In: Hori, T., Ridge, R.W., Tulecke, W., Del Tredici, P., Trémouillaux-guiller, J. and Tobe, H. (eds), *Ginkgo biloba* — A Global Treasure. Springer-Verlag, Tokyo, pp.173-182.

Hasler, A., 2000, Chemical constituents of *Ginkgo biloba*. In: van Beek T.A. (ed.), *Ginkgo biloba*, Harwood Academic Publishers, pp.109-142.

Hasler, A. and Meier, B., 1992, Determination of terpenes from *Ginkgo biloba* L. by capillary gas chromatography. Pharm. Pharmacol. Lett., 2, 187-190.

Hasler, A. and Meier, B., 1993, *Ginkgo biloba*. Content of flavonoids and terpenes from leaves during the harvest time and from full extracts determined by chromatographic and biologic methods. Planta Med., 59, A632.

Hasler, A., Meier, B. and Sticher, O., 1990, Quantitative HPLC-analysis of flavonoid aglycones in different medicinal plants. Planta Med., 56, 575.

Hasler, A., Sticher, O. and Meier, B., 1992, Identification and determination of the flavonoids from *Ginkgo biloba* by high-performance liquid chromatography. J. Chromatogr., 605, 41-48.

Hirano, T., 1971, Japanese beetle and ginkgo. In: Host selection in phytophagous insects (in Japanese), Kyoritsu, Tokyo, pp.32-34.

Holt, B.F. and Rothwell, G.W., 1997, Is *Ginkgo biloba* (Ginkgoaceae) really an oviparous plant? Am. J. Bot., 84, 870-872.

Honda, H., 1996, Practical application of plant chemicals to pest control (in Japanese). Agric. Hort., 71, 548-554.

Honda, H., 1997, Ginkgo and Insects. In: Hori, T., Ridge, R.W., Tulecke, W., Del Tredici, P., Trémouillaux-guiller, J. and Tobe, H. (eds), *Ginkgo biloba* — A Global Treasure. Springer-Verlag, Tokyo, pp.243-250.

Huguet, F., Drieu, K. and Piriou, A., 1994, Decreased cerebral 5-HT_{1A} receptors during ageing: Reversal by *Ginkgo biloba* extract (Egb 761). J. Pharm. Pharmacol., 46, 316-318.

Huguet, F. and Tarrade, T., 1992, Alpha$_2$-adrenoreceptor changes during cerebral ageing. The effect of *Ginkgo biloba* extract. J. Pharm. Pharmacol., 44, 24-27.

Huh, H. and Staba, E.J., 1993, Oncogenic aspects of ginkgolide production in *Ginkgo biloba.* Planta Med., 59, 232-239.

Jeon, M.H., Sung, S.H., Huh, H. and Kim, Y.C. 1995, Ginkgolide B production in cultured cells derived from *Ginkgo biloba* L. leaves. Plant Cell Rep., 14, 501-504.

Joyeux, M., Lobstein, A., Anton, R. and Mortier, F., 1995, Comparative antilipoperoxidant, antinecrotic and scavenging properties of terpenes and biflavones from Ginkgo and some flavonoids. Planta Med., 61, 126-129.

Juretzek, W., 1997, Recent advances in *Ginkgo biloba* extract (Egb 761). In: Hori, T., Ridge, R.W., Tulecke, W., Del Tredici, P., Trémouillaux-guiller J. and Tobe H. (eds), *Ginkgo biloba* — A Global Treasure, Springer-Verlag, Tokyo, pp.341-358.

Kanowski, S., Herrmann, W.M., Stephan, K., Wierich, W. and Hörr, R., 1996, Proof of efficacy of the *Ginkgo biloba* special extract Egb 761 in outpatients suffering from mild to moderate primary degenerative dementia of the Alzheimer type or multi-infarct dementia. Pharmacopsychiat., 29, 47-56.

Kim, Y.S. and Lee, J.K., 1990, Chemical and structural characteristics of conifer needles exposed to ambient air pollution. Eur. J. For Pathol. 20, 193-200.

Kim, Y.S., Lee, J.K. and Chung, G.C., 1997, Tolerance and suceptibility of Ginkgo to air pollution. In: Hori, T., Ridge, R.W., Tulecke, W., Del Tredici, P., Trémouillaux-guiller, J. and Tobe, H. (eds), *Ginkgo biloba* — A Global Treasure, Springer-Verlag, Tokyo, pp.233-242.

Kleijnen, J. and Knipschild, P., 1992, *Ginkgo biloba.* Lancet, 340, 1136-1139.

Kochibe, N., 1997, Allergic substances of *Ginkgo biloba.* In: Hori, T., Ridge, R.W. Tulecke, W., Del Tredici, P., Trémouillaux-guiller J., and Tobe. J., (eds), *Ginkgo biloba* — A Global Treasure, Springer-Verlag, Tokyo, pp.301-307.

Komoda, Y., Nakamura, H. and Uchida, M., 1988, HPLC analysis of ginkgolides by using a differential refractometer. Iyo Kizai Kenkyusho Hokoku (Reports of the Institute for Medical & dental Engineering), 22, 83-85.

Konishi, M., 1996, knowledge for practical use of plants in Edo era. In: Yagi A. (ed.) Control of Disease and Insect Pests with Plant Extracts (in Japanese). Nobunkyo, Tokyo, pp.20-24.

Kristofikova, Z., Benesova, O. and Tejkalova, H., 1992, Changes of high-affinity choline uptake in the hippocampus of old rats after long-term administration of two nootropic drugs (tacrine and *Ginkgo biloba* extract). Dementia, 3, 304-307.

Kutz, M., Romer, S., Suire, C., Hugueney, P., Weil, J., Schantz, R. and Camara, B., 1992, Identification of a cDNA for the plastid-located geranylgeranyl pyrophosphate synthase from *Capsicum annuum*: Correlative increase in enzyme activity and transcript level during fruit ripening. The Plant J., 2, 25-34.

Laurain, D., Trémouillaux-Guiller, J. and Chénieux, J.C., 1993a, Embryogenesis from microspores of *Ginkgo biloba* L., a medicinal woody species. Plant Cell Rep., 12, 501-504.

Laurain, D., Chénieux, J.C. and Trémouillaux-Guiller, J., 1993b, Direct embryogenesis from female haploid protoplasts of *Ginkgo biloba* L., a medicinal woody species. Plant cell Rep., 12, 656-660.

Laurain, D., Chénieux, J.C. and Trémouillaux-guiller, J., 1996, Somatic embryogenesis from immature zygotic embryos of *Ginkgo biloba*. Plant Cell Tissue Organ Cult., 44, 19-24.

Laurain, D., Trémouillaux-guiller, J., Chénieux, J.C. and van Beek, T.A., 1997, Production of ginkgolide and bilobalide in transformed and gametophyte derived cell culture of *Ginkgo biloba*. Phytochem., 46, 127-130.

Laurain-Mattar, D., Gillet-Manceau, F., Buchon, L., Nabha, S., Fliniaux, M.A.and Jacquin-Dubreuil, A., 1999, Somatic embryogenesis and rhizogenesis of tissue cultures of two genotypes of *Papaver somniferum*: Relationships to alkaloid production. Planta Med., 65, 167-170.

Le Poncin-Lafitte, M.C., Rapin, J. and Rapin, J.R., 1980, Effects of *Ginkgo biloba* on changes induced by quantitative cerebral microembolization in rats. Arch. Int. Pharmacodyn., 243, 236-244.

Lichtenthaler, H.K., Schwender, J., Disch, A. and Rohmer, M., 1997, Biosynthesis of isoprenoids in higher plant chloroplasts proceeds via a mevalonate independent pathway. FEBS Lett., 400, 271-274.

Lobstein, A., Rietsch-Jako, L., Haag-Berrurier, M. and Anton, R., 1991, Seasonal variations of the flavonoid content from *Ginkgo biloba* leaves. Planta Med., 57, 430-433.

Lobstein-Guth, A., Briançon-Scheid, F. and Anton, R., 1983, Analysis of terpenes from *Ginkgo biloba* L. by high-performance liquid chromatography. J. Chromatogr., 267, 431-438.

Maillard, M.P., Camponovo, F., Wolfender, J.-L. and Hostettmann, K., 1993, Semi-quantitative determination of terpenes from *Ginkgo biloba* by LC-TSP-MS. Poster 138, International symposium of the Phytochemical Society of Europe 'Phytochemistry of Plants used in Traditional Medicine', 29 Sept.-1 Oct. 1993, Lausanne, Switzerland.

Major, R.T., 1967, The Ginkgo, the most ancient living tree. Science, 157, 1270-1273.

Major, R.T., Marchini, P. and Boulton, A.J., 1963, Observations on the production of α-hexanal by leaves of certain plants. J. Biol. Chem., 238, 1813-1816.

Major, R.T. and Tietz, H.J., 1962, Modification of the resistance of *Ginkgo biloba* leaves to attack by Japanese beetles. J. Econ. Entomol., 55, 272.

Marcocci, I., Packer, L., droy-Lefaix, M.T., Sekaki, A. and Gardes-Albert, M., 1994, Antioxidant Action of *Ginkgo biloba* extract EGb 761. Methods Enzymol., 234, 462-475.

Maruyama, M., Terahara, A., Nakadaira, Y., Woods, M.C. and Nakanishi, K., 1967, The ginkgolides. III. The structure of the ginkgolides. Tetrahedron Lett., 4, 309-313.

Masood, E., 1997, Medicinal plants threatened by over-use. Nature, 66, 570.

Matsumoto, T. and Sei, T., 1987, Antifeedant activities of *Ginkgo biloba* L. components against the larva of *Pieris rapae crucivora*. Agric. Biol. Chem., 51, 249-250.

Maurer, K., Ihl, R., Dierks, T. and Frölich, L., 1998, Clinical efficacy of *Ginkgo biloba* special extract Egb 761 in dementia of the Alzheimer type. Phytomedicine, Vol. 5(6), 417-424.

Melzheimer, V. and Lichius, J., 2000, *Ginkgo biloba* L. : Aspects of the systematical and applied botany. In: van Beek (ed.), *Ginkgo biloba*. Harwood Academic Publishers, Singapore, pp.25-47.

Michel , P.F., 1985, *Ginkgo biloba* — L'arbre qui a vaincu le temps. Editions du Félin, Paris.

Murashige, T. and Skoog, F., 1962, A revised medium for rapid growth and bioassay with tobacco tissue cultures. Physiologia Plantarum, 28, 473-497.

Murashige, T. and Tücker, D.P., 1969, Growth factor requirements of *Citrus* tissue cultures. In: Chapman, J.V. (ed.), Proceedings of the First International *Citrus* Symposium, University of California, Riverside, pp.1155-1161.

Nakanishi, K., 1988, Ginkgolides — Isolation and structural studies carried out in the mid 1960's. In: Ginkgolides — Chemistry, Biology, Pharmacology and Clinical Perspectives, Braquet, P. (ed.), J.R. Prous Science Publishers, Barcelona, pp.27-36.

Neau, E., Cartayrade, A., Balz, J.P., Carde, J.P. and Walter, J., 1997, Ginkgolide and bilobalide biosynthesis in *Ginkgo biloba*. II: Identification of a possible intermediate compound by using inhibitors of cytochrome P-450-dependent oxygenases. Plant Physiol. Biochem., 35, 869-879.

Ni, Y., Zhao, B., Hou, J. and Xin, W., 1996, Preventative effect of *Ginkgo biloba* extract on apoptosis in rat cerebral neuronal cells induced by hydroxyl radicals. Neurosci. Lett., 214, 115-118.

Nozeran, R., 1986, Le mouvement morphogénétique spécialement chez les végétaux supérieurs pérennes. Naturalia Monspeliensia. Colloque International sur l'Arbre. pp.415- 428.

Oehrle, S.A., 1995, Analysis of ginkgolides and bilobalide by capillary electrophoresis. J. Liq. Chromatogr., 18, 2855-2859.

Oh, J.W., 1986, Effects of simulated acid rain on the growth of trees and soil. Master's thesis, Kyung Hee University, Seoul.

Oh, J.W., Kim, Y.G. and Chae, J.S., 1987, Studies on the periodical changes of air pollution at several forest stands. In: Effects of air pollution and acid precipitaion on the forest ecosystem (in Korean, summary in English). Ministry of Science and Technology, Seoul, 49-69.

Okabe, K., Yamada, K., Yamamura, S. and Takada, S., 1967, Ginkgolides. J. Chem. Soc.(c), pp.2201-2206.

O'Reilly, J., 1993, *Ginkgo biloba* — Cultivation, extraction, and therapeutic use of the extract. In: van Beek, T.A. and Breteler H. (eds), Phytochemistry and Agriculture, Clarendon Press, Oxford, pp.253-270.

Parliment, T., 1995, Characterization of the putrid aroma compounds of *Ginkgo biloba* fruits. In: R. Rouseff and M. Leahy (eds.). Fruit Flavors : Biogenesis, Characterization, and Authentication, Am. Chem. Soc. Symp. Ser., 596, 276-279.

Petit, A., David, C., Dahl, G.A., Ellis, J.G. and Guyon, P., 1983, Further extension of opine concept: plasmids in *Agrobacterium rhizogenes* cooperate for opine degradation. Mol. Gen. Genet., 190, 204-214.

Pietta, P.G., Gardana, C. and Mauri, P.L., 1997, Identification of *Ginkgo biloba* flavonol metabolites after oral administration to humans. J. Chromatogr., 693B, 249-255.

Pietta, P.G., Gardana, C., Mauri, P.L., Maffei-Facino, R. and Carini, M., 1995, Identification of flavonoid metabolites after oral administration to rats of a *Ginkgo biloba* extract. J. Chromatogr., 673B, 75-80.

Pietta, P., Mauri, P. and Rava, A., 1988, Reversed-phase high-performance liquid chromatographic method for the analysis of biflavones in *Ginkgo biloba* L. extracts. J. Chromatogr., 437, 453-456.

Pincemail, J., Dupuis, M., Nasr, C., Hans, P., Haag-Berrurier M., Anton R. and Deby C., 1989, Superoxide anion scavenging effect and superoxide dismutase activity of *Ginkgo biloba* extract. Experientia, 45, 708-712.

Prehn, J.H.M. and Krieglstein, J., 1993, Platelet-activating factor antagonists reduce excitotoxic damage in cultured neurons from embryonic chick telencephalon and protect the rat hippocampus and neocortex from ischemic injury in vivo. J. Neurosci. Res., 34, 179-188.

Ramassamy, C., Girbe, F., Pincemail, J., Christen, Y. and Costentin, J., 1994, Modifications of the synaptosomal dopamine uptake and release by two systems generating-free radicals: Ascorbic acid/Fe^{2+} and arginine/NADPH. Ann. N. Y. Acad. Sci., 738, 141-151.

Ramassamy, C., Naudin, B., Christen, Y., Clostre, F. and Costentin, J., 1992, Prevention by *Ginkgo biloba* extract (Egb 761) and trolox C of the decrease in synaptosomal dopamine or serotonin uptake following incubation. Biochem. Pharmacol., 44, 2395-2401.

Rensselaer, M., 1969, The remarkable Ginkgo. Plants Garden, 25, 50-53.

Rigney, U., Kimber, S. and Hindmarch, I., 1999, The effects of acute doses of standardized *Ginkgo biloba* extract on memory and psychomotor performance in volunteers.

Phytotherapy Research, 13, 408-415.

Sangwan-Norreel, B.S., Sangwan, R.S. and Pare, J., 1986, Haploïdie et embryogenèse provoquée *in vitro*. Bull. Soc. Bot. France, 4, 7-39.

Santamour, F.S., He, S. and McArdle, A.J., 1983, Checklist of cultivated Ginkgo. J. Arboric., 9, 88-92.

Sakabe, N., Takada, S. and Okabe, K., 1967, The structure of ginkgolide A, a novel diterpenoid trilactone. Chem. Commun., pp.259-261.

Schennen, A., 1988, Neue Inhaltsstoffe aus den Blättern von *Ginkgo biloba* L. sowie Präparation ^{14}C-markierter Ginkgo-Flavonoide. Thesis, Universität Marburg.

Schwarz, M., 1994, Terpen-Biosynthese in *Ginkgo biloba*. Eine überrraschende Geschichte. PhD thesis, Eidgenössischen Technischen Hochschule Zürich, Switzerland (ETH Nr. 10951).

Schwender, J., Seemann, M., Lichtenthaler, H.K. and Rohmer, M., 1996, Biosynthesis of isoprenoids (carotenoids, sterols, prenylside-chain of chlorophylls and plastoquinone) *via* a novel pyruvate/glyceraldehyde-3-phosphate non-mevalonate pathway in the green algae *Scenedesmus obliquus*. Biochem. J., 316, 73-80.

Seward, A.C., 1938, The story of the maidenhair tree. *Sci. Prog.*, 32, 420-440.

Seymour, A.B., 1929, Host index of the fungi of North America. Harvard University, Cambridge, p. 28.

Shunan, L., Tianen, S. and Genbao, L., 1997, Transformation of Ginkgo hairy root and establishment of its suspension culture clone. Wuhan Univ. J. Nat. Sci., 2, 493-495.

Smith, P.F., Maclennan, K. and Darlington, C.L., 1996, The neuroprotective properties of the *Ginkgo biloba* leaf: A review of the possible relationship to platelet-activating factor (PAF). J. Ethnopharmacol., 50, 131-139.

Sohier, C., 2002, La biotechnologie végétale, recherche d'avant-garde pour un arbre ancestral: le *Ginkgo biloba*. Ann. Pharm. Fr., 60, 22-27.

Song, Y., 1986, Chemical composition and utilization of *Ginkgo biloba* L. Chem. and Ind. of Forest Prod., 6, 1-4.

Spinnewyn, B., 1992, *Ginkgo biloba* extract (Egb 761) protects against delayed neuronal death in gerbil. In: Christen, Y. Costentin, J. and Lacour, M. (eds), Effects of *Ginkgo biloba* Extract (Egb 761) on the Central Nervous System, Elsevier, Paris, pp.113-118.

Spinnewyn, B., Blavet, N. and Clostre, F., 1986, Effects of *Ginkgo biloba* extract on a cerebral ischaemia model in gerbils. Press. Med., 15, 1511-1515.

Steinke, B., Müller, B. and Wagner, H., 1993, Biologische Standardisierungsmethode für Ginkgo-Extrakte. Planta Med., 59, 155-160.

Sticher, O., Meier, B. and Hasler, H., 2000, The analysis of ginkgo flavonoids. In: *Ginkgo biloba*, van Beek, T.A. (ed.), Medicinal and Aromatic Plants — Industrial Profiles, Harwood Academic Publishers, pp.179-202.

Strode III, J.B.T., Taylor, L.T. and van Beek, T.A., 1996, Supercritical fluid chromatography of ginkgolides A, B, C and J and bilobalide. J. Chromatogr., 738 A, 115-122.

Stumpf, K.-H., 1997, Pharmaceutical quality and extract specification of Egb 761, Proceeding of the International Seminar of Ginkgo 1997, Organizing Committee for 1997 International Seminar on Ginkgo, Beijing, pp.39-56.

Taylor, J.E., 1986, Binding of neuromediators to their receptors in rat brain. Presse Med., 15, 1491-1493.

Taylor, J.E., 1990, In vitro inhibition of [^{3}H]5-HT, [^{3}H]norepinephrine (NE), and [^{3}H]dopamine (DA) uptake by an extract of the *Ginkgo biloba* plant. Proc. Soc. Neurosci. Annual Meeting, St. Louis (Abstract. N° 32.11).

Teng, B.P., 1988, Chemistry of ginkgolides. In: Braquet, P. (ed.), Ginkgolides — Chemistry, Biology, Pharmacology and Clinical Perspectives, Vol. 1, J.R. Prous Science Publishers, Barcelona, pp.37-41.

Tralau, H., 1968, Evolutionary trends in the genus *Ginkgo*. Lethaia, 1, 63-101.

Trémouillaux-Guiller, J., Laurain, D. and Chénieux, J.C., 1996, Microspore and protoplast culture in *Ginkgo biloba*. In: Jain, S.M., Sopory, S.K., and Veilleux, R.E. (eds), In vitro Haploid Production in Higher Plants, Volume 3, Kluwer Academic Publishers, The Netherlands, pp.277-295.

Tsumura, Y., Motoike, H. and Ohba, K., 1992, Allozyme variation of old *Ginkgo biloba* memorial trees in western Japan. Can. J. For. Res. 22, 939-944.

van Beek, T.A., 2000, Chemical analysis of ginkgo terpene trilactones. In:*Ginkgo biloba*, T.A. van Beek (ed.) Medicinal and Aromatic Plants — Industrial Profiles, Harwood Academic Publishers, Singapore, pp.151-178.

van Beek, T.A., Bombardelli, E., Morazzoni, P. and Peterlongo F., 1998, *Ginkgo biloba* L. Fitoterapia, Vol. LXIX, N°3, pp.195-244.

van Beek, T.A. and Lelyveld, G.P., 1992, Concentration of ginkgolides and bilobalide in *Ginkgo biloba* leaves in relation to the time of year. Planta Med., 58, 413-416.

van Beek, T.A. and Lelyveld, G.P., 1993, Thin layer chromatography of bilobalide and ginkgolides A, B, C and J on sodium acetate impregnated silica gel. Phytochem. Anal., 4, 109-114.

van Beek, T.A., Scheeren, H.A., Rantio, T., Melger, W.Ch. and Lelyveld, G.P., 1991, Determination of ginkgolides and bilobalide in *Ginkgo biloba* leaves and phytopharmaceuticals. J. Chromatogr., 543, 375-387.

Van Beek, T.A., van Veldhuizen, A., Lelyveld, G.P., Piron, I. and Lankhorst, P.P., 1993, Quantitation of bilobalide and ginkgolides A, B, C and J by means of nuclear magnetic resonance spectroscopy. Phytochem. Anal., 4, 261-268.

van Dongen, M.C.J.M., van Rossum, E. and Knipschild, P., 2000, Efficacy of *Ginkgo biloba* special extracts — Evidence from randomized clinical trials. In: van Beek, T.A. (ed.), *Ginkgo biloba*. Harwood Academic Publishers, Singapore, pp.385-442.

Vanhaelen, M. and Vanhaelen-Fastre, R., 1988, Countercurrent chromatography for isolation of flavonol glycosides from *Ginkgo biloba* leaves. J. Liq. Chromatogr., 11, 2969-2975.

Wagner, H., Bladt, S., Hartmann, U., Daily, A., and Berkulin, W., 1989, *Ginkgo biloba*. DC- und HPLC-Analyse von Ginkgo-Extrakten und Ginkgo-Extrakte enthaltenden Phytopräparaten. Dtsch. Apoth. Ztg., 129, 2421-2429.

Weinges, K. and Bähr, W., 1969, Bilobalid A, ein neues Sesquiterpen mit tert.-Butyl-Gruppe aus den Blättern von *Ginkgo biloba* L. Liebigs Ann. Chem., 724, 214-216.

Weinges, K., Hepp, M. and Jaggy, H., 1987, Isolierung und strukturaufklärung eines neuen Ginkgolides Liebigs Ann. Chem., 742, 521-526.

Wettstein, A., 1999/2000, Cholinesterase inhibitors and Ginkgo extracts — Are they comparable in the treatment of dementia? Phytomedicine, Vol. 6(6), 393-401.

Xie, P., Yan, Y. and Qian, H., 1997, Determination of terpenoid lactones in *Ginkgo biloba* extract and its preparations by means of quantitative fluorophotometric TLC analysis. Proceedings of the international Seminar of Ginkgo 1997, Organizing Committee for 1997 International Seminar on Ginkgo, Beijing, pp.174-177.

Yagi, A., 1996, Control of disease and insect pests with plant extracts (in Japanese). Nobunkyo, Tokyo, p. 137.

Yao, W.X., Cheng, X.M. and Zhang, Y.Q., 1995, Determination of flavonoid compounds in *Ginkgo biloba* leaves supercritical fluid extraction and high performance liquid chromatography (SFE/HPLC). Chin. Chem. Lett., 6, 589-592.

Zhang, G. and Li, Z., 1981, Preliminarly study on *Pammene* sp. (in Chinese with English summary). J. Nanjing Tech. Coll. For Prod., 4, 83-89.

Zheng, W., 1983, Ginkgo L. In: Comm. *Sylva Sinica* (ed.) (in Chinese). Peking, pp.154-158.

CHAPTER 9

Ginseng

Kwang-Tae Choi

INTRODUCTION

Traditionally, ginseng, *Panax ginseng* C.A. Meyer, has been considered to be a medicinal plant of mysterious powers and has more than 5,000 years of history among Oriental people. Especially in Korea, China and Japan, ginseng has been recognized as the most prized medicine among all herbal curatives. Therefore, the Oriental people have traditionally used ginseng roots and extracts for geriatric, stomachic, aphrodisiac treatment, and as tonics. From the pharmacological studies of animal and human beings for more than 30 years, the efficacy of ginseng, especially tonic effects, has been gaining popularity even in western countries. The term 'tonic' refers to a drug meant for maintaining a normal physical tone or restoring a diseased state to normal. Ginseng has been known to have a tonic effect and it is the general opinion of many investigators (Bittles, 1979; D.H. Zhuo, 1982; Brekhman and Dardymov, 1969) that this plant ginseng has the effect of normalization of physical conditions, that is, maintaining individual homeostasis. The Korean workers, Oh et al. (1969) and Hong et al. (1969), have reported that in mice, the saponin fractions potentiate nembutal hypnosis, retard the onset of cocaine-induced convulsions, reduce body temperature, and enhance the process of sexual behavior.

Recently, a number of scholars have paid attention to elucidate the efficacy of ginseng scientifically. Today, I have a considerable amount of information about cultivation, breeding, biotechnology, composition, pharmacology properties, clinical demonstrations, and molecular biology. Now, it is widely accepted as a health food allover the world.

Nowadays, Korea is the largest *Panax ginseng* producing and exporting country in the world. In 1999, about 25,000 households were engaged in the cultivation of ginseng over a total area of over 11,800 ha. The estimated amount of cultivated ginseng crops reached 13,000 tons; of this, 2,930 tons were exported to more than 60 countries.

Therefore, ginseng, *Panax ginseng* C.A. Meyer, is one of the economically important plants in the world.

BIOLOGY

Taxonomical view of ginseng plants

The systematics of ginseng plants is as follows:

Phylum	Embryophyta
Subphylum	Angiospermae
Class	Dicotyledoneae
Subclass	Anchichlamyeae
Order	Umbellifloreae
Family	Araliaceae
Genus	*Panax*

The scientific name of the Korean ginseng plant is *Panax ginseng* C.A. Meyer is used on a worldwide basis. C.A. Meyer, who was a Russian, named the plants in 1843. *Panax* is a compound Greek word and (denotes all, axos means cure). Therefore *Panax* denotes cure all, and the word 'ginseng' originated from the Chinese pronunciation of ginseng plants.

Botanical characteristics of cultivated Korean ginseng plants

Ginseng is a perennial herb with flesh roots, an annual stem bearing a whorl of palmate compound leaves and a terminal simple umbel. Ginseng plants are basically self-pollinated plants. However, cross-pollination can also be found. Flowers that are wrapped prior to blooming show over 95% self-pollination, and the percentage of cross-pollination is thought to be low.

Flowers start to bloom when the plant is 3 years old. Ginseng flowers in the middle of May. It is general practice to collect the seeds once from 4-year-old plants, and the flower buds are nipped off for better growth of the roots. The ripe red fruits are gathered two or three times in the middle of July (Fig. 9.1). The mature fruits usually contain two whitish-yellow seeds.

It is believed that the berries that fall on the ground naturally are the most desirable for propagation. The ginseng seeds must be stratified with fine sand to promote embryo growth artificially because the embryo is immature when the seed is gathered. The optimum season for sowing on the nursery bed is early in November. In general, the seedlings grown on the nursery bed are transplanted in late March and early April, and harvested 3 to 5 years after transplanting the seedlings, i.e. 4- or 6-year-old roots are harvested (Fig. 9.2).

The roots are corpulent and composed of a taproot (or main root), two to five lateral roots, and fine roots. The roots are light yellowish-white in color. The size and shape of the root vary as per the soil property, fertilizer, water content in the soil, method of transplantation, climate, and germplasm.

Fig. 9.1 Ripe red fruits of Korean ginseng.

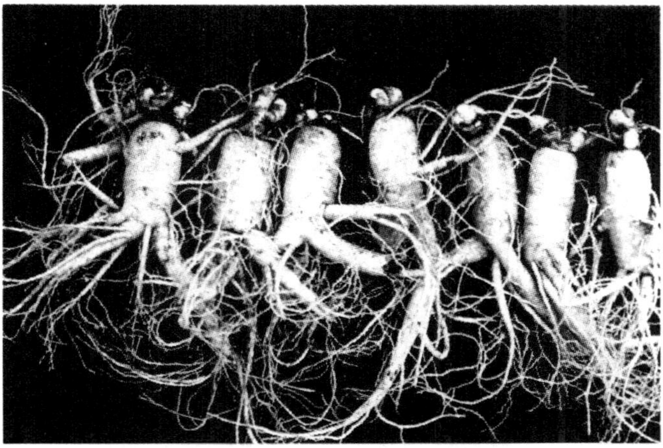

Fig. 9.2 Six-year-old roots of Korean ginseng.

Particularly, the growth of the root and its shape differ according to the plant's age. The roots are harvested when the plants are 4- to 6-year-old, but younger roots do not exhibit complete maturity in the growth of the main and lateral roots. Only a six-year-old plant of ginseng has the rhizome, main and lateral roots grown with balance to a shape that resembles the human body. At 6 years, the rhizome becomes fat and the length of the main root is about 7 to 10 cm; the diameter of the main root is about 3 cm, and in addition, it has many fat lateral roots, the total length being about 35 cm. The weight of this mature root is 70 to 100 g, and some have been found weighing 300 to 500 g per root.

Korean ginseng is a perennial plant which sprouts buds from its roots each spring and whose leaves and stems wither and die off each fall. At this time each year, the trace of an insertion is left on the rhizome, the head of the root, and its shape varies from mortar shape to a hollow shape. This rhizome is an important factor in the quality assortment process of ginseng's manufactured goods. Therefore, if this part is missing, it is judged to be inferior quality or becomes a low-priced item. For this reason, special care is taken in handling this product. In particular, this rhizome is the special feature of Korean ginseng and it is used as a manufacturing index to distinguish Korean ginseng from ginsengs of other countries.

A Korean ginseng seedling, on sprouting, has two cotyledons (primary leaves). Between these cotyledons, a small stem supports three tiny leaves. Leaves of the ginseng plants are palmately compound with a little longer petiole whorled on the tip of the stem. The number of leaves varies with the age of the plants and cultivation factors. It is usually related to the age of the plants, i.e. one for yearlings, two for two-year-old plants, three for three-year-old plants, and so on. Each compound leaf has three to five leaflets. Yearlings have three leaflets, and two-year plus plants have five leaflets in each compound leaf.

Ginseng is a semi-shade plant and the aerial part above ground is fairly weak. Therefore, if it is exposed to direct sunlight, chlorophyll in the leaves is destroyed, leaving that part dried up and dead. For that reason, it is good to set up a shade of sorts during the high temperatures of summer so

Fig. 9.3 Korean ginseng field.

as to create cooler conditions. Moreover, ginseng's growth is very slow as compared to common plants and its fertilizer tolerance is very weak, so using chemical fertilizers often hinders the growth of the plants. Therefore, nutrition management should focus on improving the soil with decomposed leaves. Another consideration is ginseng's severe unsuitability for consecutive planting—thus, it is desirable to wait for usually 10 to 15 years before planting the same field with ginseng again. Reasons to explain this include blight and damage from harmful insects, the lack of particular components in the soil, and the accumulation of toxic components. At present, the most likely theory involves blight and insects.

Given that ginseng has various botanical characteristics which demand great care, high production cost, and hard labor in its cultivation, it is understandable that the end-product is expensive.

Related species

The genus *Panax* was established by Linnaeus. Depending on who is doing the taxonomy, there are anywhere from six to eight species of the *Panax* genus.

The six species about which there is little debate are as follows:

Panax ginseng C.A. Meyer

Panax ginseng C.A. Meyer is found in Korea, North China, Manchuria, Siberia, and Japan, and is usually referred to as Korean or Asian ginseng. *Panax ginseng* is a long-peduncled plant, with the peduncle 3 to 6 times longer than the petiole. The chromosome number of *P. ginseng* is 2n=48.

Panax quinquefolium L.

Panax quinquefolium L. was found in America and Canada, and commonly called American ginseng, Canadian ginseng, or North American ginseng. This species is a short-peduncled plant. Its flowering peduncles are equal to, shorter than, or only slightly exceeding the petioles of the same plant. The chromosome number of *P. quinquefolium* is 2n=48.

Panax trifolium L.

Panax trifolium L. was found in North America, and commonly called dwarf ginseng or groundnut. The plants of *Panax trifolium* L. are all sexually specialized (Hu, 1976). The male plants bear white flowers with slender long pedicels that are 5 or 6 times longer than the obconic floral tube. The female plants bear pink flowers with short stout pedicels.

Panax japonicum C.A. Meyer

Panax japonicum C.A. Meyer was found only in Japan, and called Japanese ginseng or Chikusetsuninjien. This species is a long-peduncled plant.

Panax notoginseng Burkill

Panax notoginseng Burkill was found in China, and commonly called San Chi or Tien Chi ginseng.

Panax pseudoginseng Wallich

Panax pseudoginseng Wallich was found in India, Nepal, Bhutan and Sikkim (Wallich, 1829).

Of these six ginseng species, *Panax ginseng* C.A. Meyer and *Panax quinquefolium* L. are thought to have exceptional curative properties and, therefore, have the greatest commercial value.

Distribution and cultivation

Natural ginseng, *Panax ginseng* C.A. Meyer, is distributed in East Asia of 30-48 latitude north in areas such as Korea, the northeast of China, and the Russian Far East because this plant enjoys the cooler climate conditions of the temperate zone.

Ginseng grows all over Korea at 33 to 46 north latitude, especially in the Tae-Baek mountain range and the northern district of Korea. And it is distributed in a range of 100-800 m above sea level (Fig. 9.3). Natural ginseng, or, wild ginseng has been disappearing gradually while its demands for medicinal use, trade, and political exchange, have all increased. Therefore, people began cultivating it artificially. The early methods of artificial cultivation involved sowing the seeds or planting the seedlings of wild ginseng in the forest. It was later moved gradually from mountainous to flat areas, and eventually to today's shaded cultivation method. At present, ginseng cultivation prevails throughout almost all of Korea.

The environmental conditions suitable for Korean ginseng cultivation include an average annual temperature of 0.9°C to 13.8°C and summer average of 20°C to 25°C. Rainfall should average 1,200 mm annually, with comparatively little snow. Direct sunlight is harmful, diffused light being preferred in amounts 1/8 to 1/13 of the total sunlight capacity. Sandy loam is good for the surface soil, but for subsoil, it is desirable to have somewhat more clay. Old soil—neither too fertile nor damaged by blight or insects, and with a pH of 5.5 to 6.0—is very suitable.

It is extremely difficult to adapt and raise ginseng in places where the climate or soil is incompatible with its natural surroundings. Even if it is possible to cultivate this species, it is recognized that ginseng, if unsuitably grown differs in shape, quality, and pharmacological effect.

Biotechnological approaches for micropropagation and production of active principle

MICROPROPAGATION Korean ginseng plants (*Panax ginseng* C.A. Meyer) are perennial herbaceous plant, which grow very slowly. Cultivation of ginseng plants requires a period of more than three years in order to produce seeds. At the time of seed harvest, zygotic embryos of Korean ginseng are still in an immature globular stage; thus the seeds require stratification and cold treatment for several months. Therefore, tissue culture procedures could contribute to clonal propagation and genetic breeding in ginseng. However, it has been accepted that plant regeneration from somatic embryos of ginseng has been very recalcitrant. In many reports of Korean ginseng tissue cultures, structurally-abnormal somatic embryos such as multicotyledonary or multiple embryos frequently formed (Arya et al., 1991; Butenko et al., 1968; Chang et al., 1980; Lee et al., 1990). In addition, regenerated plantlets from somatic embryos developed into multiple shoots either without roots or with inadequate roots (Chang et al., 1980; Choi et al., 1998; Shoyama et al., 1998). Root development of plants is important for field habituation after transfer. However, little is known about the main problem of plant regeneration in ginseng.

In general, exogenous growth regulators are the most important components for somatic embryo induction. In *Panax ginseng*, numerous research concerning somatic embryogenesis were conducted from the callus culture. Lately, direct somatic embryogenesis of ginseng on a hormone-free medium can be applied to a technique for plant regeneration (Choi et al., 1994, 1997). This method has some advantages due to the rapid propagation and to the low frequency of genetic variation compared to calli-derived embryogenesis. Choi et al. (1997, 1998) tried to introduce an efficient plant regeneration protocol using direct somatic embryogenesis on a hormone-free medium.

When the cotyledon explants were cultured on MS agar medium with 5% sucrose, somatic embryos were developed directly near the basal excised region of cotyledon explants (Figs. 9.4 and 9.5). The frequency of somatic embryo formation was the highest (93%) in the cotyledon explants from midmature zygotic embryos and the competence for embryo formation decreased to 66% as the zygotic embryos matured and no embryos formed at seedling stage (60 mm in length) (Table 9.1). Somatic embryos were developed into multiple or single state (Table 9.1). The frequency of multiple and single embryo formation differed markedly, according to the degree of maturity of the zygotic embryos. On cotyledons from midmature zygotic embryos, most of somatic embryos (96% of total embryos) developed as a multiple state, while on cotyledon explants from a mature one, 13% of somatic embryos developed into single state (Table 9.1). In midmature cotyledons, multiple embryos (Fig. 9.4 A, B, D, E) formed on the basal portion of cotyledons, which developed into numerous multiple shoots (Fig. 9.4 B, C, E, F). For a while, in fully-mature cotyledons, single embryos formed from single epidermal cells of cotyledons and eventually developed into single cotyledonary embryos (Fig. 9.5 A-C).

Table 9.1 Frequency of somatic multiple or single embryo formation from cotyledon explants of Korean ginseng zygotic embryos at different stages of maturity on hormone-free MS medium after 2 months of culture

Stage of explant	Size (mm)	Explants forming somatic embryos (%)	Proportion of embryos Multiple	Single
Immature cotyledon	1	0	0	0
Midmature cotyledon	3	94±7	96	4
Mature cotyledon	6	66±5	13	87
Seedling	60	0	0	0

Single and multiple cell origin of somatic embryo were reported to depend on the differentiated state of maternal ginseng cotyledon (Choi et al., 1994). We compared the developmental pattern and plant conversion rate of multiple and single embryos. In the culture of immature cotyledons, somatic embryos derived from multiple cells developed into abnormal multiple embryos, fused each other and then fused to maternal explants; these embryos then regenerated into only multiple shoots (Fig. 9.4 A-F). While, in cotyledons from fully-matured zygotic embryos, somatic embryos originated from single cells and developed into single independent state (Fig. 9.5 A-C). These single embryos developed into normal plantlets with both shoots and roots (Fig. 9.5 D, E). Therefore, direct single embryogenesis derived from single cells is important for normal plant regeneration in ginseng. However, to induce single embryos, only cotyledons from fully-matured zygotic embryos should be sampled as an explant.

Recently, Langhansova et al. (2004) reported improvement from 31% to 70% in maturation and regeneration of *P. ginseng* somatic embryos by treatment of polyethylene glycol and abscisic acid.

Herbicide-resistant transgenic *Panax ginseng* plants were produced by introducing the phosphinothricin acetyl transferase gene that confers resistance to the herbicide Basta through *Agrobacterium tumefaciens* co-cultivation. Integration of transgene into the plants was confirmed by polymerase chain reaction and Southern analysis. Transgenic plants showed strong resistance against Basta in field conditions (Choi et al., 2003).

In conclusion, mass production of ginseng plants was accomplished via direct single somatic embryogenesis from preplasmolyzed ginseng cotyledons, and this protocol may be applied to the micropropagation of Korean ginseng plants.

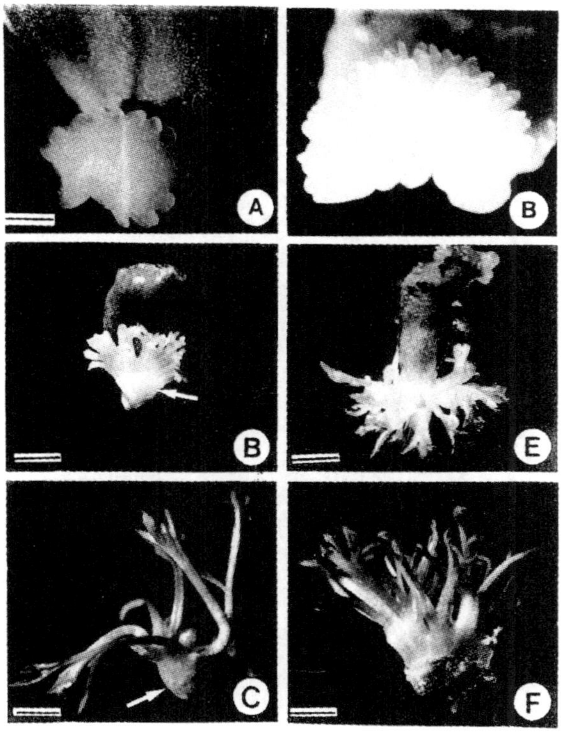

Fig. 9.4 Multiple shoot formation from somatic multiple embryos of ginseng. A, D ; Somatic multiple embryos formed directly from the basal surface of midmature cotyledons on MS basal medium after 1 month of culture (bars: 830 μm). B, E; Greening of cotyledonary multiple embryos just after treatment with 3 × 10^{-5} M GA_3, for 2 weeks (bars: 2.3 cm). C, F; Multiple shoots formed from somatic multiple embryos on half-strength MS medium (bars: 3.2 cm). Arrows in B and C indicate radical region.

Production of Ginsenosides by in vitro Culture of Ginseng Cells In general, ginseng cultivation requires large facilities and manpower as well as unusually long cultivation times. Therefore, the production of ginsenosides by in vitro culture of ginseng cells or tissues has been studied by a number of investigators (Kim et al., 1980; Choi et al., 1990). Large-scale plant cell cultures may be cost-effective and also enable one to produce a higher amounts of ginsenosides. The production of ginsenosides and other secondary metabolites from cell and tissue cultures of *Panax ginseng* cell lines have been patented by this laboratory (Korea Patent No. 045232, 045233, 045234 and 047779). For commercial availability of ginsenoside production from ginseng cell cultures, significant amounts of ginsenosides must be produced in a relatively short period of time. This

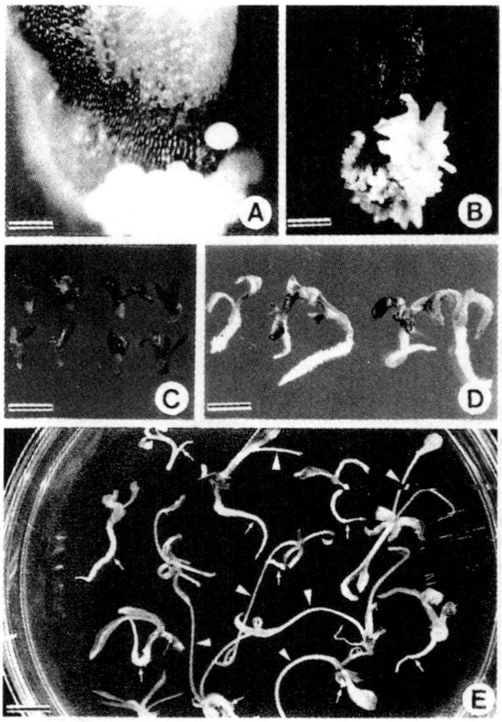

Fig. 9.5 Plant regeneration with both shoots and roots from somatic single embryos of Korean ginseng. A; Somatic embryos formed directly from cotyledons of mature zygotic embryos on MS basal medium after 1 month (bar: 300 mm). B; Cotyledonary embryos after 2 months (bar: 920 mm). C; Somatic single embryos with well-developed cotyledons and radicles after separation from the cotyledon explants (bar: 1.2 cm). D; Germinating somatic single embryos with well-developed roots on MS medium with 3×10^{-5} M GA_3 (bar: 2.2 cm). E; Plantlets regenerated from single embryos on half-strength MS basal medium with both shoots and roots (arrows roots, arrowheads shoots, bar: 1.36 cm).

can be accomplished with a slowly-growing cell line that produces relatively large amounts of ginsenosides or with a rapidly-growing cell line that produces relatively small amounts of ginsenosides. The idea, of course, is to select a rapidly-growing cell that produces large amounts of ginsenosides.

Choi et al. (1994) studied the effects of auxins on the ginsenoside-biosynthesis in ginseng cells cultured in liquid media. There have been numerous reports on the conditions of phytohormones to enhance the yields of secondary metabolites, but it is obvious that there is no general rule in most cases of plant cell cultures. The critical factors of secondary metabolite formation is the phytohormone regime. Since 2,4-D and IAA was clarified as the good phytohormone for the growth of ginseng cells cultured in the liquid medium, experiments were done in 10^{-4} M of IAA and 3 mg/l of 2,4-D in order to clarify the effects of IAA and 2,4-D on the ginsenoside biosynthesis in ginseng cells suspension-cultured.

The amounts of total ginsenoside of KGC2 cultured in 10^{-4} M of IAA and 3 mg/l of 2,4-D were 12.29 mg and 6.51 mg per g dry weight, respectively. In general, ginseng cells were grown faster in 2,4-D than in IAA. However, the total ginsenoside was more highly synthesized in 10^{-4} M of IAA than in 3 mg/l of 2,4-D. Also, amounts of ginsenoside fractions in 10^{-4} M of IAA were higher than those in 3 mg/l of 2,4-D. Specifically, amounts of ginsenosides Rb_2 and Re in 10^{-4} M of IAA were increased to 2.3 and 3.2-fold of 3 mg/l 2,4-D. However, the amount of ginsenoside Rg_1 in 3 mg/l of 2,4-D was increased to 2.2 fold of 10^{-4} M IAA. Choi et al.(1994) investigated the growth of ginseng cells, ginsenoside content and ginsenoside productivity in ginseng cell lines cultured on the MS liquid medium containing 10^{-4} M of IAA for 34 days. Table 9.2 shows the growth and amounts of total ginsenoside and ginsenoside productivity of ginseng cell lines at 34 days after culture. In general, the ginsenoside amount was maximized at 34 days after the liquid culture. Especially, KGC 13 was selected as a cell line having the highest total ginsenoside, 65.92 mg per g dry weight, among the other cell lines, such as KGC 2, KGC 15 and KGC 16. The fresh weight of KGC 13 was 1.22 g per 15 ml culture solution. Although KGC 13 produced the highest amount of total ginsenoside, KGC 13 was grown slowly in the liquid medium. The amount of total ginsenoside in KGC 15 was 22.45 mg per g dry weight. However, KGC 15, of which the fresh weight was 3.18 g per 15 ml culture solution, was the most rapidly grown among the other cell lines. In general, the productivity of ginsenoside in the cells is dependent on the capability of the ginsenoside biosynthesis and the growth of individual cells. In this experiment, the productivities of total ginsenoside in KGC 2, KGC 13, KGC 15 and KGC 16 were 12.69 mg, 80.42 mg, 71.39 mg, and 56.04 mg per 15 ml culture solution, respectively.

Expressed sequence tags (ESTs) provide a valuable tool that can be used to identify genes in secondary metabolite biosynthesis. Jung et al. (2003) sequenced 11636 ESTs from five ginseng libraries in order to create a gene resource for biosynthesis of ginsenosides. ESTs involved in ginsenoside biosynthesis were identified by a key word BLASTX results and a domain search of ginseng ESTs. They identified 4 oxidosqualene cyclase candidates involved in the cyclization reaction of 2,3-oxidosqualene, 9 cytochrome P450 and 12 glycosyltransferase candidates, which may be involved in modification of the triterpene backbone. Attempt has been made to identify the changes in protein profile during transformation (Kim et al., 2003). The proteomic analysis of cultured hairy root of *P. ginseng* on 2-D polyacrylamide gel electrophoresis was performed and protein spots were analysed and identified by peptide finger printing and internal amino acid sequencing by matrix assisted laser desorption/ionization-time of flight mass spectrometry (MALDI-TOF MS) and electrospray ionization quadrupole-time of flight mass spectrometry (ESI Q-TOF MS), respectively. Major protein spots (159) were analysed by peptide finger printing or de novo sequencing and the function of 91 of these proteins were identified. Protein identification was achieved using the EST database from *P. ginseng* and protein database of *Arabidopsis thaliana* and *Oryza sativa*. Internal amino acid sequences were recorded as better markers for identification. These results suggested that protein patterns of normal and hairy root were different and some proteins were expressed exclusively in hairy root cultures (Kim et al., 2003a).

Table 9.2 Fresh weight and ginsenoside content of ginseng cell lines cultured in the liquid medium containing 10^{-4} M of IAA for 34 days

Cell lines	FW (g/15 ml)	Ginsenoside (mg/g dry weight)	Productivity (mg/15 ml media)
KGC 2	1.03	12.32	12.69
KGC 13	1.22	65.92	80.42
KGC 15	3.18	22.45	71.39
KGC 16	1.85	30.29	56.04

Initial inoclum size — 200 mg
Productivity — Amount of cells (g/15 ml of culture media) × total ginsenoside (mg/g dry weight).

In conclusion, Choi et al.,(1994) selected two ginseng cell lines, KGC 13 and KGC 15, having higher productivity of total ginsenoside. In particular, the amount of total ginsenoside in KGC 13 cell line in vitro cultured was the highest reported so far, for any ginseng tissues and was much higher than that in 6-year-old native root. KGC 15 cell line was the most rapidly grown among ginseng cell lines selected. Therefore, the ability of ginseng to produce a medicinal compound by a cultured cell line is apparently higher than that of intact plants.

CHEMISTRY

Active components

Ginsenoside (ginseng saponin)

Chemical studies of ginseng components were first reported by Garriques (1854) of the United States, who isolated a blend of amorphous glycoside and called it 'panaquilon', and claimed it as a peculiar component of ginseng. Further studies of ginseng were restimulated when Brekhman and Dardymov (1969) suggesting saponin as the active principle contained in ginseng. He found that ginseng exhibits antifatigue effects, increases work efficiency, and can either stimulate or pacify the central nervous system.

In 1960's, Shibata et al. (1963, 1965) isolated 13 kinds of saponins from Korean ginseng and identified their structures, differentiating them as to their order of moved distance on thin-layer chromatography, ginsenoside Ro, Ra, Rb, Rc, Rd, Re, Rf, Rg, and Rh_1.

Plant saponin gets its name from the fact that it forms a soapy foam when dissolved in water. In general, it is any of a group of glycosides and forms a rather highly polarized compound which has a destructive effect on red blood cells and a toxicity towards fish. In addition, it is able to join with cholesterol in the blood to form a complex. Given this, what is so different about ginseng saponin when compared to that in other plants that makes it effective while not causing the toxic effects which may lead to problems such as hemolysis?

First, ginseng saponin is a glycoside structured with dammarane from triterpenoid. However, this is not the case in general herbal pharmaceutical saponin. Secondly, while the saponin originating from other plants has polarity with effects such as hemolysis, ginseng saponin is a neutral glycoside which has hardly any toxicity. Thirdly, ginseng saponin's pharmacological action is remarkably different from other herbal medicinal saponins.

Hydrolysis of glycoside mixtures with acid results in the formation of carbohydrates and their aglycones. Panaxadiol, panaxatriol, β-sitosterol and oleanolic acid are known to be aglycones of ginseng glycosides. Glycosides of panaxadiol and panaxatriol are distributed in ginseng and related plants only, and their physiological activities have attracted much academic interest since they had essentially the same effects as in public experience.

Glycoside groups containing protopanaxadiol or protopanaxatiol as genuine aglycone are called dammarane triterpene glycoside. First of all, terpenes are carbons whose number is a multiple of 5, that is, 5n (n=2), and in the view of herbal synthesis it is the "n" number of isoprene or isopentane; $[CH_2=CH-C(CH_3)=CH_2]$. Terpene is classified according to the number of carbons, and when n=6, triterpene is indicated with 30 carbons. The basic structure of dammarendiol, with its four rings, is dammarane.

As shown in Fig. 9.6, the combination in the position of R_1, R_2, and R_3, in dammarane's basic structure determines the kind of ginsenoside.

Ginseng sapoin is thus called 'ginsenoside', which is a combination of the words ginseng and glycoside. Korean ginseng, *Panax ginseng* C.A.

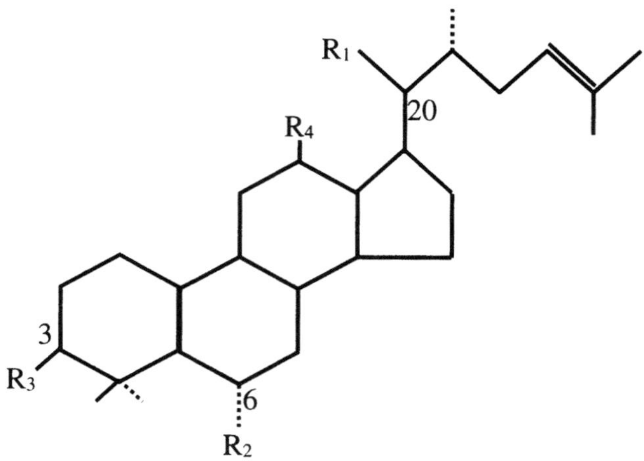

Ginsenosides	R_1	R_2	R_3	R_4
Ginsenoside-Ra_1	-O-Glu^2-Glc	-H	-O-Glc-Ara (fur)-xyl	-OH
Ginsenoside-Ra_2	-O-Glu^2-Glc	-H	-O-Glc-Ara (pyr)-xyl	-OH
Ginsenoside-Rb_1	-O-Glu^2-Glc	-H	-O-Glc^6-Glc	-OH
Ginsenoside-Rb_2	-O-Glu^2-Glc	-H	-O-Glc^6-Ara (fur)	-OH
Ginsenoside-Rb_3	-O-Glu^2-Glc	-H	-O-Glc^6-Xyl	-OH
Ginsenoside-Rc	-O-Glu^2-Glc	-H	-O-Glc^6-Ara (pyr)	-OH
Ginsenoside-Rd	-O-Glu^2-Glc	-H	-O-Glc	-OH
Ginsenoside-Re	-OH	-O-Glu^2-Rha	-O-Glc	-OH
Ginsenoside-Rf	-OH	-O-Glu^2-Glc	-OH	-OH
20-Gluco Ginsenoside-Rf	-OH	-O-Glu^2-Glc	-O-Glc	-OH
Ginsenoside-Rg_1	-OH	-O-Glc	-O-Glc	-OH
Ginsenoside-Rg_2	-OH	-O-Glu^2-Rha	-OH	-OH
Ginsenoside-Rh_1	-OH	-O-Glc	-OH	-OH

	R_1	R_2
Ginsenoside-Ro	-O-$GlcUA^2$Glc	-O-Glc

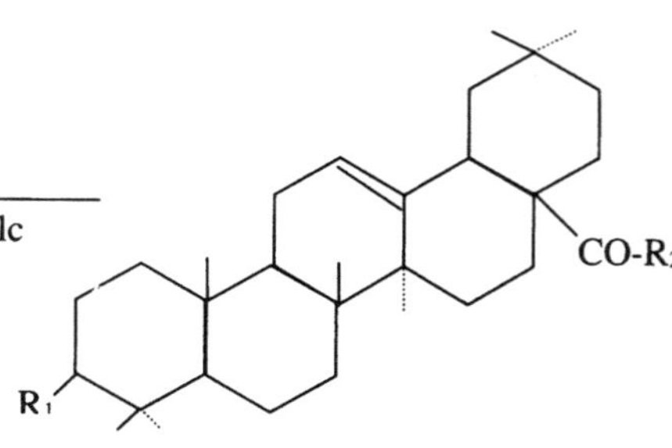

Fig. 9.6 Structure of ginsenosides

Meyer, is known to contain more than 20 kinds of ginsenosides, but as shown in Table 9.3, only half this number is found in related species. Considering that each of these ginsenosides has different pharmacological actions, it becomes apparent that Korean ginseng, *Panax ginseng* C.A. Meyer, might have a pharmacological effectiveness superior to any other.

Table 9. 3 The composition of ginsenosides in Korean ginseng (*Panax ginseng* C.A. Meyer), American ginseng (*Panax quinquefolium L.*), and Chinese Tienchi ginseng (*Panax notoginseng* Burkill)

Components	P. ginseng	P. quinquefolium	P. notoginseng
Total number of ginsenosides	22	13	14
Panaxadiol group	14	8	5
Panaxatriol group	7	4	9
Oleanolic acid group	1	1	0

Reversed-phase HPLC was used to determine the natural variation in the levels of ginsenosides Rg1, Re, Rf, Rb1, Rc, Rb2 and Rd in *Panax quinquefolius* and total contents in different plant parts. Though variation was recorded in individual ginsenoside content, no difference was recorded between wild and cultivated plants of *Panax quinquefolius* in terms of total contents (Assinewe et al., 2003).

Other active components

There are certain very important substances in ginseng. Among them are components that demonstrate anticancer, antioxidation, antidiabetes, and anti-inflammation effects, as well as those which exhibit hematopoiesis.

Polyacetylene compounds (Ahan et al., 1988 ; Kim et al., 1989) exhibit an anticancer effect, and phenol compounds (Han et al., 1981, 1992) exhibit antioxidation and, thus, anti-aging effects through the inhibition of the formation of lipid peroxide in the body.

Among refined oil constituents, pinene acts as an analgesic, and ocinene has an antiphlogistic effect, antibiotic tendency and also inhibits overfermentation in the intestinal tract. Moreover, through the discovery of antidiabetic constituents such as adenosine, acidic peptide, and a manganese-containing substance, along with the discovery of other non-saponin constituents which have effective pharmacological action, the importance of these non-saponin ingredients was recognized (Han et al., 1981; Okuda, 1990). Meanwhile, the importance of the anti-inflammatory constituents, panaxynol and linoleic acid, was emphasized. In addition, 12% to 16% of nitrogen-containing compounds such as protein, nucleic acid, as well as many other components including essential amino acids, essential fatty acids are contained in Korean ginseng (*Panax ginseng* C.A. Meyer).

Therefore, Korean ginseng is spotlighted not only for pharmacological use, but also as a health food.

Pharmacological properties

Effects of ginseng on hepatic diseases

When ginseng was administered to rats who had been given a two-thirds partial hepatectomy, the rate of liver regeneration was higher by 34% than that in the control (Igarashi et al., 1975). Liver disease is usually hepatitis, especially acute viral hepatitis type B by nature, and when this becomes chronic, it turns into cirrhosis of the liver, and may progress further to liver cancer. Therefore, the treatment of hepatitis is vital so as to prevent it from passing into a chronic state. Professor Koo (1981) observed that ginseng, when administered to 201 patients of acute viral hepatitis-B, not only promoted early recovery but also prevented the disease from becoming chronic. Ginseng reduces tissue damage such as the death of liver cells induced by acutely toxic chemicals like carbon tetrachloride or phenacetin (Hahn, 1978). Moreover, it has been revealed that ginseng promotes RNA synthesis in the liver (Igarashi et al., 1975).

Many people have heard that ginseng is good for a hangover, and probably many others have experienced this themselves. Joo et al. (1984) studied the effect of ginseng saponin fraction on enzymes involved in alcohol metabolism such as alcohol dehydrogenase (ADH) and the microsomal ethanol oxidizing system (MEOS). Electron micrographs of the liver from rats receiving 12% ethanol and 12% ethanol supplemented with ginseng saponin were also compared. A moderate amount of ginseng saponin fraction stimulated ADH, ALDH and MEOS in vivo as well as in vitro. It appears that ginseng saponin fraction enhances the removal from the liver of the toxic aldehyde produced during ethanol oxidation. Furthermore, in the presence of saponin fraction, the excess hydrogen generated from the oxidation of ethanol is efficiently mobilized for biosynthetic processes and, thus, protects the liver against alcohol intoxication. Electron microscopic observation revealed that the hepatocytes of rats treated with 12% ethanol for 6 days were severely damaged, whereas no damage was observed in hepatocytes of ginseng saponin-treated animals. Such results suggest that the ginseng fraction protects the liver against ethanol intoxication. Yu et al. (1982) measured the activity of glutamic oxaloacetic transaminase (GOT), glutamic pyruvic transaminase (GPT) and alkaline phosphatase (ALP) in serum to study the effect of ginseng on alcohol-intoxication. Alcohol administration induced fat degeneration, the accumulation of intracellular lipid, the thickening of venular wall, cellular swelling, congestion and bile pigmentation in the Kuffer cell and also increased the GOT, GPT, and ALP activities. Combined treatment with ginseng saponin and a moderate amount of ethanol, however, did not induce any histological changes and furthermore, the levels of GOT, GPT and ALP were lower than the control group receiving alcohol treatment alone.

Effects of ginseng on Diabetes mellitus

Chinese healers known for a very long time that ginseng increases stamina, prevents collapse, increases the secretions of body fluids, and quenches thirst. This means that ginseng can counter the special characteristics of diabetes, such as thirst and body weakness. Also recent research reports reveal ginseng's effect on the treatment of diabetes as well as its prevention. Ginseng not only acts directly to lower the level of blood sugar, but also acts indirectly and improves the symptoms that accompany diabetes and its complications. Okuda (1978), Ando et al. (1979, 1980) and Sekiya et al. (1981) isolated insulin-like compounds from water extracts of Korean ginseng and identified them as acidic peptides, ginsenoside—Rb_1 and Re, adenosine and Mn-containing compounds. Recently, it was found that ginsenoside $Rb_{2,}$ which is one of the components of ginseng saponin, also exhibits the effect of lowering blood sugar levels. Meanwhile, Professor Kimura (1980, 1981) reported that he had isolated DPG3-2, which is a blood sugar depressive component, and found this component to function in stimulating insulin release.

Anticarcinogenic Effects of Ginseng

Effects to develop the use to cancer chemotherapy have intensified over the last several decades. However, many types of cancer remain difficult to cure. It is obvious that the war against cancer is still far from being won. The studies of Dr Yun (1997) have focused on the anticarcinogenicity of *Panax ginseng* C.A. Meyer, known as a mysterious tonic, against various chemical carcinogens since 1978. The prolonged administration of red ginseng extract inhibited the incidence and also proliferation of pulmonary tumors induced by DMBA, urethane and aflatoxin by modified long term in vivo experiments. A 9-week medium-term anticarcinogenicity test using pulmonary adenoma induced by benzo(a)pyrene was established. Statistically-significant anticarcinogenic effects were observed in powder and extract of 6-year fresh ginseng, 5- and 6-year white ginseng, and 4-, 5- and 6-year red ginseng. Comparative studies on the anticarcinogenicity between *Panax ginseng* C.A. Meyer and *Panax notoginseng* using the 9 week's medium-term mouse system (Yun's test) were carried out. In *Panax ginseng*, both the water extract and the ethanol soluble fraction showed a significant reduction of lung tumor incidence as compared to the benzo(a)pyrene alone group. However, in the case of *Panax notoginseng*, only the highest dose of ethanol soluble fraction showed evidence of anticarcinogenicity. The ethanol insoluble fractions of both ginsengs did not show any significant decreases of lung adenoma incidence.

An epidemiological study to confirm whether ginseng has anticarcinogenic effect in human being was conducted in relation to two case-control studies on patients and a cohort study on a population of a ginseng cultivation area. In case-control studies, odd ratios of the cancer of lip, oral cavity and pharynx, larynx, lung, esophagus, stomach, liver, pancreas, ovary and colorectum were significantly reduced. In a cohort study with seven year's follow-up conducted in a ginseng cultivatiion area, ginseng intakers had a decreased relative risk (RR : 0.38) compared with non-intakers. On the types of ginseng, the RRs of cancer were 0.52 for fresh ginseng extract intakers and 0.47 for multiple combination intakers. Among 24 red ginseng intakers, there were no reports of cancer deaths. The RRs of ginseng intakers were 0.35 in gastric cancer and 0.23 in lung cancer. There was a decrease in risk with rising frequency of ginseng consumed, showing a statistically significant dose-response relationship. These findings strongly suggest that *Panax ginseng* C.A. Meyer has a non-organ specific cancer preventive effect against various cancers. In 1984, potential cytotoxic activity of the petroleum ether extract of *Panax ginseng* root (crude GX) and its partially purified fraction (7 : 3 GX) on tumor cells were compared with that of 5-fluorouracil (5-FU). Administration of crude GX normalized the elevated level of white blood cells in mice caused by inoculation with S-180 cells. A decrease in Hb value in rats inoculated with Walker 256 was also restored to the normal level by GX-treatment. Administration of the 7 : 3 GX fraction increased the survival time of the mice challenged with S-180 cells by 1.5 to 2 times that of the control group. These results demonstrate that the inhibitory effects of ginseng on the proliferation of cancer cells are attributed to the lipid-soluble fraction of ginseng.

Effects of ginseng on the cardiovascular system

Ginsenosides reduced the mean arterial blood pressure in anesthetized rats. The studies of Kim (1999) have focused on characterizing the endothelium-dependent and independent relaxation elicited by ginsenosides, a mixture of saponin extracted from *Panax ginseng*, in isolated rat aorta. Ginsenosides stimulated the formation of nitric oxide (NO) in endothelial cells which caused relaxation of blood vessels by enhancing the production of cyclic GMP in rings of rat and rabbit aorta and femoral and coronary artery of dogs. Ginsenosides increased the permeability of the vascular smooth muscle cell membrane to K^+, resulting in hyperpolarization, which indirectly relaxed the blood vessel by decreasing the opening of voltage-dependent Ca^{2+} channels. Ginsenoside Rg_3 was the most potent vasodilator among the ginsenosides examined. These results indicated that ginsenosides-induced relaxation might be regulated by dual mechanisms. Ginsenosides-induced relaxation in rat aorta was caused by the release of endothelium-derived nitric oxide and direct inhibition of vascular smooth muscle via the activation of K^+ channels. Endothelium-dependent relaxation to acetylcholine was impaired in the thoracic aorta isolated from hypercholesterolemic rabbits. Dietary supplementation of rabbits with ginsenosides improved the endothelium-dependent relaxation to acetylcholine in the aortic blood vessels in cholesterol-fed rabbits, providing evidence that ginsenosides increased the production of nitric oxide by endothelial cells. Ginsenosides reduced the blood pressure of spontaneous hypertensive rats (SHRs). Acetylcholine caused endothelium-dependent contraction in the aorta of SHRs. Ginsenosides prevented endothelium-dependent contraction to acetylcholine in the isolated thoracic aorta of SHR. Prostaglandin endoperoxide (PGH2)- and oxygen-derived free radical-induced contractions of the isolated aorta of SHR were significantly inhibited by ginsenosides. These results indicate that the hypotensive effect of ginsenosides in SHR may be associated with decreased formation of PGH2 and inhibition of superoxide anion in the aorta. Ginsenosides reduced rat and human platelet aggregation induced by collagen or thrombin, possibly by inhibiting the formation of TXA2. These results indicate that ginsenosides may be useful in endothelium-dependent relaxation, vasorelaxation via activation of potassium channels, hypertension, platelet aggregation, atherosclerosis, and a positive inotropic effect.

Antifatigue, Antistress and Anti-Aging Effects of Ginseng

The ginsenoside Rg_1 exerted an antifatigue effect in mice subjected to exhaustive exercise (Saito et al., 1974). Takagi (1974) conducted pharmacological studies of active components extracted from ginseng. Water extracts, Rg_1 and the G-5 fraction exhibited antifatigue effects. When seven ginsenosides, Rb_1, Rb_2, Rc, Rd, Re, Rf, and Rg_1, were individually administered to mice, they all showed antifatigue effects (Kaku et al., 1975).

Brekhman (1969) studied the pharmacological effect of ginseng and presented the concept of ginseng's tonic effect as a phenomenon that generally increases resistance, and said that ginseng improves defense ability nonspecifically by increasing the resistance against harmful environmental conditions in the body. This effect of ginseng was named "adaptogen effect". Dr Petkov (1978) found that Korean ginseng accelerates the release of hormones related to the defense system against stress and thus exhibits an antistress effect. The effects of ginsenosides (Rb_1, Rb_2, Rc, Re, and Rg_1) isolated from *Panax ginseng* and their partial hydrolyzates, prosapogenins, on specific cold stress (SART) and water immersion stress in mice were studied (Kita et al., 1981). A single intraperitoneal administration of ginsenosides or prosapogenins did not affect the pentobarbital-induced sleeping time in non-stressed mice. But the inhibition of normal body weight increase in stressed mice was significantly prevented by the administration of ginsenosides and prosapogenins for 5 days at a dose of 2.5 mg/kg body weight. Writhing, caused by the administration of acetic acid, was significantly inhibited by ginsenosides and prosapogenins in non-stressed and SART-stressed mice. Studies conducted on the effect of red ginseng powder and extract on the sex cycle of the adult -CS strain female mice exposed to chronic hanging stress were conducted by Bao et al. (1984). Administration of red ginseng powder and extract prevented disturbances on the sex cycle and a decrease in body temperature induced by chronic stress.

The non-saponin fraction of ginseng inhibited the harmful action of oxygen radicals such as O_2^-, $H_2O_2^-$, OH^-, and $^1O_2$, thus, peventing tissue degeneration, which causes aging (Park et al., 1984). The mechanism of the antioxidant effect of ginseng was also investigated. The non-saponin fraction of ginseng prevented the inhibition of skin collagen gelation induced by xanthine and xanthine oxidase. Total ginseng extract and its non-saponin fraction exhibited a quenching effect on $^1O_2$. These results suggest that the collagen alteration induced by oxygen radicals was prevented by the active components present in the non-saponin fraction which have a free radical ($^1O_2$) quenching property. Also, Cho et al. (1984) investigated the effect of ginseng on age-related enzynes such as isocitrate dehydrogenase, 6-phosphogluconate dehydrogenase, glucose-6-phosphate dehydrogenase, glutachione peroxidase, and glutathion reductase. Ginseng extract enhanced the activities of those age-related enzymes in vitro and also retarded a decline in their activity occurring in relation to aging. Such results indicate that ginseng has the potential to retard aging processes.

In two recent studies, potential neuroprotective action of ginseng extract have been established. Oral administration of ginseng provided protection against neurotoxicity experimentally created by parkinsonism-inducing neurotoxin (MPTP, 1-methyl-4-phenyl-1,2,3,6-tetrahydropyridine) in rodent model of Parkinson disease (Van Kampen et al., 2003). Pretreatment of neuronal PC12 cells with ginseng radix prior to exposure with MPP+ (1-methyl-4-phenylpyridine) showed a decrease in the apoptotic features as evident by flow cytometry, DNA fragmentation assay, reverse-transcription polymerase chain reaction, Western blotting and other assays (Kim et al., 2003b).

REFERENCES

Ahan, B.Z., Kim, S.I., Kang, K.S., and Kim, Y.S., 1988, The action of cytotoxic components of Korean ginseng against L1210 cells and their structure-activity relationship. Proc. 5th. Int'l Ginseng Symp., Korea Ginseng and Tobacco Research Institute, Korea:19-24.

Ando, T., Muraoka, T., Yamasaki, N. and Okuda, H., 1979, Preparation of insulin-like peptides from *Panax ginseng*. Proc. Symp. Wakan Yaku 12:15.

Ando, T., Muraoka., T., Okuda, H. and Yamasaki, N., 1980, Preparation of an antilipolytic substance from *Panax ginseng*. Planta Medica 38:18.

Arya, S., Liu, J.R., and Eriksson, T., 1991, Plant regeneration from protoplasts of *Panax ginseng* (C.A. Meyer) through somatic embryogenesis. Plant Cell Rep. 10:277-281.

Assinewe, V.A., Baum, B.R., Gagnon, D. and Arnason, J.T.,2003, Phytochemistry of wild populations of *Panax quinquefolius* L. (North American Ginseng). J. Agric. Food. Chem.51(16): 4549-4553.

Bao,T.T. and Saito, H., 1984, The effects of red ginseng, vitamins and their preparations (IV): Effects on the sex cycle in stressed female mice. Japanese Pharmacology & Therapeutics 12(4):37.

Bittles, A.H., Fulder, S.J., Grant, E.C., and Nichills, M.R., 1979, The effect of ginseng on lifespan and stress response in mice. Gerontology 25:125.

Brekhman, I.I. and Dardymov I.V., 1969, New substances of plant origin which increase non-specific resistance. Ann. Rev. Pharmacol. 9:419-430.

Brekham, H. and Dardymov, I.V., 1969, Pharmacological investigation of glycosides from ginseng and Eleutherococcus. Lloydia, 32: 46-51.

Butenko, R.G., Brushwitzky, I.V., and Slepyan, L.I., 1968, Organogenesis and somatic embryogenesis in the tissue culture of *Panax ginseng* C.A. Meyer. Bot. Zh. 7: 906-913.

Chang, W.C., and Hsing, Y.I., 1980, Plant regeneration through somatic embryogenesis in root-derived callus of ginseng (*Panax ginseng* C.A. Meyer). Theor. Appl. Genet. 57:133-135.

Cho, Y.D., Koo, B.S. and Lee, S.J., 1984, Studies of ginseng extract on age-related enzymes. Proc. 4th Int'l. Ginseng Symp.:37.

Choi, K.T., Park, J.C. and Ahn, I.O., 1990, Saponin production in tissue culture of ginseng (*Panax ginseng* C.A. Meyer). Korean J. Ginseng Sci. 14:107-111.

Choi, K.T., Lee, C.H., Ahn, I.O., Lee, J.H. and Park, J.C., 1994, Characteristics of the growth and ginsenosides in the suspension-cultured cells of Korean ginseng (*Panax ginseng* C.A. Meyer). Proc. Int'l. Ginseng Conference, Vancouver: pp. 259-268.

Choi. Y.E., and Sho, W.Y., 1994, Origin of somatic embryos induced from cotyledons of zygotic embryos at various developmental stages of ginseng. J. Plant Biol. 37:365-370.

Choi. Y.E., Yang, D.C., Kim, H.S., and Choi, K.T., 1997, Distribution and changes of reserve materials in cotyledon cells of *Panax ginseng* related to direct somatic embryogenesis and germination. Plant Cell Rep. 16:841-846.

Choi, Y.E., Yang, D.C., Park, J.C., Soh, W.Y., and Choi, K.T., 1998, Regenerative ability of somatic single and multiple embryos from cotyledons of Korean ginseng on hormone-free medium. Plant Cell Rep. 17:544-551.

Choi, Y.E., Jeong, J.H., In, J.K. and Yang, D.C. 2003, Production of herbicide-resistant *Panax ginseng* through the introduction of the phosphinothricin acetyl transferase gene and successful soil transfer. Plant Cell Rep. 21(6): 563-68.

Garriques, S. 1854, On panaquilon, a new vegetable substance. Ann. Chem. Pharm.90:231.

Hahn, D.R., 1978, Pharmaco-biological action of the ginsenosides Rb1, Rg1, and Re. Proc. 2nd Int'l. Ginseng Symp.:135.

Han, B.H., Park, M.H. and Han, Y.N., 1981, Studies on the antioxidant components of Korean ginseng(III). Identification of phenolic acid. Arch. Pharm. Res. 4(1):53-58.

Han, B.H., Park, M.H., Han, Y.N., and Suh, D.Y., 1992, chemical and biochemical study on non-saponin constituents of Korean ginseng. Korean J. Ginseng Sci. 16(3):228-234.

Hong, S.A., Cho, Y.Y. and Hong, S.K., 1969, Influence of each fraction from *Panax ginseng* on the hypothermia in mice elicited by reserpine, nembutal and chlorpromazine. Korean J. Pharmacol, 5:19-29.

Hwang, W.I. and Oh, S.K., 1984, Study on the anticancer activities of lipid soluble ginseng extract and ginseng saponin derivatives against some cancer cells. Korean J. Ginseng Sci. 8(2):153.

Igarashi, S., Oka, Y., Oda, T. and Oura, H., 1975, Effect of saponin extract from *Panax ginseng* on protein and amino acid metabolism in liver. Proc. Symp., Waken Yaku 9:63.

Joo, C.N., 1984, The preventive effect of the saponin fraction of *Panax ginseng* C.A. Meyer against ethanol intoxication of rat liver. Proc. 4th Int'l. Ginseng Symp.:63.

Jung, J.D., Park, H.W., Hahn, Y. Hur, C.G., In, D.S., Chung, H.J. Liu, J.R. and Choi, D.W., 2003, Discovery of genes for ginsenosides biosynthesis by analysis of ginseng expressed sequence tags. Plant Cell rep., 22(3): 224-30.

Kaku, T., Miyata, T., Uruno, T., Sako, I. and Kinoshita, A., 1975, Chemico-pharmacological studies on saponins of *Panax ginseng* C.A. Meyer. Arzenim.-Forsch. (Drug Res.) 25(4):539.

Kim, M.W., Choi, K.T., Bae, H.W. and Kang, Y.H., 1980, Effect of 2,4-D and kinetin on saponin content of callus derived from ginseng root. Korean J. Botany 23:91-98.

Kim, N.D., 1999, Cardiovascular actions of ginsenosides. Proceedings of '99 Korea-Japan Ginseng Symposium:1-15.

Kim, S.I., Lee, Y.H., and Kang, K.S., 1989, 10-Acetylpanaxytriol, a new cytotoxic polyacetylene from *Panax ginseng*. Yakhak Hoeji 33:118.

Kimura, M., 1980, Hypoglycemic components in ginseng radix and its insulin release. Proc. 3rd Int'l Ginseng Symp.:37.

Kimura, M., Waki, I., Tanaka, O., Nagai, Y. and Shibata, S., 1981, Pharmacological sequential trials for the fractionation of components with hypoglycemic activity, in alloxan diabetic mice, from ginseng radix. J. Pharm. Dyn. 4:402.

Kim, S.I., Kim, J.Y., Kim, E.A., Kwon, K.H., Kim, K.W., Cho, K., Lee, J.H., Nam, M.H., Yang, D.C., Yoo, J.S. and Park, Y.M.. 2003a, Proteome analysis of hairy root from *Panax ginseng* C.A. Meyer using peptide fingerprinting, internal sequencing and expressed sequence tag data. Proteomics 3 (12):2379-92.

Kim, E.H., Jang, M.H., Shin, M.C., Shin, M.S. and Kim, C.J., 2003b, Protective effect of aqueous extract of ginseng radix against 1-methyl-4-phenylpyridinium-induced apoptopsis in PC12 cells. Biol. Pharm Bull.,26(12):1668-73.

Kita, T., Hata, T., Kawashima, Y., Kaku, T., and Itoh, E., 1981, Pharmacological actions of ginseng saponin in stress-mice. J. Pharm. Dyn. 4:387.

Koo, K.H., 1981, Clinical observation on the preventive effect on *Panax ginseng* against liver cirrhosis and its treatment at an early stage. Ann. Rept. Korean Ginseng Res., Korea Ginseng & Tobacco Research Institute.

Langhansova, L., Konradova, H. and Vanek, T., 2004, Polyethylene glycol and abscisic acid improve maturation and regeneration of *Panax ginseng* somatic embryos. Plant Cell Rep., 21: in press.

Lee, H.S., Liu, J.R., Yang, S.G., Lee, Y.H., and Lee, K.W., 1990, In vitro flowering of plantlets regenerated from zygotic embryo-derived somatic embryos of ginseng. HortSci 25:1652-1654.

Oh, J.C., Park C.W. and Moon, D.Y., 1969, Effect of *Panax ginseng* on the central nervous system. Korean J. Pharmacol., 5:123-28.

Okuda, H., 1978, Studies on insulin-like substances in *Panax ginseng*. Proc. 2nd Int'l Ginseng Symp.: 75.

Okuda, H., and Lee, S. D., 1990, Biological activities of non-saponin compounds isolated from Korean red ginsneg. Proc. Int'l. Symp. on Korean ginseng. The Society for Korean Ginseng, Seoul, Korea:15-19.

Park, C.W., Lim, J.K., Lee, J.S. and Chung, M.H., 1984, Effects of ginseng components on the actions of oxygen radicals to the gelation of skin collagen. The Seoul J. Med. 25(1):45.

Petkov, V.D., 1978, Effect of ginseng on the brain biogenic monoamines and 3',5'-AMP system: Experiments on rat. Arzeim Forsch. 28:388-393.

Saito, H., Yoshida, Y. and Takagi, K., 1974, Effect of *Panax ginseng* root on exhaustive excercise in mice. Japanese J. Pharmacol. 24:119.

Sekiya, K. and Okuda, H., 1981, Purification of an anti-lipolytic (insulin-like) substance from *Panax ginseng*. Proc. Symp. Wakan Yaku 14:133.

Shibata, S., Fujita, M., Itokawa, H., Tanaka, O., and Ishii. T., 1963, Panaxadiol, a sapogenin of ginseng roots (I). Chem. Pharm. Bull. 11:759-761.

Shibata, S., Tanaka, O., Soma, K., Iita, Y., Ando, T., and Nakamura, H., 1965, Studies on saponins and sapogenins of ginseng. The structure of panaxatriol. Tetrahedron Lett. 3:207-213.

Shoyama, Y., Kamura, K., and Nisioka, I., 1998, Somatic embryogenesis and clonal multiplication of *Panax ginseng*. Planta Med. 54:155-156.

Takagi, K., 1974, Pharmacological studies on ginseng. Proc. 1st Int'l Ginseng Symp.:119.

Van Kampen, J., Robertson, H., Hagg, T. and Drobitch, R.,2003, Neuroprotective actions of the ginseng extract G115 in two rodent models of Parkinson's disease. Exp. Neurol. 184:521-29.

Wallich, N., 1829, An account of the Nepal ginseng. Trans. Med. Phys. Soc. Calcutta 4:115-120.

Yu, H.Y., Kim, C.M. and Koo, K.H., 1982, An experimental study on the effect of ginseng saponin upon alcoholic liver injury. J. Hanyang Med. Coll. 2(2):287.

Yun, T.K., 1997, Anticarcinogenic and cancer preventive effects of Korean ginseng. Proceedings of '97 Korea-Japan Ginseng Symposium:1-31.

Zhao, X.Z., 1990, Antisenility effect of ginseng rhizome saponin. Chung Hsi Chieh Ho Ta Chih 10(10):486-589.

Zhuo, D.H., 1982, Preventive geriatrics: An overview from traditional Chinese medicine. American J. Chinese Med. 10(1-4):32.

CHAPTER **10**

Nardostachys jatamansi DC.— The Indian Nard

A.K. Kukreja and *A.K. Singh*

INTRODUCTION

Medicinal plants constitute a group of industrially-important crops which hold a great value for domestic use as well as for export to many other countries. Plant-based drugs are being increasingly preferred in medical sciences with a growing tendency worldwide towards a shift from synthetic to natural-based products. Utilizing plant parts of several drug-based plants in curing specific ailments has been in vogue from ancient times in Ayurvedic, Unani and Siddha—the indigenous system of medicines in India. Apart from India, these systems of medicines are also prevalent in China, Korea and many other West and Southeast Asian countries. India is also fortunate to have a rich heritage of information available from tribal areas in the form of folklore medicine.

However, at present, many of these valuable and useful groups of plants are threatened with extinction because of over-exploitation and habitat destruction. Hence, there is immediate need to identify the natural plant wealth; developing agrotechniques for cultivation and production of these plants and their genetic improvement; and also to strengthen the research and development activities for developing new herbal products and their production technology.

The present chapter provides a comprehensive and condensed information on the genus *Nardostachys*, found and grown in different parts of India, China, Nepal and Bhutan covering the various important and interesting aspects.

BIOLOGY

Description of the plant

Nardostachys DC. (Family: Valerianaceae) is a small genus of two species (Rao et al., 1977; Shah, 1983; Thakur et al., 1989). *Nardostachys jatamansi* DC., native to India, is popularly known as 'Spikenard' or 'Indian nard' (Vernacular names: Sanskrit—*Jatamansi,* Hindi—*Jatamansi, bal - chir. Nardostachys grandiflora* DC. was kept as synonym under *N. jatamansi* for a long time. However, recently, both these varieties have been fixed as two distinct species (Rastogi and Mehrotra, 1998). Another species found in China and Japan and in use for a very long time in the Oriental system of medicine as the crude drug 'Chinese spikenard' has been chritstened as *Nardostachys chinensis,* commonly known as 'Kanshoko' in Japan. *Nardostachys* (Fig. 10.1) is an aromatic perennial herb (Satyawati et al., 1987). The part of the plant stem partly underground has a long thickened stout woody root stock generally enclosed with fibers (Fig. 10.2) originating from the petiole of the dead and withered leaves. The plant can grow upto 60 cm in height. The upper portion of the stem is hairy. Leaves develop from both the root stock and stem, differing from each other. Leaves, developing from the root stock, are entire, elongate, spathulate, 15 to 18 cm long and 1.5 to 2.5 cm wide, longitudinally veined and have a petiole, whereas those arising from the stem are in 1 or 2 pairs approximately 7-8 cm long and 2-3 cm wide, oblong or ovate in shape and sessile.

Rhizomes are short, thick and dark gray in color. Each rhizome is crowned with the reddish-brown coloured tufted fibrous remains of petioles of rootstock leaves. The inflorescence may have terminal capitate cluster frequently with broad ovate bracts or 3 to 7 terminal and axillary clusters. The flowers (Fig. 10.1) are rosy, pale pink or blue in dense cymes. The fruit is small with while hairs parallel to the main axis and is crowned by clayx teeth (Weberling, 1975), while the seeds are obovate and compressed.

Geographical distribution

Nardostachys generally grows in alpine Himalayas at an altitude of 2200 to 5000 m and in the geographical area from Punjab to Sikkim and Bhutan. It is also located in Tehri Garhwal (Uttaranchal, India) in the western regions of the Himalayan mountains to Yunnan and southwest Szechuan in the east and Tibet in the north, covering a large area of Nepal, Sikkim and Bhutan (Anonymous, 1970; Kihara, 1955; Kitamura, 1954; Nayar, 1964; Shah, 1994; Tiwari and Joshi, 1974; Weberling, 1975). In Nepal, the plant is located through the eastern and western regions of the Himalayas (Amatya and Staphit, 1994). The plants grow in dry open pine forests among dwarf *Rhododendron* and Juniper scrub, and on open stony and grassy slopes in alpine meadows or small depressions. Dripping moss-laden rocks and

Fig. 10.1 *Nardostachys jatamansi* in natural habitat.

moist boulders are the most preferred habitats of this plant (Airi et al. 2000).

Plant propagation and cultivation

The plant is propagated by cuttings of underground parts and occasionally also by seeds. The rhizomes are collected from wild plants. Collection of the plant in the wild is highly dependent on snowfall. Early snowfall and late spring snowfalls generally affect rhizome collection. Rhizomes can be harvested from wild and cultivated plants from June to October. However, at higher altitudes, the possible collection in the wild frequently lasts for only a few days due to heavy rains and snowstorms. Domestication from wild sources are being tested and the details of cultivation and actual economics are currently being worked out.

In vitro approaches in *Nardostachys*

In vitro studies on family Valerianaceae have concentrated mainly on cell culture for secondary metabolite production. Regeneration from callus and suspension culture has been reported for *Valeriana wallichi* (Mathur and Ahuja, 1991 and references cited therein; Mathur, 1992). However,

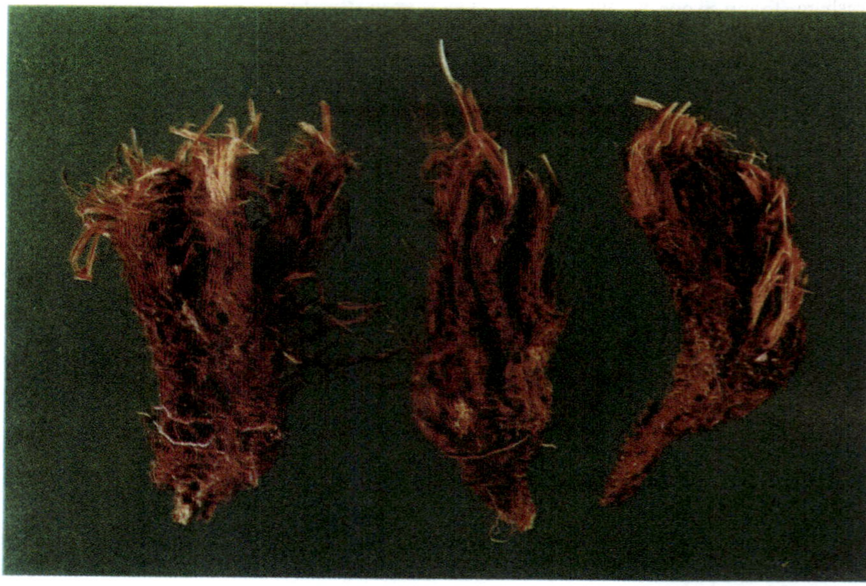

Fig. 10.2 Root stock of *N. jatamansi.*

there have been only a few tissue culture studies in *Nardostachys* (Mathur, 1992, 1993). Both the genera suffers from low seed viability and rely heavily on vegetative means of propagation. However, inspite of high potential for vegetative reproduction, large-scale uprooting of plants from the natural habitat in order to meet the domestic and international demand of pharmaceutical industry, both the genera have attained the status of endangered species and have been enlisted in Red data book of India (Jain and Sastry, 1980). Following the successful application of in vitro propagation techniques for a large number of threatened plants, the ex situ conservation strategy for *Nardostachys* includes tissue culture methods which hold a great promise. Mathur (1992, 1993) explored the regeneration potential of *N. jatamansi* roots to evolve a rapid propagation protocol for this plant species.

Callus initiation and regeneration of shoot buds from roots

Callus cultures of *N. jatamansi* DC. were initiated from petiole explants on Murashige and Skoog's medium containing 3.0 mg l^{-1} of α-naphthalene acetic acid (NAA) and 0.25 mg l^{-1} of kinetin (Kn). Profuse root formation was observed when petiole-derived callus was subcultured on 0.25-1.0 mg

l^{-1} of indole acetic acid (IAA) and indole-3-butyric acid (IBA). Root tips, approximately 1-2 cm long and 0.2-0.3 mm in diameter from callus-derived roots and from callus-regenerated somatic embryos (0.25-0.5 cm from apex) were subsequently placed on MS medium containing 2-6 mg l^{-1} Kn produced shoot buds on each media tested. The mode of regeneration, however, depends on the level of Kn present in the medium. Histological studies revealed that shoot buds originate either directly from superficial cortical layers or endogenously. Creamy white coloured root apices turned green within 10 days when cultured on medium containing 0.5-3.0 mg l^{-1} Kn and developed numerous protrusions on the surface mainly near the shrivelled root apex which eventually developed into organized shoot buds and, after 30 days, cluster of shoots were visible. Adventitious roots developed from the base of these regenerated shoots either on the same medium or when cultured on 1.0 mg l^{-1} IAA or IBA supplemented medium. Occasionally, shoot and root primordia developed from a common meristematic zone exhibiting axiate relationship. These bipolar structures differed from somatic embryos as they did not have cotyledonary leaves or any single root and could be easily excised and grown into complete plantlets. Such observations of shoot bud initiation response to exogenous cytokinin application have also been reported earlier in many instances, suggesting that the application of cytokinin can alter the polarity of shoot bud formation along the root axis. Root apex segments at higher levels of Kn (4.0-6.0 mg l^{-1}) swelled and formed fist like structures within 30 days of incubation and surface of explant showed numerous protrusions and in nearly 65% of the cultures, shoot clusters emerged from the explant surface. Adventitious roots were absent from such buds and complete plantlets could not be obtained from them. Endogenous shoot buds have, however, failed to produce roots either on a MS basal medium or on any other medium containing auxins and only reverted back to callusing.

Root segments excised from somatic embryos showed shoot bud regeneration response along with secondary somatic embryogenesis on the low Kn containing medium. Shoot bud regeneration was achieved only at higher concentrations > 4.0 mg l^{-1} to 6.0 mg l^{-1}. The secondary embryoids developed through typical embryogenic stages, lending credence to earlier views regarding secondary somatic embryogenesis which may be due to a weakening ionic polarity along the primary embryoid axis. About 15-20 shoot buds per root were obtained through direct regeneration from the cortical region of the explant and developed into complete plantlets on being transferred to a medium containing 1.0 mg l^{-1} IAA or IBA.

Histological observations revealed ill-defined vasculature in the in vitro regenerated endogenous shoot buds. Vascular connections with the original stele were missing from the cortical buds which terminated in the nests of tracheids in the explant cortex. The difference in origin of the shoot buds from similar explants upon being placed on media differing in their relative cytokinin content may be attributed to the degree of differentiation. Relatively superficial and less differentiated cells of the root cortex respond to growth regulator treatments of a low order while higher concentrations of the same growth regulator are required to elicit a similar response from more differentiated cells in the stelar region.

Somatic embryogenesis in callus cultures

Somatic embryogenic response of the cultured somatic cells is controlled by in vitro and in vivo environmental variables. The role of balance between the type and concentration of auxins and cytokinins at different stages of embryoid developments has been adequately demonstrated in numerous earlier investigations. Substantiating further the above inferences, Mathur (1993), in his study, highlighted the importance of sequential and gradual changes in the auxin:cytokinin ratio in the medium for achieving plant regeneration in *N. jatamansi.* Such gradual shifts in exogenously-supplied auxin:cytokinin has resulted in complete differentiation of plantlets in otherwise recalcitrant systems such as *Vitis, Asparagus, Panax,* etc. However, similar sequential treatments given to *Valeriana wallichii* callus obtained from petiole explants on the same levels of growth regulators could not elicit any embryogenic response.

Petiole explants cultured on MS medium containing Kn (1.16-2.32 μm) and NAA (2.69-2.15 μm) showed callusing response form the cut ends. Presence of kinetin in the medium though not essential for callus initiation, 1.16 μm Kn along with 16.1 μm NAA (hereafter referred as stock culture medium) in the medium supported optimal callusing from petiole and its subsequent proliferation. Lower levels of NAA (<16.1 μm) in the medium induced root formation in the callus.

Three distinct morphogenic responses were clearly evident when callus from the stock culture medium was cultured on various media combinations of auxins (IAA, IBA or NAA) and cytokinins (Kn, BA, 2ip, zeatin): (a) continuous unorganized growth of callus; (b) callus growth accompanied with rhizogenesis; and (c) development of dark green nodular structures 'embryoid' on the callus surface. Direct placement of callus on a range of auxin:cytokinin media elicited either rhizogenesis or differentiation of morphogenically-arrested green nodular structures, whereas sequential treatments resulted in directing the competent cells into a complete expression of their regenerative potential. Sequential transfer of callus over a gradual series of auxin:cytokinin ratio was highly effective in achieving regeneration in callus. Graded series of decreasing levels of NAA in combination with an increasing level of Kn were most effective than NAA-BA and NAA-2iP series. Green nodular structure differentiating on 5.37 μm NAA and 4.60 μm Kn developed into globular structure upon further transfer to medium containing 2.69 μm NAA and 4.60 μm Kn, which subsequently developed into somatic embryos with well-defined radicular and plumular ends on a medium supplemented with 1.35 μm NAA and 6.95 μm Kn. A further increase in cytokinin level to 9.30 μm allowed the development of complete plantlets from these somatic embryos at a frequency of 35-40% only. No regeneration response was elicited when the callus from stock cultures was directly transferred to media that provided somatic embryogenesis when used in sequence or to the basal medium. Transplantation of these somatic embryo-derived plantlets to the soil has not yet been achieved.

CHEMISTRY OF *NARDOSTACHYS*

The first report about the chemistry of *N. jatamansi* is about the isolation of a water insoluble crystalline acid named jatamanshic acid [1] and essential oil (Choudhary et al., 1951). In continuation, they further reported the structure of jatamanshic acid and chemical examination of essential oil (Choudhary et al., 1958 a, b). During the same period, isolation of a bicyclic sesquiterpene ketone named jatamansone [2] was also reported (Govindachari et al., 1958). Structure of jatamansone and another compound valeranone, isolated and characterized (Djerasi et al., 1961; Govindachari et al., 1959, 1960; Krepinsky et al., 1962; Kulkarni et al., 1964), was the subject of discussion until Klyne et al. (1964) established the absolute chemistry and found both the compounds to be the same. This compound was also present in *N. jatamansi* and *Valeriana officinalis* (Horster et al., 1977). The presence of at least 6 hydrocarbons in essential oil of Chinese spikenard subsequently resulted in elucidation of two compounds. β-maaliene [3] and calarene [4] (Buchi et al., 1962). The hydrocarbon calarene had also been isolated from sweet flag oil (*Acorus calamus*). Calarene was later on found to be a mixture of two compounds β-gurjunene [5] and aristolene [6] (Vrkoc et al., 1963). The presence of all these three compounds in *N. jatamansi* was further confirmed in a separate investigation (Pesnell and Ourisson, 1963). From the solvent extracted oil from the roots by refrigeration and elaborate column chromatography a new optically-active terpenic, coumarin jatamansin [8] and other identical to coumarin oroselol [7] were isolated from a grayish brown variety of Indian nard growing in Jammu and Kashmir (Shanbhag et al., 1964). The isolation of oroselol has also been reported earlier (Mitsuhashi and Itoh, 1962). In continuation of the examination of the oil, Shanbhag et al. (1965) further reported several hydrocarbons, a new oxide, alcohol jatamansinol [9], a polyoxygenated crystalline compound named angelicin [10] along with other known compounds, i.e. α-pinene, β-pinene, carene, β-eudesmol, elemol, β-sitosterol, angelecin and jatamansin. Jatamansinol (3'-hydroxy-3'-4'dihydroseselin) is also obtained by the hydrolysis of jatamansin with alkali. It is identical with lomatin obtained from *Lomatium nuttallii* (Soine and Jawad, 1964) and can be converted into dihydrojatamansin. Jatamansin is closely related to active vasodilatory drugs (visnadin, samidin, dihydrosamidin, pteryxin and suksdorfin). Detailed investigation of the essential oil of roots and rhizomes of *N. jatamansi* revealed the presence of isovaleric acid, hexacosanylarchidate, calarene, n-hexacosane, n-hexacosanyl-isovalerate, valeranone, valeranal, n-hexacosanol, β-sitosterol along with three new sesquiterpene nardol [11], calarenol [12] and nardostachone [13] (Pai et al., 1971; Rucker et al., 1978; Sastry et al., 1966, 1967 a,b,c; Saunders and Pinder, 1977). Calarenol was a new monoethynoid tricyclic tertiary alcohol named thus because of its similarity to calarene. Calarenol was found to be a mixture of two inseparable structural isomers [12]. The presence of a new ketone nardosinone [14] was reported from the essential oil of *N. chinensis* (Rucker, 1970 and references cited therein; Schulte et al., 1965, 1967) along with the previously reported ketone jatamansone from the ketonic fraction and 1(10) aristolene [15], 9-aristolene [16] in the hydrocarbon fraction. Maaliol [17], β-Maaliene [3], and jatamanshic acid were not present in the *N. chinensis* oil, while a sample of commercial oil showed their presence, indicating that it is a mixture of oils of *N. jatamansi* and *N. chinensis*. Rucker and his group (1971) identified many new aristolane type [18] sesquiterpene ketones, i.e. 1(10) Aristolene-2-one [19] 1,8,9,10 tetradehydroaristolan-2-one [20], 9 hydrodroxy aristolene-2-one (debilon) [21], 9-aristolene-1-α-ol [22] and 1,2,9,10 tetrahydroarist olene [23], along with β-ionone and previously- reported compounds; many of these compounds reported from *N. chinensis* were also reported from *N. jatamansi* (Rucker, 1968, 1969, 1971 and references cited therein; Rucker and Kretzschmar, 1971). A new trisnorsesquiterpenic diketoalcohol narchinol A ($C_{12}H_{14}O_3$) [24] was isolated in the form of yellow needles from methanol extract of roots and rhizomes of *N. chinensis* (Hikino et al., 1972), while gas chromatographic study led to the isolation of patchouli alcohol [25] and patchoulene [26] (Rucker and Kretzschmar, 1972), along with nine other constituents (Rucker et al., 1976). The presence of 9 aristolene [16] was confirmed, whereas valepotriates were not present. Nardosinone present in the oil of *N. chinensis* was converted into deoxonarchinol [29], a C12 compound (Rucker and Eberhard, 1974) and many other compounds present in the plant having C15 skeleton. Later norseychelanone, patchouli alcohol [25], seychellen [27] and α-and β-patchoulene were reported from *N. jatamansi* (Rucker et al., 1976). Seychellene [27], a unique sesquiterpene hydrocarbon reported to be present in *Pogostemon patchouli*, was isolated along with two epimeric forms of seychelane (dihydroseychellene) in *N. jatamansi* (Maheshwari and Saxena, 1974). An alkaloid actiridine [28] was isolated from the acetone extract of *N. jatamansi*, while this was reported to be absent in *N. chinensis* (Hirose et al., 1978). The presence of 9-aristolene 1-α-ol (nardostachnol) and 1(10) aristolene-2-one in the petroleum ether extract of Indian jatamansi was confirmed (Maheshwari et al., 1980). Five compounds—nordostachnol, 9-aristolene-1-α-ol [22], 9-10 aristolene, 1(10) aristolene, β-maaliene and 1,2,9,10 tetradehydro aristolene were isolated from both the species although their content varied (Han Dang et al., 1980). Work on fresh underground parts of *N. chinensis* growing in the province of Sichuan (China) has led to the isolation of a new instable aristolane type sesquiterpene ketone gansongone [29] (Shide et al., 1987 a,b,c) in addition to nordosinone [14], nordosinondiol [30], deoxonarchinol A [31], β-sitosterol, oleanolic acid and ethyl-β-D glucopyranoside on the basis of 2D NMR spectroscopy and they could revise the structure of previously known 9-aristolene -1 α-ol [22] to 1(10) aristolene 9-β-ol [22-A] which, on oxidation, converted into gansongone. They further reported a new sesquiterpene nardonoxide [32] from *N. chinensis*. In exhaustive investigations carried out on the chemistry of rhizomes of *N. chinensis*, Bagchi et al. (1988 a,b,c,d,e) isolated and reported several new compounds from dichloromethane extract: nardostachin ($C_{20}H_{32}O_6$) an iridoid [33], together with known triterpenoid ursolic acid, two new sesquiterpenoids kanshone A [34] ($C_{15}H_{22}O_2$) as colorless oil and kanshone B [35] ($C_{15}H_{22}O_4$) as colorless needles, highly oxygenated aristolene type sesquiterpenoid kanshone C ($C_{15}H_{20}O_3$) [36]. Further reinvestgation of the methylene chloride extract of this plant material has resulted into two minor constituents: kanshone D [37] ($C_{15}H_{22}O_4$) and kanshone E [38] ($C_{15}H_{20}O_4$) in addition to major constituent nardosinone, isonardoisonone [39], nardiosnonediol and nardofuran [40]. The advancement of structures for kanshone D and E were mainly based on spectral data and chemical transformation. The last two of these were earlier reported as reaction products of nordosinone. As an extension of work on valerianaceous plants, phytochemical investigation of the underground parts of Indian spikenard *N. jatamansi* revealed the presence of many new compounds: one new neolignan [41] together with two known neolignans [42, 43], two lignans (Bagchi et al., 1991). One of the neolignans was new erythro-1-(3,4-dimethoxy phenyl)-2-(2 methoxy-4(E) propenyl phenoxy)propan-1-ol [41], which was diasteromer of virolin [42], another neolignan also present in the plant. The third neolignan identified was erythro-1-(4-hydroxy-3-methoxyphenyl)-2-(2 methoxy-4(E) propenylphenoxy-propan-1-ol [43]. Acyclic bisphenyl propanoids were till date isolated exclusively from the plants belonging to the family Myrsticaceae. It is worth mentioning that valerianaceous plant *N. jatamansi* is another new possibility to obtain this class of compounds together with lignans. The lignan were (+) pinoresinol [44] and (+)-1-hydroxypinoresinol [45]. A sesquiterpene ($C_{15}H_{26}O$) having novel spiranic skeleton spirajatamol [46] and two eudesmane sesquiterpene jatamols A

[47] (viscous oil $C_{15}H_{24}O$) and B [48] (viscous oil; $C_{20}H_{32}O_3$) along with valeranone, patchouli alcohol were also isolated from dichloromethane extract of the same plant species (Bagchi et al., 1998 b, 1990, 1991). During the survey of novel cytotoxic antitumour compounds from medicinal plants, methanolic extract of *N. chinensis* exhibited cytotoxic activity against P-388 cells. Chromatographic purification of the compounds exhibiting cytotoxicity led to the isolation of some new sesquiterpenes: desoxonarchinol A [49], nardosinone, debilon, nardosinonediol and kanshone A (Rucker, 1970, 1971), all having cytotoxic activity to P-388 cells (Itokawa et al., 1993). Nardosinone acts as a novel enhancer of nerve growth factor in neurite outgrowth from PC 12 D cells (Li et al., 1999). The steric structure of [49] was determined by nuclear overhauser effects (NOEs) and excitation chirality method. One new sesquiterpene possessing eudesmane structure eudesm-11-en-2,4 diol [50] also having cytotoxic activity to P-388 cells was isolated from *N. jatamansi.* Three new quaiane type sesquiterpenoids nardoperoxide [51], isonardoperoxide [52] and nardoxide [53] have been isolated from *N. chinensis* (Takaya et al., 1998). Amongst these three compounds, the first two showed antimalarial activity against *Plasmodium falciparum.* A novel compound BR-606 [54] patented by Kawashima et al. (1994) was found to be bone sorption inhibitor which is useful for oesteoporosis and hypercacemia. Nardosaldehyde, a new peroxide sesquiterpene [55], was isolated and structure was determined by 2D NMR (Shide et al., 1997).

Takaya et al. (2000 a,b) isolated and elucidated nine novel guaiane endoperoxides, Nardoguaianone A-I from the roots of *Nardostachys chinensis* and also evaluated the antinociceptive and antimalarial activities of constituents of *N. chinensis,* including some of these novel compounds. Further, the same group isolated two more guaiane-type compounds, nardoguaianone J and K and two aristolane-type compounds kanshone F and G (Tanitsu et al., 2002).

Essential oil

On hydrodistillation, roots and rhizomes of *N. jatamansi* yield an essential oil (Spikenard oil) with a peculiar pleasant, sweet and spicy aroma, suggestive of patchouli and *Valeriana.* The oil flavour is warm and spicy with a bitter burning power. The site and time of collection of plant material, type and maturity of rhizomes greatly affect the yield, aroma character and quality of the essential oil (Amatya and Staphit, 1994). Plant material collected at different locations differs in oil content. Oil yield varies greatly from 0.5% to 1.9% of rhizome dry weight. The oil resinifies on being exposed to air. Low yield occurs because of the loss of the fine rhizome hairs during harvest, drying and storage. The volatile oil cells are generally located on the fine fibrous hairs of the rhizome. Rupture of some

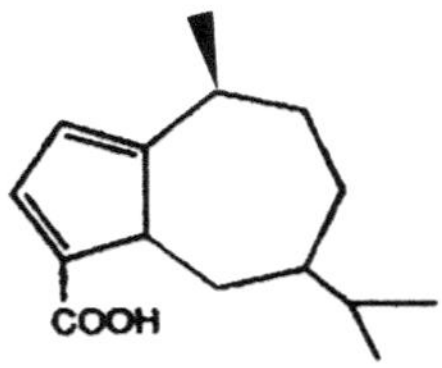

Jatamanshic acid [1]

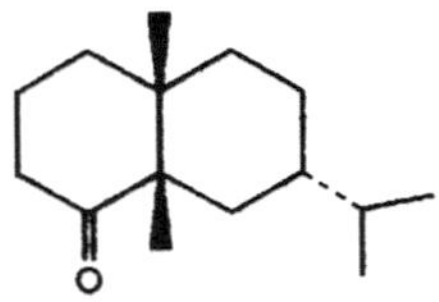

Valeranone (Jatamansone) [2]

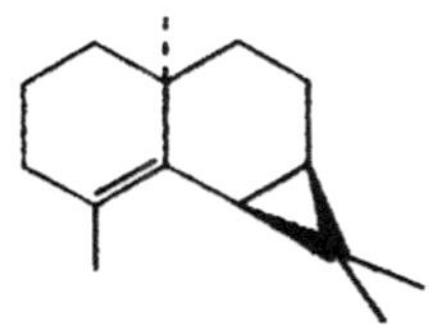

β Maaliene [3]

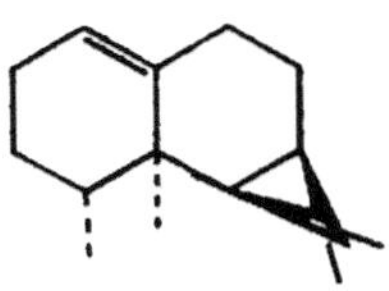

Calarene [4]

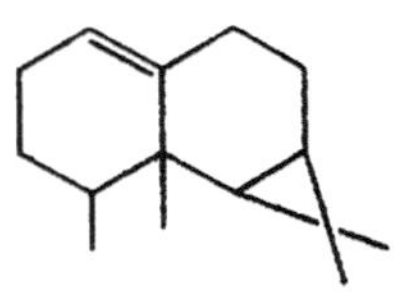

b Gurjunene [5]

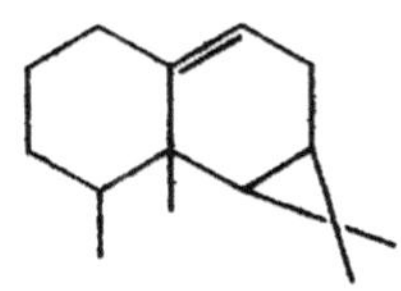

Aristolene [6]

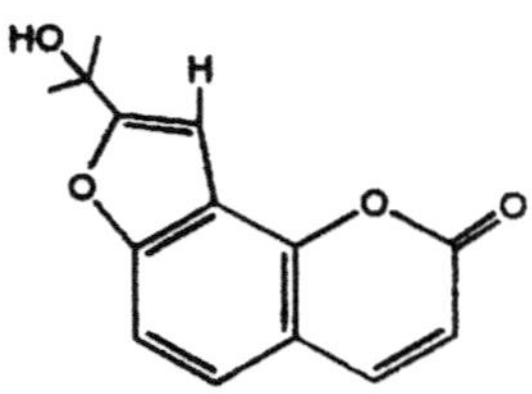

Oroselol [7]

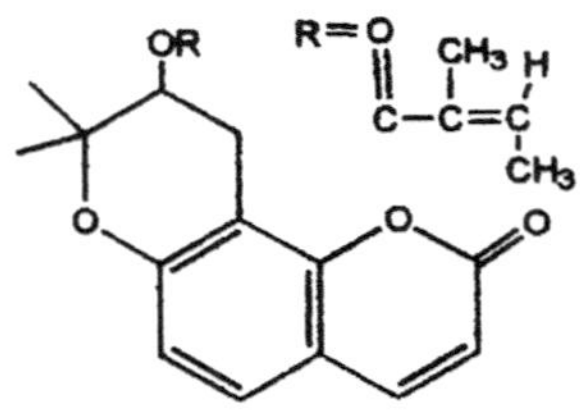

Jatamansin [8]

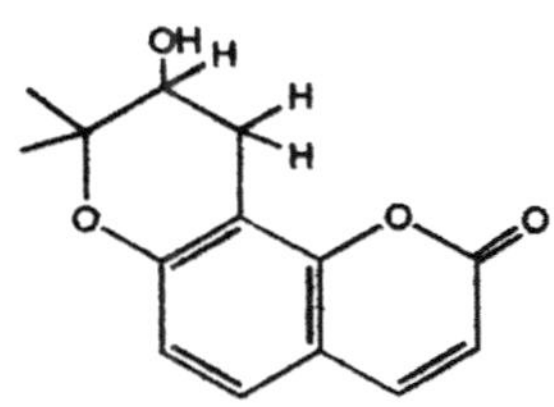

Jatamansinol [9]

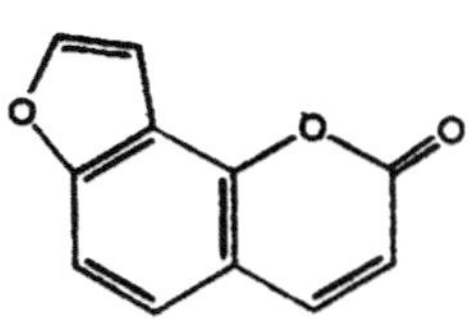

Angelicin [10]

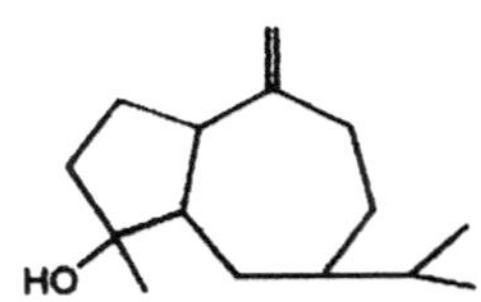

Nardol [11]

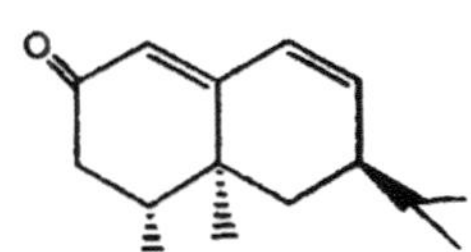

Nardostachone [13]

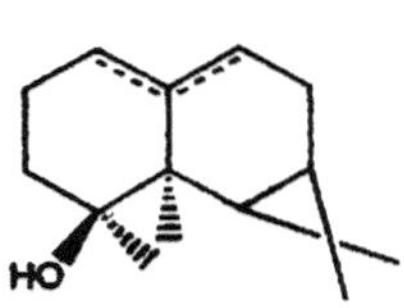

Calarenol [12]

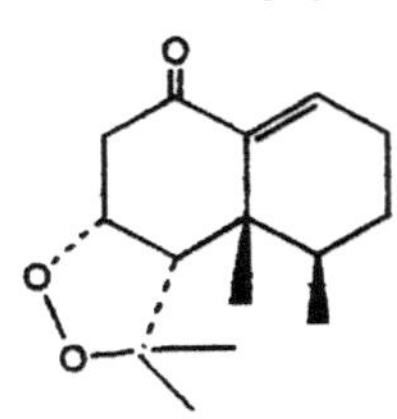

Nardosinone [14]

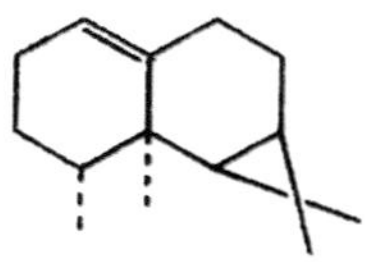

1 (10)Aristolene [15]

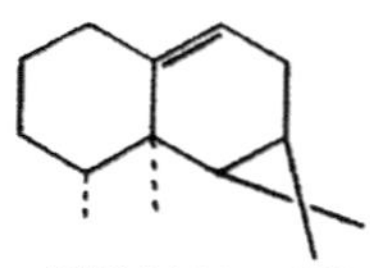

9(10) Aristolene [16]

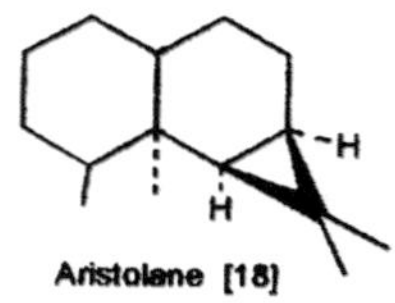

Aristolane [18]

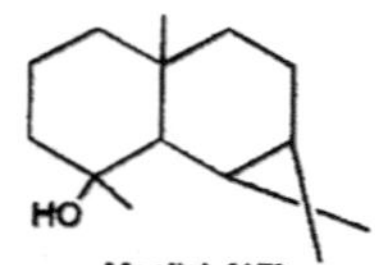

Maaliol [17]

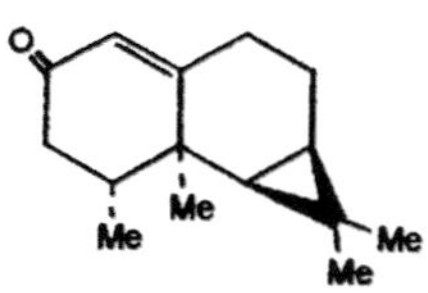

1(10)Aristolene 2-one [19]

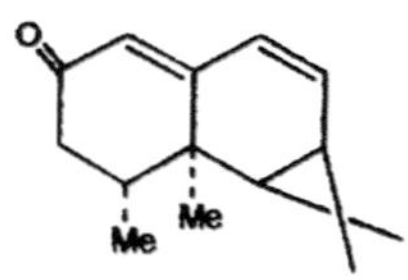

1,8,9,10-Tetradehydro aristolene-2-one [20]

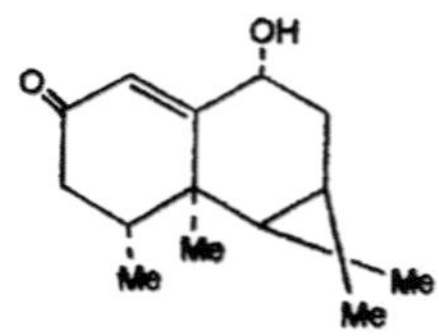

9 Hydroxy-$\Delta^{1(10)}$ aristolene- 2 one (debilon) [21]

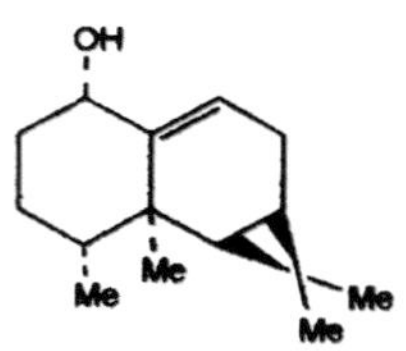

9-Aristolene-1-α-ol (Nardostachnol) [22]

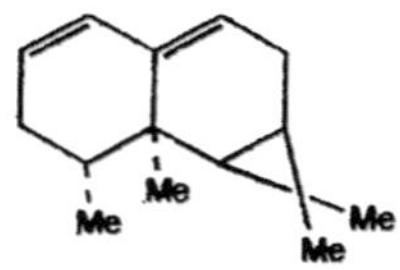

1,2,9,10 Tetradehydro aristolene [23]

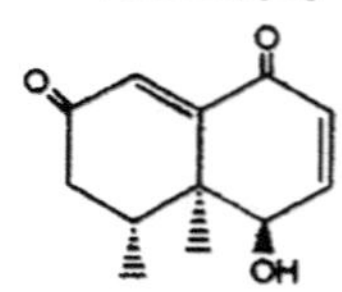

Narchinol A [24]

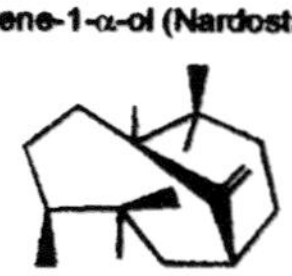

Seychellene [27]

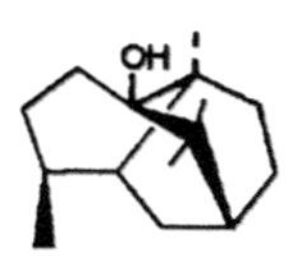

Patchouli alcohol [25]

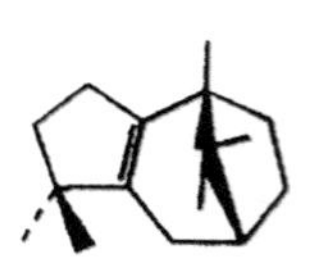

β Patchoulene [26]

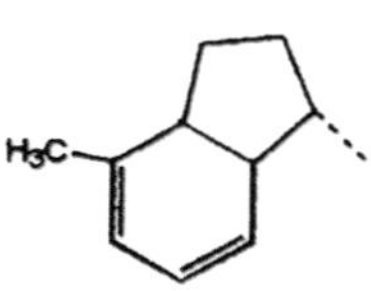

Actiridine [28]

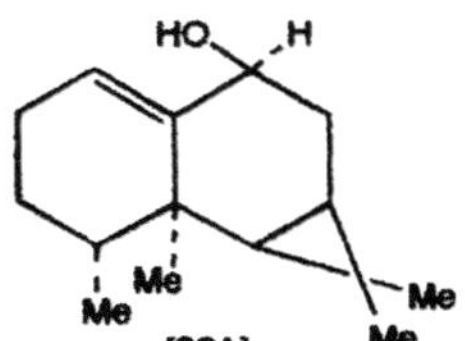

[22A]

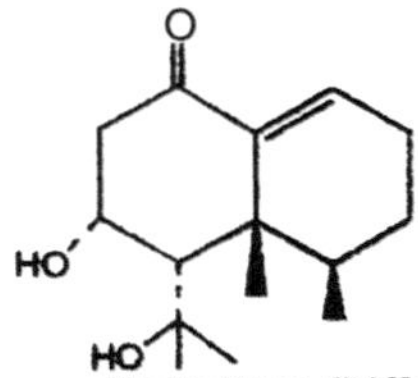

Nardosinone diol [30]

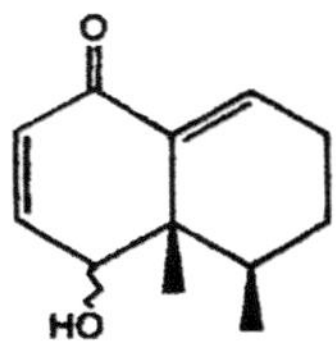

Deoxo narchinol-A [31]

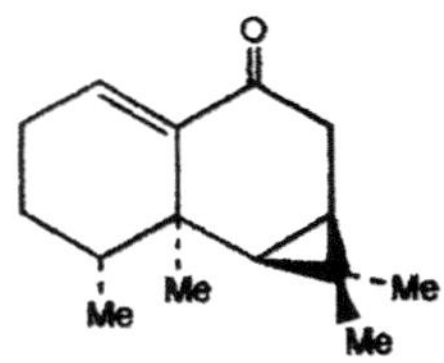

Gansongone [29]

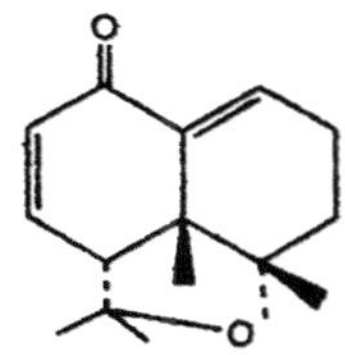

Nardonoxide [32]

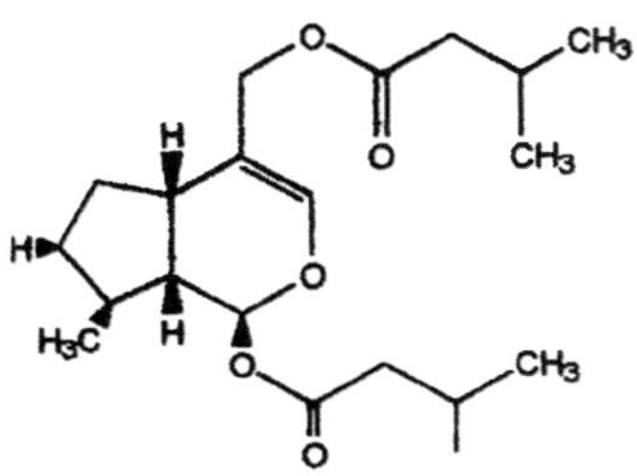

Nardostachin [33]

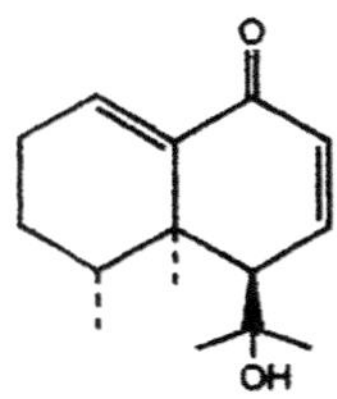

Kanshone A [34]

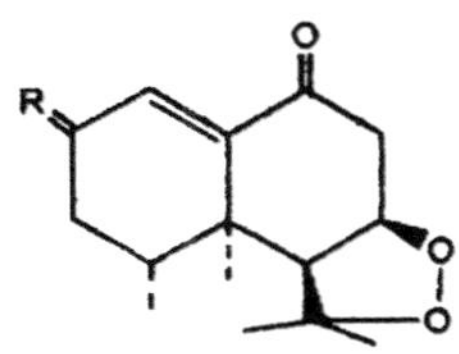

R=α-H, β-OH

Kanshone B [35]

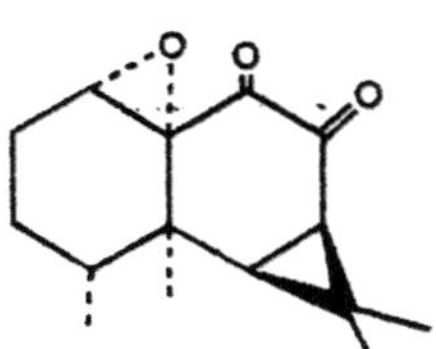

Kanshone C [36]

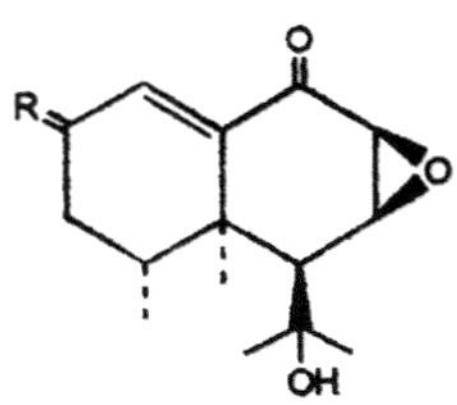

R = α-H, β-OH
Kanshone D [37]

R= O, Kanshone E [38]

R = H_2 Isonardesinone [39]

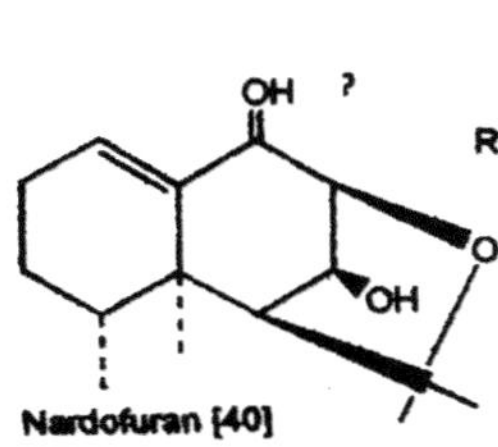

Nardofuran [40]

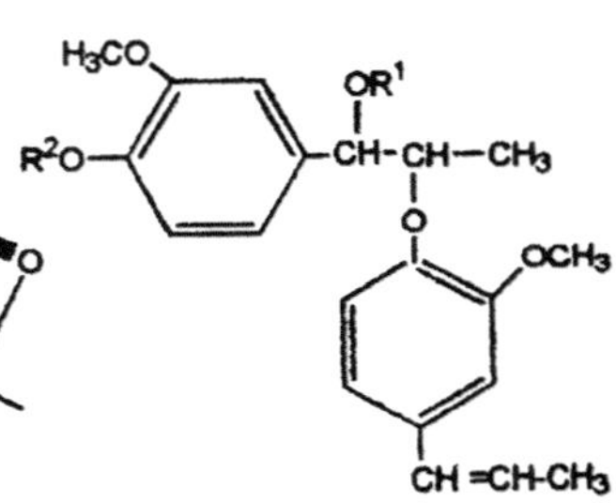

R1 = H, R2 = CH3 (erythro) [41]

R1 = H, R2 = CH_3, virolin. (threo) [42]

R =H, R2 = H (erythro) [43]

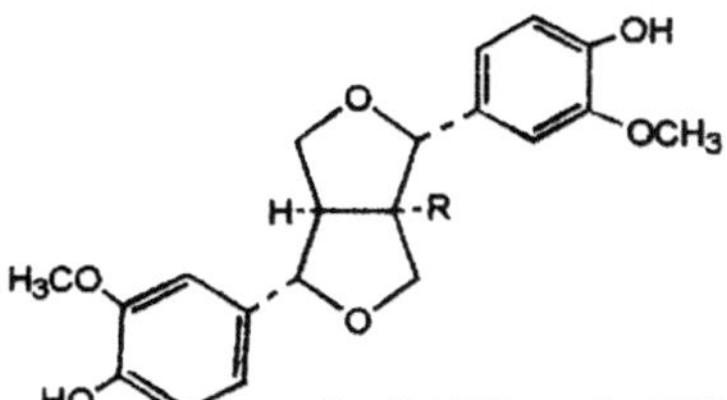

R = H (+) Pinoresinol [44]
R = OH 1-hydroxypinoresinol [45]

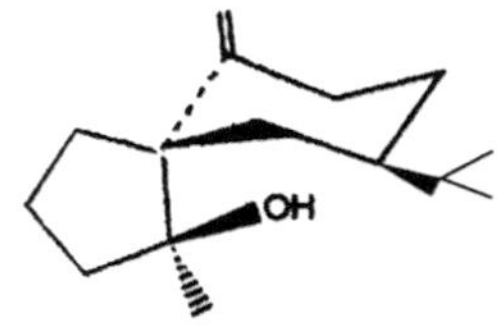

Spriojatamol [46]

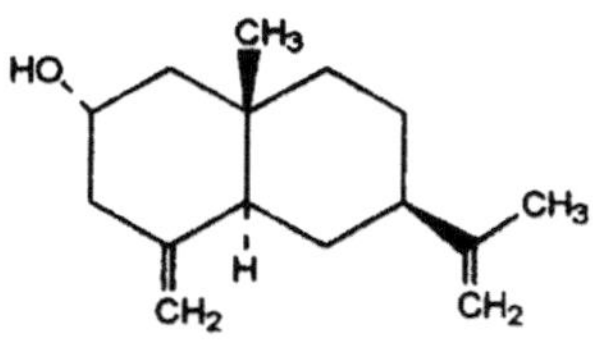

R =H = Jatamol A [47]

R = OH- $C(CH_3)_2$-CH_2 - C=O Jatamol B [48]

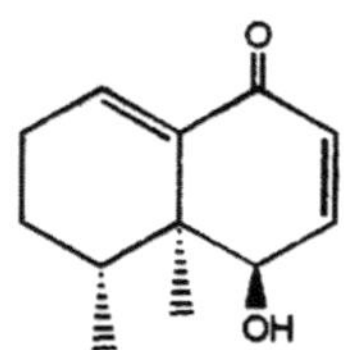

Desoxo narchinol A [49]

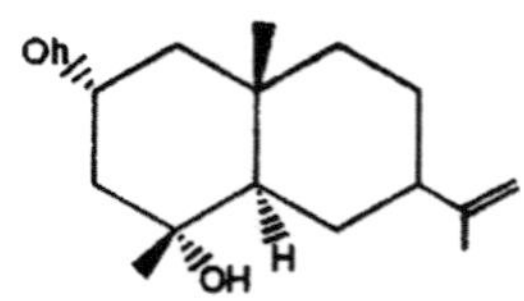

Eudesm-11-en 2,4 α diol [50]

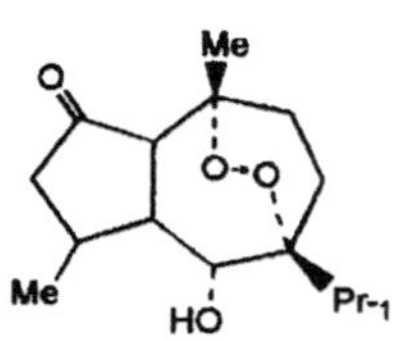

Nardoperoxide [51]

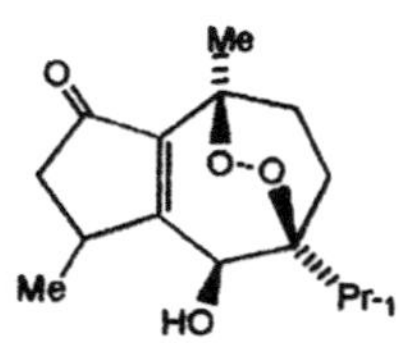

Iso nardoperoxide [52]

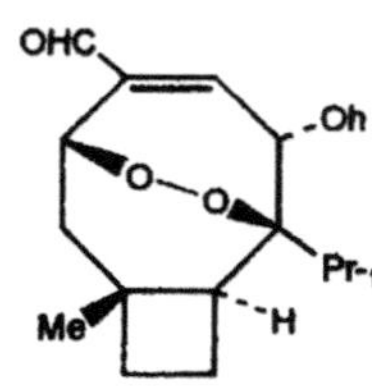

Nardosaldehyde [55]

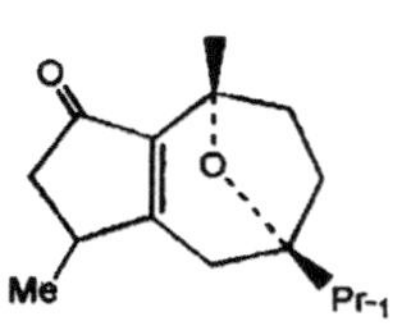

Nardoxide [53]

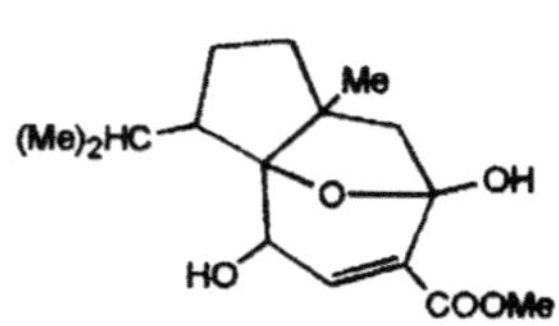

BR-606 [54]

of these oil cells/glands with a consequent loss of volatile oil during the preparation of drug also results in low oil yield. However, under controlled conditions, oil yields greater than 2% have also been reported. Tests in a pilot processing plant indicate that extractable volatile oil can reach a maximum of 2.9% of dry weight of rhizome over a 15-hour distillation period. Rhizomes harvested at later stages of maturity (2-3 year-old plants) possess a relatively higher essential oil content as compared to younger plants. Oil colour (yellow to light amber, greenish blue-blue) varies with the site of collection and maturity and nature of rhizomes. Use of copper distillation apparatus promotes brownish green oil colour, while distillation in glass yields a blue coloured oil and distillation in stainless steel makes a greenish blue coloured oil. No commercial or pharmaceutical standards for the essential oil of *N. jatamansi* have been prescribed. The trade relies entirely on the basis of conformance to traditionally-accepted physico-chemical variables (Table 10.1) and on an acceptable characteristic aroma and flavour. The Indian oil is d-rotatory, while oil from Japan is reported to be l-rotatory. The unavailability of a uniform quality has always created uncertainty in the international market (Okuda, 1991).

Table 10.1 Physico-Chemical properties of spikenard oil

Colour	light yellow to greenish blue
Acid value	2.16
Ester value	17.35
Ester value after acetylation	32.76
Cabonyl content	22.76%
Specific gravity at 17°C	0.9649
Optical rotation at 17°C	20° 20°
Refractive index at 17°C	1.5078
Solubility in alcohol	0.15

Extracts of oil from rhizomes of *N. jatamansi* have proven to be the source of many sesquiterpenes. The major constituent of oil are d-nardostachone and valaronone, while the minor constituents were identified as α-pinene, β-pinene, limonene, 1,4-cineol, linallool, α-gurjunene, β-gurjunene, *trans* pinocarveol, β-bisabolene, α-curcumene, β-ionone, and humulene oxide, iso-valeric ester of alcohol, jatamansi acid. GC-MS analysis of oil of two different locations with two different instruments reported — patchoulene, aromadendrene, and δ-cadinene in addition to jatamansone and d-nardostachone. Other reported constituents are aristolene, 8-hydroxypinoresinol, jatamol A & B, lomatin, maaliol, nordol, pinoresinol, selinidin, spirojatamol. Recently, Qim et al. (1999) have reported the extraction of essential oil and an analysis of its chemical constituents. They identified 48 components amounting to about 82% of the total quantity of the oil extracted from *N. chinensis* while Mahalwal and Ali (2002) identified

41 constituents comprising 70% of oil by capillary GC and GC-MS from steam distilled oil of Indian nard (*N. jatamansi*).

Spikenard oil possesses fungitoxic (Misra et al., 1995), nematicidal (Saxena et al., 1987), antiarrhythmic (Arora and Madan, 1956) activity with possible therapeutical usefulness in cases of auricular flutter, it is less effective and simultaneously less toxic than quinidine. Jatamansone is more potent than the oil. The oil exerts a hypotensive effect (Arora et al., 1958) and in moderate doses, it has a distinct depressent action on the central nervous system. Lethal doses cause deep narcosis and ultimately, death within a few hours.

THERAPEUTICS IN FOLK AND TRADITIONAL MEDICINES: DRUG AND BIOLOGICAL ACTIVITIES

The drug, commercially known as jatamansi or Bal chhad, consists of rhizomes and roots from the plant and has an agreeable aroma with a bitter aromatic taste. The underground parts, i.e., roots and rhizomes of *N. jatamansi* have been used in Indian traditional medicines such as Ayurveda and Unani (Nadkarni, 1954; Satyawati et al., 1987; Ahmad et al., 1993) as a bitter tonic, which acts as a stimulant for the cardiac, respiratory and nervous systems, antispasmodic, carminative, stomachic, laxative, diuretic, antiseptic and emmenagouge. The decoction of the roots of this plant have been used for centuries in Ayurveda for the treatment of epilepsy, hysteria, convulsions, mental disorders and diseases of blood and circulatory systems (Debelmas and Hache, 1976; Hamied et al., 1962; Mukherjee, 1961; Nadkarni, 1954). In addition, the plant has been used as a curative for dysmenorrhoea, flatulence, excessive urination, jaundice and several other disorders of the digestive system. In traditional systems of medicine, this drug is used alone or in combination with other drugs. When used as an anthelmintic, the plant material is prepared with a purgative. Extracts are generally prepared as an infusion in doses of 1-2 ounces 3 times per day for medical applications. Besides, *N. jatamansi* is also valued for its cosmetic uses. It is well known as a hair tonic and is an ingredient of many washes and hair oils. Pulverized roots of *N. jatamansi* along with extracts of walnut and certain other ingredients have been used for a herbal formulation for treatment of human nails (Bindra et al. 2001). Jatamansi is occasionally used as a substitute for valerian (*Valeriana officinalis* Linn.). The roots of *Valeriana* and *Cymbopogon schoenanthus,* a aromatic grass are often mistaken for jatamansi. Roots and rhizomes of *Selenium vaginatum* C.B. Clarke have been found to be adulterants of jatamansi.

The roots and rhizomes of *Nardostachys chinensis* Batalin, commonly known as 'Kanshoko' in Japan have been used in the crude drug Chinese spikenard in the Oriental system of medicine for sedative, aromatic, and stomachic purposes. A literature survey has shown that so far, mainly the isolation and structure elucidation of sesquiterpenoids from the rhizomes and roots of the plant have been reported. A new iridoid named Nardostachin, isolated from dichloromethane extract, was screened for its possible biological activity and it was found that at the dose of 0.1-1.0 mg/ml in the culture medium, it exerted a weak antihepatoxic activity in the carbon tetrachloride-induced cytotoxicity model system employing primary cultured rat hepatocytes, whereas at similar doses, it exhibited a pronounced protective activity against D-galactosamine produced liver damage in the rat hepatocytes (Kiso et al., 1983 a, b).

In the early stages, sesquiterpenoids kanshones A and B isolated from the methylene chloride extract of rhizomes and roots of Japanese Kanshoko were examined for their possible biological activities and it was found that kanshone B at the dose of 0.01-1 mg/ml in the culture medium exerted a weak antihepatoxic activity in the complement-mediated cytotoxicity model system employing primary cultured mouse hepatocytes (Kiso et al., 1987). However, kanshone C, at a dose of 0.1-1.0 mg/ml in the culture medium, showed a remarkable protective activity against D-galactosamine induced liver damage in rat hepatocytes. Eudesm-11-en-2,4-diol isolated from methanolic extract of *N. chinensis* showed moderate cytotoxic activity against P-388 cells with an IC 50 value of 12 μg ml^{-}1(Masuyama et al., 1993). A scientific and systematic evaluation (Prabhu et al., 1994) of the effect of *N. jatamansi* extract on the activity of the rat brain demonstrated that the acute and subchronical administration has brought about changes in monoamines as well as inhibitory amino acids, suggesting that a balanced concentration between different neurotransmitters is essential to increase the threshold against maximal electroshock seizures (MES).

Roots and rhizomes of *N. jatamansi* DC on stem distillation yield a light yellow peculiar smelling essential oil. This volatile oil possesses less antiarrhythmic activity than jatamansone. The active ketonic sesquiterpenes have reserpine and chlorpromazine-like activity and pronounced and prolonged hypotensive activity (Arora, 1965). This essential oil is also known to have antibacterial, fungitoxic (Girgune et al., 1978), and insect repellent activity (Dixit et al., 1992). Besides, it is also used in the perfumery industry (Han Dong et al., 1980) and has a tranquillizing effect when massaged onto the head. It is also applied on leprous wounds.

REFERENCES

Ahmad, R., Sheikh, T.U., Ahmad, A. and Ahmad, M., 1993, Medicinal importance of essential oils. Hamdard Medicus, 36:101-115.

Airi, S., Rawal, R.S., Dhar, V. and Purohit, A.N., 2000, Assessment of availability and habitat preference of jatamansi - a critically endangered medicinal plant of west Himalaya. Current Science 79: 1467-1471.

Amatya, G. and Staphit, V.M., 1994, A note on *Nardostachys jatamansi.* Journal of Herbs, Spices and Aromatic Plants, 2:39-47.

Anonymous, 1997, Medicinal plants of Nepal. No. 3. Department of Medicinal Plants. H.M.G. of Nepal. pp.54-55.

Arora, R. B. and Madan B.R., 1956, Antiarrhythmics III. Antiarrhythmic activity of *Nardostachys jatamansi* (an Indian indigenous drug). Indian J. Med. Res. 44:259-269.

Arora, R.B., Singh, K.P., Das, P.K. and Mistry, P.N., 1958, Prolonged hypotensive effect of the essential oil of *Nardostachys jatamansi.* Arch. Inter. Pharmacodynamic. 113:367-376.

Arora, R.B., 1965, *Nardostachys jatamansi:* A chemical, pharmacological and clinical appraisal. Indian Council of Medical Research, New Delhi.

Bagchi, A., Oshima, Y. and Hikino, H., 1988a, Nardostachyn, an iridoid of *Nardostachys chinensis.* Planta Medica, 54:87-88.

Bagchi A., Oshima, Y. and Hikino, H., 1988b, Jatamols A and B: sesquiterpenoids of *Nardostachys jatamansi* roots. Planta Medica, 57:282-283.

Bagchi, A., Oshima, Y. and Hikino, H., 1988c, Kanshones A and B, sesquiterpenoids of *Nardostachys chinensis.* Phytochemistry, 27:1199-1201.

Bagchi A., Oshima, Y. and Hikino, H., 1988d, Kanshone C, a sesquiter penoid of *Nardostachys chinensis* roots. Phytochemistry, 27:2877-2879.

Bagchi, A., Oshima, Y. and Hikino, H., 1988e, Kanshone D and E sesquiterpenoids of *Nardostachys chinensis* roots. Phytochemistry, 27:3667-3669.

Bagchi, A., Oshima, Y. and Hikino, H. 1990, Spirojatamol, a new skeletal sesquiterpenoid of *Nardostachys jatamansi* roots. Tetrahedron, 46:1523-1530.

Bagchi, A., Oshima, Y. and Hikino, H., 1991, Neolignans and lignans of *Nardostachys jatamansi* roots. Planta Medica, 57:96-97.

Buchi, G., Greuter, F., and Tokoroyama, T., 1962, Terpenes XVIII. Structure of calarene and stereochemistry of aristolone. Tetrahedron. Lett., 18:827-833.

Bindra, R.L., Singh, A.K., Shawl, A.S., and Kumar, S., 2001, Antifungal herbal formulation for treatment of human nails fungus and process thereof. US Patent 6,296,838.

Choudhary, G.R., Sharma, V.N., and Salimuzzaman Siddiqui., 1951, Chemical constituents of *Nardostachys jatamansi* I. isolation of a crystalline acid and an essential oil. Jour. Sci. Ind. Res., 10B:48.

Choudhary, G.R., Sharma, V.N. and Kaul, K.N., 1958a, Chemical examination of the essential oil from *Nardostachys jatamansi.* Jour. Sci. Ind. Res. 17B:159-162.

Choudhary, G.R., Sharma, V.N. and Kaul, K.N., 1958b, Chemical examination of the essential oil from *Nardostachys jatamansi.* Jour. Sci. Ind. Res., 17B:473-474.

Debelmas, A.M. and Hache, J., 1976, Pharmacological study of some medicinal plants of Nepal. Acute toxicity behavioural study and effect on the central nervous system. Plantes Medicinales et Phytotherapie, 10:128-138.

Dixit, V., Chaturvedi, R.V. and Tripathy, S.C., 1992, Evaluation of some essential oils against pulse beetle (*Callosobruchus chinensis*). Nat. Acad. Sci. Lett., 15 : 255-257.

Djerasi, C., Govindachari, T.R., Pai, B.R. and Purushothaman, K.K., 1961, Determination of the absolute configuration of jatamansone (valeranone) by the axial α-haloketone rule. Tetrahedron Lett. 226.

Girgune, J.B., Jain, N.K. and Garg, B.D., 1978, Antifungal activity of some essential oils-II. Current Science, 47: 454-455.

Girgune, J.B., Jain, N.K. and Garg, B.D., 1978, Antifungal activity of some essential oils-II. Indian Drugs 16:24.

Govindachari T.R., Rajadurai, S., and Pai, B.R., 1958, Structure of jatamansone. Chem. Ber. 91: 56-58.

Govindachari,T.R., Pai, B.R., Purushothaman, K.K. and Rajadurai, S., 1959, Jatamansone. Tetrahedron Lett., 5.

Govindachari, T.R., Pai, B.R., Purushothaman, K.K. and Rajadurai, S., 1960, Structure of jatamansone. Chem. Ind., 1059.

Hamied, K.A., Bakshi V.M. and Aghara L.P. 1962, Pharmacological investigation of *Nardostachys jatamansi.* Jour. Sci. Ind. Res. 21 C:100-103.

Han Dong Sun, Ding, Jing-Kai, Zhong-Wen, and Che Fang Rong., 1980, Study on the chemical constituents of the essential oil of *Nardostachys grandiflora* DC and *N. chinensis* Batalin and their uses on the perfume. Yun nanchih Wu Yen Chin 2:213-223.

Hikino, H., Hikino, Y., Koakutsu, S. and Takemoto, T., 1972, Structure and absolute configuration of Narchinol A. Phytochemistry, 11:2097-2099.

Hirose, Y., Yonemitsu, R. and Sonada, T., 1978, Chemical studies on the components of *Nardostachys jatamansi* de Candolle. Shoyakugaku Zasshi 32:121-122.

Horster, H., Rucker, G. and Tautges, J., 1977, Valeranone content in the roots of *Nardostachys jatamansi* and *Valeriana officinalis.* Phytochemistry, 16:1070-1071.

Itokawa, H., Masuyama, K., Morita, H. and Takeya, K. 1993, Cytotoxic sesquiterpenes from *Nardostachys chinensis.* Chem. Pharm. Bull., 41:1183-1184.

Jain, S.K. and Sastry, A.R.K., 1980, Threatened Plants of India—A state of the Art Report. Botanical Survey of India, Calcutta.

Kawashima, A., Kishmito, M., Morimoto, S., Akyama, T., Maegima, A. and Kawada, I., 1994, Extraction of a novel compound BR-606 from *Nardostachys jatamansi* roots as bone sorption inhibitor for treatment of osteoporosis and hypercalcemia. Jpn. Kokai Tokkyo Koho JP 0841068.

Kihara, H., 1955, Fauna and flora of Nepal Himalayas. Scientific result of the Japanese expedition to Nepal Himalayas 1952-53. Fauna and Flora Research Society, Kyoto University, Kyoto, Japan, pp. 235-236.

Kiso, Y., Tohkin, M. and Hikino, H., 1983a, Assay method for antihepatotoxic activity using carbon tetrachloride induced cytotoxicity in primary cultured hepatocytes. Planta Medica, 49:222.

Kiso, Y., Tohkin, M. and Hikino, H., 1983b, Assay method for antihepatotoxic activity using galactosamine induced cytotoxicity in primary cultured hepatocytes. Jour. Nat. Prod. 46:841-847.

Kiso, Y., Kawakami, Y., Kikuchi, K. and Hikino, H., 1987, Assay method for antihepatotoxic activity using complement mediated cytotoxicity in primary cultured hepatocytes. Planta Medica, 53:241-247.

Kitamura, S., 1954, New species from Nepal collected by Sasuke nakao, a member of the Himalayan expedition III. Acta Phytotax. et Geobot. 15:129-134.

Klyne, W., Bhattacharya, S.C., Paknikar, S.K., Narayan, C.S., Kulkarni, K.S., Krepinsky, J., Romanuk, M., Herout, and V., Sorm F., 1964, The absolute stereo chemistry of the sesquiterpene ketone valeranone. Tetrahedron Lett., 23:1443-1448.

Krepinsky, V., Romanuk, M., Herout, H., and Sorm, F., 1962, On terpenes. CXLII. Structure of the sesquiterpene ketone valerone. Colln. Czech. Chem. Commun, 27:2638.

Kulkarni, K.S., Paknikar, S.K., and Bhattacharya, S.C., 1964, Structure and stereochemistry of hydroxy valernone and acetylhydroxy valeranone. Tetrahedron. Lett., 23:1289-1300.

Li, P., Matsunaga, K., Yamamoto, K., Yoshikawa, R., Kawashima, K., and Ohizumi, Y., 1999, Nardosinone, a novel enhancer of nerve growth factor in neutrite outgrowth from PC12 D cells. Neurosci. Lett., 273:53-56.

Mahalwal, V. and Mohamad A., 2002, Volatile constituents of the rhizomes of *Nardostachys jatamansi.* Journal of Essential Oil bearing Plants 5: 83-89.

Maheshwari, M.L. and Saxena, D.B., 1974, Seychellene and seychellone from *Nardostachys jatamansi.* Indian J. Chem., 12:1221.

Maheshwari, M.L. and Saxena, D.B., 1980, 9-aristolen-1-α–ol and 1(10)-aristolen-2-one from *Nardostachys jatamansi* D.C. Indian Perfumer 24:171-174.

Masuyama, K., Morita, H., Takeya, K. and Itokawa, H., 1993, Eudesm-11 -en-2,4α-diol from *Nardostachys chinensis*. Phytochemistry, 34:567-568.

Mathur, J. and Ahuja, P.S., 1991, Plant regeneration from callus of *Valeriana wallichii* DC. Plant Cell Reports, 9:523-526.

Mathur, J., 1992, In vitro morphogenesis in *Nardostachys jatamansi* DC. Shoot regeneration from callus derived roots. Annals of Botany. 70:419-422.

Mathur, J., 1993, Somatic embryogenesis from callus cultures of *Nardostachys jatamansi*. Plant Cell Tissue and Organ Culture, 33:163-169.

Mishra, D., Chaturvedi, R.V. and Tripathy, S.C., 1995, The fungitoxic effect of the essential oil of the herb *Nardostachys jatamansi* DC. Tropical Agriculture, 72:48-52.

Mitsuhashi, H. and Itoh, T., 1962, Constituents of Umbelliferae plants VI. Constituents of *Angelica edulis*. Chem. Pharm. Bull. (Japan), 10:514-518.

Mukerjee, B., 1961, Pharmacology and chemistry of *Nardostachys jatamansi* Pakistan Jour. Sci. Ind. Res., 4:277-279.

Nadkarni, A.K., 1954, The Indian Materia Medica, New Delhi. Vol. I, 840 pp.

Nayar, S.L., 1964, Medicinal plants of commercial importance found wild in Uttar Pradesh and their distribution. Jour. Bombay Nat. Hist. Soc., 61:651-661.

Okuda, I., 1991, Analysis of *Nardostachys jatamansi* oil. Final report. Screening of Plants Bearing Essential Oil. Department of Forestry and Plant Research, Thapathalli, Kathmandu, Nepal., pp.21-28.

Pai, P.P., Chakravarti, K.K. and Paknikar, S.K., 1971, Revised structure for nardostachone: identify with 1,8,9,10-tetrahydroaristlane-2-one. Indian Jour. Chemistry, 9:729.

Pesnell, P. and Ourisson, G., 1963, Aristolene isolation in Nard essence. Bull. Soc. Chim. Fr., pp.912-913.

Prabhu, V.M., Karanth, S.K. and Rao, A., 1994, Effects of *Nardostachys jatamansi* on the biogenic amines and inhibitory amino acids in the rat brain. Planta Medica, 60:114-117.

Qim, Q., Tinli, L., Zhaogic, C., Liancuer, G. and Yi, Z., 1999, Extraction of essential oil and analysis of its chemical components. Shandeng Daxu Xuebao Ziram Kexueban, 34:192-197.

Rao, P.S., Sethi, H., Randhawa, G.S. and Badhwar, R.L., 1977, Aromatic plants of India. Part XVIII. Family LX — Valerianaceae. Ind. Jour. of Horticulture 34 : 323-333

Rastogi, R.P. and Mehrotra, B.N., 1998, Compendium of Indian Medicinal Plants. Vols. I to IV, CDRI and NISCOM, New Delhi.

Rucker, G., 1968, 1(10) aristolenon-(2) und 1,8,9,10-tetradehydro-aristolanon-(2), neue sesquiterpen-ketone vom aristolan-typ. Liebigs Ann. Chem., 717:221.

Rucker, G., 1969, Compounds of commercial Chinese spikenard oil *Nardostachys jatamansi*. Planta Medica 17:32-34.

Rucker, G., 1970, Structure of Nardosinone. IV. The constitution of isonardosinone. Liebigs Ann. Chem., 733:152-158.

Rucker, G. and Kretzschmar, U., 1971, 9-aristolene-1-α–ol and 1,2,9,10 tetradehydroaristolene— a new aristolane type sesquiterpene. Liebigs Ann. Chem., 748:214.

Rucker, G. 1971, 9 hydroxy-1(10) aristolene-2-one(debilon) from *Nardostachys chinenesis*. Planta Medica, 19:16-18.

Rucker, G. and Kretzeschamar, U., 1972, Patchouli alcohol and patchoulen aus *Nardostachys jatamansi*. Planta Medica, 21:1-4.

Rucker, G. and Dyck Eberhard., 1974, Formation of deoxynarchinol A from Nardosinone. Phytochemistry, 13:1907-1909.

Rucker, G., Tautges, J., Maheshwari, M.L. and Saxena, D.B., 1976, Norseychelanone, α and β patchoulenes and patchouli alcohol from *Nardostachys jatamansi*. Phytochemistry, 15:224.

Rucker, G., Tautges, J., Sieck, A., Wenzel, H. and Graf, E., 1978, Studies on isolation and pharmacodynamic activity of the sesquiterpene valeranone from *Nardostachys jatamansi* DC. Arzneim Forsch, 28:7-13.

Sastry, S.D., Maheshwari, M.L. and Bhattacharya, S.C., 1966, Terpenoids LXXXVII. The structure of nardol. Tetrahedron Lett., pp.1035-1042.

Sastry, S.D., Maheshwari, M.L., Chakravarti, K.K. and Bhattacharya, S.C., 1967a, Terpenoids CXV. Chemical constituents of *Nardostachys jatamansi*. Perfum. Essent. Oil. Rec., 58:154-158.

Sastry, S.D., Maheshwari, M.L., Chakravarti, K.K. and Bhattacharya S.C., 1967b, Terpenoids.CVI. Structure of calarenol. Tetrahedron, 23:1997-2000.

Sastry, S.D., Maheshwari, M.L., Chakravarti, K.K. and Bhattacharya, S.C., 1967c, Terpenoid. CVII. The structure of nardostachone. Tetrahedron, 23:2491-2493.

Satyawati, G.V., Gupta, A.K., and Tandon, N., 1987, Medicinal Plants of India. Vol II. 312 pp. Indian Council of Medical Research, New Delhi, India.

Saxena, D.B., Goswami, B.K. and Tomar, S.S. 1987, Nematicidal activity of some essential oils against *Meloidogyne incognita*. Indian Perfumer 31:150-154.

Saunders, W.D. and Pinder, A.R. 1977, Further studies relating to structure of nardostachone. Tetrahedron Lett., pp. 1687-1690.

Schulte, K.E., Rucker, G. and Glauch, G., 1965, Nardosinone a new constituent of *Nardostachys chinenesis*. Tetrahedron Lett., 35: 3083-3084.

Suhulte, K.E., Rucker, G. and Glauch, G., 1967, Planta Medica 15: 274-281.

Shah, N.C., 1983, Endangered medicinal and aromatic plants of UP Himalayas. In:An Assessment of Threatened Plants of India, pp. 40-49. Jain S.K. and Rao R.R. (eds), Botanical Survey of India, Kolkata.

Shah, N.C., 1994, Cultural and religious aspects of jatamansi, *Nardostachys grandiflora* DC. in the Kumaon Himalayas, India, In: IV International Congress of Ethnobiology: Ethnobiology in Human Welfare, Lucknow, India, 17-21 Nov.

Shanbhag, S.N., Mesta, C.K., Maheshwari, M.L., Paknikar, S.K. and Bhattacharya, S.C., 1964, Terpenoids LII. Jatamansi, a new terpenic coumarin from *Nardostachys jatamansi*. Tetrahedron, 20:2605-2615.

Shanbhag, S.N., Mehta, C.K., Maheshwari, M.L. and Bhattacharya, S.C. 1965, Terpenoids. LXXXV. Constituents of *Nardostachys jatamansi* and synthesis of (±) dihydrosemidin and visnadin from jatamansin. Tetrahedron, 21:3591-3597.

Shide, Luo., Mayer, R. and Rucker, G., 1987a, Nardonoxide, a new skeletal sesquiterpenoid of *Nardostachys jatamansi*. Tetrahedron, 46:1523-1530.

Shide, Luo., Olbrich, A., Mayer, R. and Rucker, G., 1987b, Gansongone — a new aristolane ketone from *Nardostachys chinensis* and structure revision of an aristolenol. Planta Medica, 53:556-558.

Shide, Luo., Mayer, R. and Rucker, G., 1987c, Nardonoxide, a new nardosinane type sesquiterpene ether from *Nardostachys chinensis*. Planta Medica, 53: 332-334.

Shide, Luo., and Zhang, Q., 1997, Nardosaldehyde: new peroxide-sesquiterpene. Tianvan Chanwu Yanjiu Yu Kaifa Bianjibu 9:7-9.

Soine, T.O. and Jawad, F.H. 1964, Structure of lomatin, A new coumarin, J. Pharm. Sci. 53:990.

Takaya, Y., Kurumada, K., Takeuji, Kim HS., Shibata, Y., Ikemoto, N., Watsya, Y. and Oshima, Y., 1998, Novel antmalarial quaiane type sesquiterpenoids from *Nardostachys chinensis* roots. Tetrahedron Lett., 39:1361-1364.

Takaya, Y., Takeuji, Y., Akasaka, M., Nakagawasai, O., Tadeno, T., Kisara, K., Kim, H.S., Wataya, Y., Niwa, M. and Oshima, Y., 2000. Novel guaiane endoperoxides, Nardoguaianone A-D, from *Nardostachys chinensis* roots and their antinociceptive and antimalarial activities. Tetrahedran 56: 7673-7678.

Takaya, Y., Akasaka, M., Takeuji, Y., Tanitsu, M., Niwa, M. and Oshima, X., 2000, Novel guaianoids, Nardoguaianone E-I, from *Nardostachys chinensis* roots. Tetrahedran 56: 7679-7683.

Tanitsu, M., Takaya, Y., Akasaka, M., Niwa, M., and Oshima, Y., 2002, Guaine- and aristolane type sesquiterpenoids of *Nardostachys chinensis* roots. Phytochemistry 59:845-849.

Thakur, R.S., Puri, H.S. and Husain, A., 1989, Major medicinal plants of India. Central Institute of Medicinal and Aromatic Plants, Lucknow, India. 361.pp.

Tiwari, K.C. and Joshi, H.C., 1974, Economics of medicinal herbs cultivation in hills of Kumaon and Uttarkhand. J. Res. Ind. Med. 9:73-78.

Vrkoc, I., Krepinsky, J., Herout, V. and Sorm, F,. 1963, The nature of sesquiterpenic hydrocarbon calarene and structure of β-gurjunene. Tetrahedron Lett., pp. 225-225.

Weberling, F., 1975, On the systematics of *Nardostachys* (Valerianaceae). Taxon, 24: 443-452.

CHAPTER 11

Podophyllotoxin and Related Lignans Produced by Plants

*Albert Koulman, Wim J. Quax** and *Niesko Pras*

INTRODUCTION

Lignans comprise a group of plant secondary metabolites receiving a great deal of attention in the field of natural product chemistry since the discovery of podophyllotoxin (PT). Lignans are defined as ββ-linked dimers of coniferyl alcohol, and are widely spread throughout the plant kingdom. The present interest for this group of natural products is based on its application in the fields of pharmacy and nutrition.

One group of lignans—those with an enterolactone-like structure—has drawn the attention as food additives. These lignans may have a preventive role in breast cancer, due to their phytooestrogenic activity. Different studies claim this product's cancer preventive activity based on in vitro as well as animal experiments (Cho et al., 1999) and epidemiological data (Adlercreutz, 1999). While other studies cannot proof this cancer preventive activity (Kilkkinen et al., 2004).

The attention of pharmacy for lignans and podophyllotoxin in particular is due to the pronounced cytotoxic activity of a number of these compounds. Several different plant species accumulate the 2,7'-cyclolignan lignan PT. This lignan is used as the starting material for the production of three semisynthetic anticancer drugs. Most commonly used for its antineoplastic properties is Etoposide, which was discovered in 1966 and approved by the FDA in 1983. In 1996, a phosphate analogue Etopophos, produced by Bristol-Meyers Squibb Co., was FDA approved. The third derivative is Tenoposide, and is less frequently used for chemotherapy in comparison with Etoposide (Hande, 1998). PT is one of the success stories in medicinal plant research. However, a better resource of PT is still required. The

*Corresponding author

isolation of PT from the rhizomes of *Podophyllum peltatum* and *Podophyllum hexandrum* (Berberidaceae) plants is not a very ideal production system. The supply of *P. hexandrum* rhizomes, which contain ca. 4% of PT on a dry weight basis, becomes increasingly limited due to both intensive collection and lack of cultivation (Choudhary et al., 1998; Rai et al., 2000). *Podophyllum peltatum* contains lower amounts of PT.

This supply problem forms the motivating a large number of scientists to search for alternative sources of PT. The stereo-specific chemical synthesis of PT and its related lignans is possible but is not economically feasible because of its low yield. An ideal resource would be a fast-growing, easy to cultivate plant, with a high lignan content, preferably in the leaves. A biotechnological approach might also solve the supply problem. The third option would be producing of lignans by organic chemistry in combination with plant enzymes that are able to catalyse difficult reaction steps.

In this chapter we will try to give an overview of the current state of research dealing with plants producing PT and related lignans. The chemical structure of PT and a number of important related lignans have been depicted in Fig. 11.1. The subjects under discussion range from

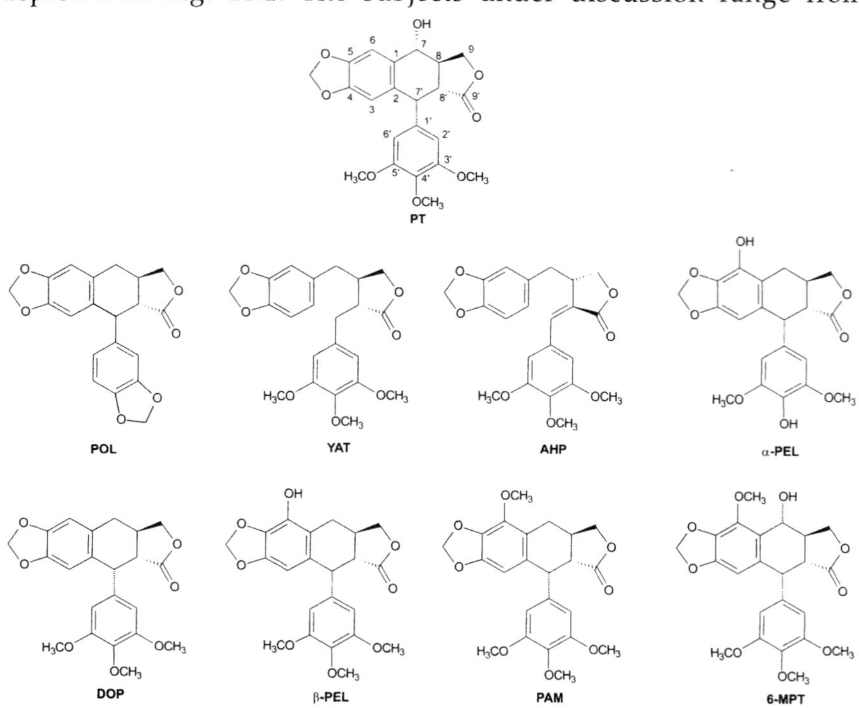

Fig. 11.1 Podophyllotoxin (PT) and its related lignans polygamain (PGM), yatein (YAT), anhydropodorhizol (AHP), α-peltatin (α-PEL), deoxypodophyllotoxin (DOP), β-peltatin (β-PEL), β-peltatin-A-methylether (PAM), 6-methoxypodophyllotoxin (6-MPT).

pharmacological activity, analysis, lignan-producing plants including chemotaxonomy, the biosynthesis of lignans and their ecological role to, finally, biotechnological production systems.

PHARMACOLOGY OF PODOPHYLLOTOXIN AND DERIVATIVES

Background

Since ancient times, several lignan-producing plants containing PT and related compounds have been used for the treatment of certain malignancies. The plant species *Juniperus sabina* has the longest history with respect to treatment of cancer. As early as in the first century A.D., Pliny the Elder mentions that the smaller species of *Juniperus* could be, among other things, used to stop tumors or swelling ('tumores' in Latin) (König and Hopp, 1993). Dioscorides mentions the use of the oil of *Juniperus* species (*J. sabina, J. phoenicea* and *J. communis*) for the treatment of ulcers, carbuncles and leprosy (Gunther, 1959). Generally, dried needles, called savin, or the derived oil was applied. In 47 A.D., Scribonius Largus wrote that savin oil was used to soften 'hard female genital parts' (Sconocchia, 1983). Later references indicated the use of savin to treat uterine carcinoma, venereal warts and polyps (Hartwell and Schrecker, 1958). At present, we know that the pharmacological activity of *J. sabina* needles can be ascribed to the lignans deoxypodophyllotoxin (DOP) and PT (Hartwell et al., 1953a).

The Leech book of Bald, 900-950 A.D., an early English medicinal book, has reported the use of root of *Anthriscus sylvestris*. These roots contain several PT-related lignans like DOP and were used in ointments prepared from a large number of plant and plant extracts like savin to cure cancer (Cockayne, 1961). In Japan and China, the roots of this plant were also used as a kind of crude drug called 'Zengo' in Japan (Kozawa et al., 1982) and 'qianhu' in China (Kozawa et al., 1978).

Surprisingly, Kozawa and co-workers stated that: 'the root is soaked in water, and then crushed and pulverized in a dry atmosphere for use as food. In addition, the young aerial part of this plant is sometimes used for food' (Kozawa et al., 1982). Even though this method was still practised in the eighties, the authors do not mention any form of intoxication.

The plant species currently used for the extraction of PT are *Podophyllum peltatum* (May apple or American Mandrake) and *Podophyllum hexandrum* (syn. *P. emodi*). *P. peltatum* is indigenous to the Eastern part of the United States of America and Canada, while *P. hexandrum* is found in the higher parts of the Himalayan mountains and is also referred to as the Indian *Podophyllum*. The natives of the Himalayas as well as the American Indians independently discovered that extracts of the rhizomes possessed a cathartic action. The Indians introduced podophyllin, a resin obtained by ethanolic extraction of the *Podophyllum* roots and rhizomes, to colonists for the use as a catharic, an anthelmentic and misuse as a mortal poison. The main constituents in podophyllin are the lignans PT, 4'-demethylpodophyllotoxin, α- and β-peltatin (see Fig. 11.1). The colonists also used this resin as an emetic and cholagogue. Podophyllin was included in the first U.S. Pharmacopoeia, dating from 1820, as a cathartic and cholagogue. Due to its severe toxicity, the drug was removed from the 12th edition of this Pharmacopoeia, that appeared in 1942 (Ayres and Loike, 1990; Horwitz and Loike, 1977). In the same year, however, it was reported that venereal warts (*Condyloma acuminatum*) could be selectively destroyed by the topical application of podophyllin. Nowadays, this resin is still in use to treat this disease (Frega et al., 1997), but pure PT is also applied (Gross, 2001).

Development of Podophyllotoxin based cytostatics

It was not until 1947 that the first reports appeared dealing with the inhibition of growth of cancer in animals by podophyllin (Horwitz and Loike, 1977). Due to their high cytotoxicity, the constituents of podophyllin induce severe gastrointestinal side effects, as is also the case with their glycosides, thus limiting their clinical use (Stähelin and Von Wartburg, 1991). With the aim to find clinically-applicable drugs, related to these naturally-occurring lignans, a series of glycoside and aglycone derivatives have been prepared by chemical synthesis by Sandoz. This study resulted in the discovery of Etoposide in 1966.

This semi-synthetic entered clinical trials in 1971 and it took another 12 years (!) before Etoposide became FDA approved. By that time, Bristol-Myers took over the license for both Etoposide and Tenoposide (see Fig. 11.2). In 1996, the phosphate analogue Etopophos (see Fig. 11.2) was also approved. A detailed history of these compounds has been recently reviewed (Imbert, 1998). Etoposide is still administered, often in combination with cisplatin and for instance bleomycin, for the treatment of metastatic testicular germ-cell tumors (Flechon et al., 2001). But Etoposide alone is also used for the treatment of small cell lung cancer (Mascaux et al., 2000).

The mode of action of Podophyllotoxin

PT inhibits the formation of the microtubules. In vitro, it binds to tubulin dimers, giving a PT-tubublin complexes. This stops the further formation of the microtubules at one end but does not stop the disassembly at the other end, leading to the degradation of the microtubules. Cells treated with PT are arrested in the metaphase of the mitosis. Its mode of action is comparable to the alkaloid colchicin and such compounds are called spindle poisons. Several spindle poisons exist in clinical use. Paclitaxel and vincristine-like alkaloids have the effect that cells will enter the mitosis but

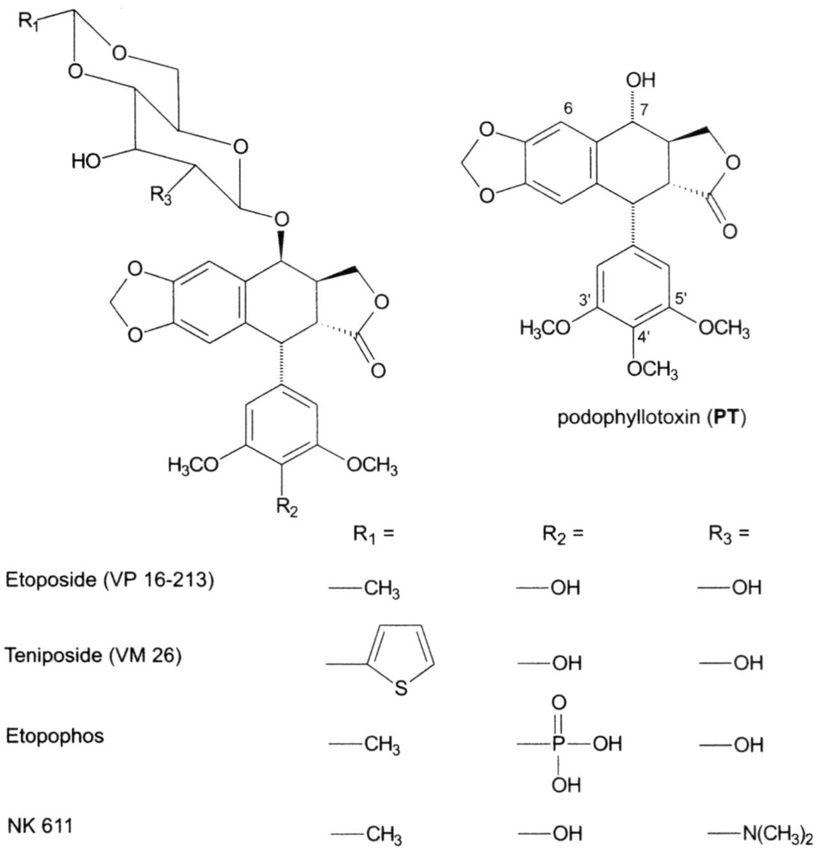

Fig. 11.2 Podophyllotoxin-derived anticancer drugs.

the duplicated chromosomes will not be seperated. In this way, the cell cannot duplicate and growth stops, but these compounds block the microtubuli in different ways (Ayres and Loike, 1990; MacRea and Towers, 1984; Stähelin and Von Wartburg, 1991).

Some structure activity relationship studies have been carried out using all kinds of PT analogues, demonstrations that the main structure of DOP is already responsible for the cytotoxicity. The extra methoxygroup (6-MPT, Fig. 11.1) on the 6 position does not significantly change the in vitro cytotoxicity compared to PT. Also, the methyl group on the 4' position of the pendent ring is of little effect on the cytotoxicity (Hadimani et al., 1996; Van Uden et al., 1992).

The mode of action of Etoposide

The clinically-applied PT-derivatives Etoposide, Tenoposide and Etopophos have a totally different mode of action. The clinically-applied derivatives of PT are all topoisomerase II inhibitors. Topoisomerase II is an enzyme that cleaves double-strand DNA and seals it again. It is heavily involved in the processes of DNA replication and repair. For its function, topoisomerase II binds covalently to the broken DNA. Etoposide and other derivatives stabilize the DNA-topoisomerase II complex in such a way that the resealing of the DNA strands is impossible. Cells that are duplicating their DNA for the mitosis are very sensitive for this mechanism. The overall effect of these anticancer drugs is the arrest of the cells in late S or early G2 phase of the cell cycle (Hande, 1998; Stähelin, 1973). The advantage of the newly-introduced Etopophos (Etoposide phosphate) is the improved solubility in water. It is a pro-drug of Etoposide. After administration, the phosphate group is hydrolysed in the human body in order to yield Etoposide with the same bioactivity. Being water soluble, it is much easier to administer (Witterland et al., 1996). Several new derivatives of Etoposide are currently in the clinical phase of studies. For instance, NK 611 (Fig. 11.2) also has an improved water solubility and still has the topoisomerase II blocking activity (Hanauske et al., 1995; Mross et al., 1996).

Recently, a Swedish company (Conpharm) began clinical trials with a new PT derived drug CPH82 (Remacon) for the treatment of rheumatoid arthritis. Reumacon is a mixture of two PT glucosides (podophyllotoxin-4,6-O-benzylidene-β-D-glucopyranoside, AS 3738 and 4′-demethyl-podophyllotoxin-4,6-O-benzylidene-β-D-glucopyranoside, AS 3739) (Carlstrom et al., 2000).

In summary, it can be concluded that nowadays, lignans are still compounds of interest for the pharmaceutical industry. PT is an important starting compound to prepare semi-synthetic cytostatics. Among the related lignans, new cytototoxic compounds can be found that may possibly lead to new cytostatics.

More detailed information on the pharmacological activity of lignans and their derivatives can be found in a number of extensive reviews (Bernasconi, 1989; Damayanthi and Lown, 1998; Hande, 1998; Imbert, 1998).

ANALYSIS OF PODOPHYLLOTOXIN AND RELATED LIGNANS

Background

There are several reasons to develop an analytical procedure for PT or its related lignans. Basically, there are two important applications. First, there is demand for quantitative and qualitative analytical procedures for the profiling of lignans present in plant material, for instance, in the cultivation of PT producing plants and the elucidation of lignan biosynthesis. The other main field of PT analysis is concentrated on the clinic, i.e. the fluids of patients treated with semi-synthetic analogues of PT.

In this chapter, we will focus on the first topic. Different techniques are available, some of which are developed for the quantitative analysis of lignans, while others are more suitable for qualitative detection.

Most commonly used are thin layer chromatography (TLC) and high performance liquid chromatography (HPLC); in lesser extent also liquid chromatography coupled to mass spectrometry (LC-MS) gas chromatography (GC) and GC coupled to mass spectrometry (GC-MS). Micellar electrokinetic capillary chromatography (MECC) can also be used for the analysis of lignans but this new method is seldom applied.

TLC

Thin layer chromatography (TLC) is a suitable method for the rapid screening of large numbers of samples, for example, during isolation of a certain compound or for the selection of high producing plants. For lignan analysis, generally, a polar stationary phase is used like silica 60. The mobile phase consists of an apolar solvent (e.g. $CHCl_3$) mixed with a more polar solvent (MeOH) in ratio of 10 to 1 or 25 to 1 (Broomhead and Dewick, 1990a; Van Uden et al., 1989; Van Uden et al., 1990a). Most lignans can be detected in these systems but the discrimination between YAT (yatein) and DOP is not possible (see Fig.11.1 for structures). The lignans can be detected using 254 nm UV light in combination with coated plates. It is possible to bath the plate in a mixture of MeOH and concentrated H_2SO_4, followed by heating at 100°C for 10 minutes. The lignans are then carbonized, leaving gray or purple spots (Wichers et al., 1991; Woerdenbag et al., 1990).

For the isolation of lignans, a preparative thin layer chromatography is still commonly practised (Jiao et al., 1998; Rahman et al., 1990a).

HPLC

High performance liquid chromatography (HPLC) is the most common and widely-applied method for quantitative and qualitative analysis of lignans. Almost all of the published methods use a reversed phase (RP-18) stationary phase. Some of the older publications mention the use of straight phase columns; the authors apply silica 60 columns with a eluents consisting of heptane, $CHCl_3$ and MeOH (Van Uden et al., 1989; Wichers et al., 1990). The routinely applied reversed phase columns, e.g. Lichrosorb or Nucleosil, are combined with an isocratic mobile phase consisting of water and MeOH or MeCN. In some cases, a gradient is used to enable more subtle changes in the H_2O/MeOH or H_2O/MeCN ratios.

UV-spectrophotometry is used to detect lignans. The UV spectra of PT or related lignans generally have a peak around 290 nm. For qualitative analysis, a diode array detection (DAD) system is very suitable. Lignans have a very distinct UV absorption spectrum. The differences in the substitutions at the aromatic ring have a clear effect on the UV spectrum. Some examples are shown in Fig. 11.3. Also, an extra double bond is easy to recognize in UV-spectrums. PT and most of the related lignans can be identified with the aid of modern software, combined with DAD and an UV-spectrum library.

A few methods have been published dealing with LC-MS for lignans. LC-MS was used for the first time to analyse Etoposide in human serum. The LC system consisted of a standard HPLC system, as described above, coupled to an atmospheric pressure ionization-electronspray inlet for the mass spectrometer (Chen and Uckun, 2000).

An other method that was initially developed for the detection of phenolic constituents in *Krameria triandra* roots should be basically applicable for the detection of PT or related lignans in plant material. This method clearly showed the presence of other phenylpropanoids like flavanoids and neolignans. Here also, a standard HPLC-system with a gradient of water/MeOH to MeCN was used. The HPLC was coupled to fast atomic bombardment MS/MS (FAB-MS/MS) using xenon as bombarding gas (Facino et al., 1997).

Recently, a study was published in which Wong and co-workers analyzed plant material with an atmospheric pressure ionization tandem mass spectrometer coupled to HPLC (API-LC/MS/MS). The authors analyzed different samples of *Podophyllum emodi*. A MeOH water gradient (from 35% MeOH to 65%MeOH over an hour) with a Hypersil BDS column was applied. It was possible to detect PT, its glycoside and several derivatives, e.g. podorhizol β-D-glycoside (Wong et al., 2000).

This system is extremely useful for the analysis of plant material, because it yields qualitative and quantitative data. Such systems can play a crucial role in the elucidation of the lignan biosynthesis.

Seidel and co-workers used LC-MS to monitor the metabolic fate of ^{13}C –labelled phenylpropanoids in cell cultures of *Linum album*. For the quantification of incorporation they used an HPLC system connected to a triple-quadrupole tandem mass spectrometer. Samples were analysed by electrospray ionisation (ESI) employing the Micromass Z-Spray source. Elution was carried out on an RP-18 Luna column (50 mm long, 2 mm i.d.; 3 µm particle size; isocratically at a flow rate of 0.2 ml min^{-1} using two solvent mixtures: A (60% water, 40% acetonitrile; v/v) and B (50% water, 50% acetonitrile; v/v). Mass spectra were recorded in the positive and/or negative mode using multiple reaction monitoring (MRM). Ion transitions for MRM (collision energy 10-15 eV) were selected by tandem mass spectrometry (MS/MS) experiments (product ion scans) with reference compounds (Seidel et al., 2002).

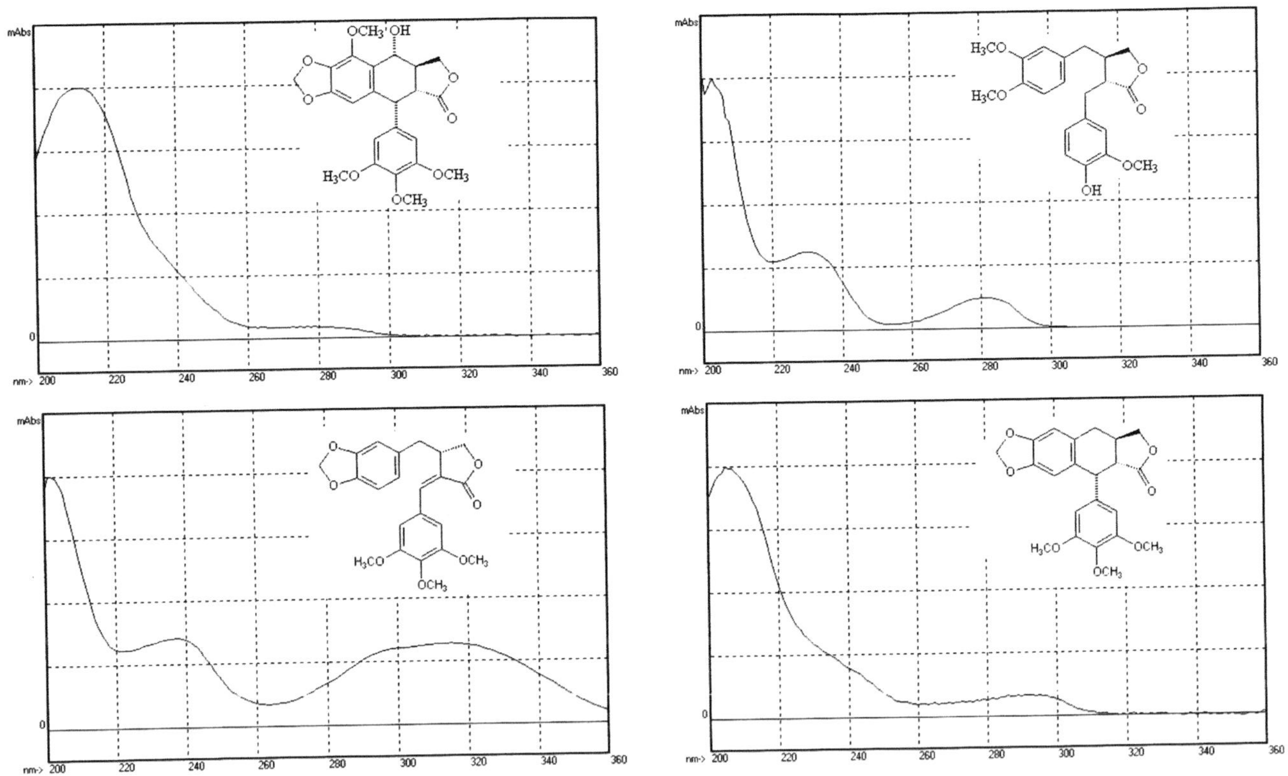

Fig. 11.3 The UV spectra of 6-MPT (top left) arcteginin (top right) AHP (bottom left) and DOP (bottom right). These spectra were taken with a Shimadzu diode array detector, with: H_2O:MeCN (60:40) eluent and a RP-18 column.

The combination of HPLC and ^{1}H-NMR is suitable for the qualitative analysis of plant material. It has been applied on leaves and needles of *Torreya jackii* that produce matairesinol derivatives (Zhao et al., 1999). The results demonstrate the usefulness of such an analytical technique. However, the advantage over LC-MS or GC-MS is limited; only some additional stereochemical information can be obtained. The electron ionization mass spectrometry (EIMS) or MS/MS deliver such unambiguous information about lignans that HPLC-^{1}H-NMR is not of advantage when new lignans are detected. For a definite structure elucidation, the compounds still have to be isolated and subjected to thorough ^{13}C and 2-dimensional NMR. Due to the high costs of the equipment, the role of HPLC-^{1}H-NMR in lignan research remains limited to a certain extent.

GC

Gas chromatography is rarely used for the detection of lignans. The main reason is that lignans as such are not really suitable for GC-analysis. Also lignans are not volatile due to the molecular mass of over 400 and the polarity. However, GC analysis is possible under certain conditions. The advantage of GC analysis is the easy combination with mass spectrometry. This makes it possible to analyse unknown lignans and to detect and interpret structural modifications after bioconversion experiments with enzyme systems or metabolic degradation.

GC-MS is used in the field of nutritional research, e.g. to investigate the metabolic fate of the phytoestrogenic lignans matairesinol and arctigenin. It is possible to analyse these compounds with a GC-MS system after derivatization with, for instance, N,O- bis(Trimethylsilyl)trifluoroacetamide (BSTFA). This will shield off the free hydroxyl groups of the lignan molecule with a trimethylsilyl moiety, making the lignan much more apolar and suitable for GC analysis. For the analysis, a silica-coated capillary column is used with a high temperature program (140°C to 245° C) coupled to electron ionization mass spectrometry (EIMS) (Morton et al., 1999).

A direct GC-MS method can be used for the analysis of less polar lignans like DOP and anhydropodorhizol (AHP). The temperature program is high (150°C to 320°C at 15°C/min.) and applied with a silica-coated capillary column. The lignans can be detected with a flame ionization detector (FID) but the real advantage of this method lays in the combination with EIMS. The EIMS spectra are easy to interpret and reveal a lot of information about the chemical structure (Koulman et al., 2001). The structure of most lignans can be confirmed using this method and, for unknown lignans, it is possible to get a basic indication of their structure (Koulman and Konuklugil, 2003). As an example, the GC-MS profiling of an *Anthriscus sylvestris* root extract is shown in Fig. 11.4. The interpretation of the mass spectra of lignans has been reviewed (Arimoto et al., 1996).

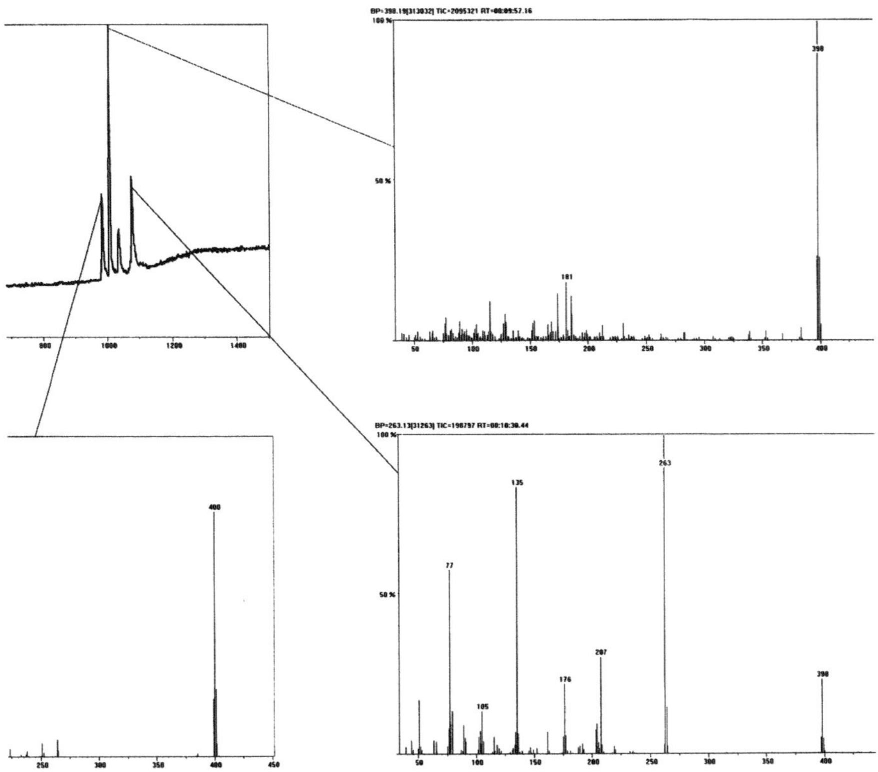

Fig. 11.4 GC-MS profiling of an apolar (CH_2Cl_2) extract of *A. sylvestris*. GC: Unicam 610/Automass 150, Column: Chrompack CP-Sil 5-CB (15 m × 0.32 mm ∅) Inj. Volume: 2 µl, Inj. Temp.: 260°C, Temp:150°–320°C, 15°C min^{-1}, Carrier gas: He, pressure: 5 psi, velocity 32 cm s^{-1}, Ionization: electron impact 70 eV (250°C), interface Temp.: 280°C, Scan speed: 2 scans s^{-1}(range 34-600 u). Top left: total ion current of the chromatogram; Top right: mass spectrum of DOP; Bottom left: mass spectrum of YAT; Bottom right: mass spectrum of AHP, see Fig. 11.1 for the chemical structures.

Capillary chromatography

For lignan analysis, there is only one published capillary chromatography method available as yet. Micellar electrokinetic capillary chromatography (MECC) was used to separate the lignans present in *Podophyllum* resin (Ganzera et al., 1999). The results are comparable with HPLC and the latter method should, therefore, be a cheaper alternative. Maybe the combination of MECC with MS will be an alternative for LC-MS with respect to the analysis of lignan glycosides in the near future (see also the sections on biosynthesis and the general discussion) but on the condition that a higher resolution can be reached in comparison with that of LC.

PLANTS PRODUCING PODOPHYLLOTOXIN AND RELATED LIGNANS

Occurrence in the plant kingdom

Presently there are least 22 different plant species cited in the literature that produce PT (Tables 11.1 and 11.2). Another 20 different species do not produce PT but only related lignans such as some *Bursera* species. These species do produce DOP and β-peltatin-a-methylether (PAM) in detectable amounts.

Some publications mention an even higher number of PT-producing plants. This goes as high as 83 different species (Konuklugil, 1996), but from this article, it is unclear to the authors on what evidence this is based. There are, however, 230 species in the *Linum* genus (Van Uden et al., 1994) and on the basis of the work of Konuklugil (1996), one might indeed assume that a high number of these species produce PT in detectable amounts, but this has still to be proven by analysis. In Tables 11.1 and 11.2 are only published results summarized.

Based on the classification of Cronquist (1988), PT can be found in the four orders of the Magnoliopsida (flowering plants: Table 11.2) and in the order Pinales of the Gymnospermae (Table 11.1). In total, there are 13 families distributed over the entire plant kingdom producing PT and related lignans. The actual number of PT-producing plants might be much higher, since only a small number of plant species has been phytochemically analyzed. During re-examination, PT was found in some species which were already studied on their lignan content (Kozawa et al., 1978; Kuo et al., 1999). A detailed survey of the vascular plants for the presence of PT could significantly extend our knowledge (Tables 11.1 and 11.2).

Chemotaxonomy

On basis of Tables 11.1 and 11.2, it is clear that lignans are widely spread in the entire plant kingdom. The study of related species in a PT-producing genus can deliver in very different results with respect to yields. A high percentage of the studied species in the *Linum* genus produce lignans, mainly 6-MPT, while the species of the *Juniperus* genus all produce PT or peltatins. On the other hand, the genus *Anthriscus* contains high lignan-producing species *Anthriscus sylvestris* that accumulates mainly DOP but also PT in small amounts. Within the same genus, the species *Anthriscus cerifolium* does not produce any detectable amounts of DOP or PT (Koulman et al., 2001). It should be remarked that only a few species of the family Apiaceae have been studied, namely, *A. sylvestris*, *A. cerifolium*, *Cicuta*

Table 11.1 The occurrence of podophyllotoxin and related lignans in the order Pinales of the gymnosperms. See Fig. 11.1 for the chemical structures.

Plant	*Plant part*[1]	DOP[2]	PT[2]	PAM[2]	6MPT[2]	*Reference*
Fam. Cupressaceae						
Callitris columellaris	L / S	0.062				(Aynehchi, 1971)
Callitris drummondii	N		1.4			(Kier *et al.*, 1963)
Juniperus chinensis	Calli		0.005			(Muranaka *et al.*, 1998)
Juniperus procera	L	X		X		(Muhammad *et al.*, 1995)
Juniperus lucayana	?		0.10			(Hartwell *et al.*, 1953a)
Juniperus phoenicae	L	X		X		(Cairnes *et al.*, 1980)
Juniperus sabina	N	X	0.20	X	X	(Arturo *et al.*, 1990; Hartwell *et al.*, 1953a; San Feliciano *et al.*, 1989)
Juniperus sabina var. *tamariscifolia*	?		0.14			(Hartwell *et al.*, 1953a)
Juniperus scopulorum	?		0.17			(Hartwell *et al.*, 1953a)
Juniperus thurifera	N	0.04		0.004		(San Feliciano *et al.*, 1992)
Juniperus virginiana	?		0.10			(Hartwell *et al.*, 1953a)
Thuja occidentalis	L	0.0014				(Chang *et al.*, 2000)
Fam. Pinacea						
Libocedrus decurrens		L/S ?	0.04			(Kupchan *et al.*, 1967)1967)

[1]: L= leaf, N= needles, S= stem, Sd= seeds, ?= plant part not known.
[2]: The concentration in weight percentage based on dry weight. YAT: yatein; DOP: deoxypodophyllotoxin; PT: podophyllotoxin; 6MPT: 6-methoxypodophyllotoxin. X: detectable amounts.

Table 11.2 The occurrence of podophyllotoxin and related lignans in flowering plants. Taxonomy based on the classification of Cronquist (Cronquist, 1988), in the four orders of Magnoliidae, Dilleniidae, Rosidae and Asteridae. See Fig. 1 for the chemical structures.

Plant	*Plant part*[1]	POL[2]	YAT[2]	DOP[2]	PT[2]	ß-PEL[2]	PAM[2]	6MPT[2]	*Reference*
				Ord. Magnoliidae					
Laurales									
Fam. Hernandiaceae									
Hernandia nymphaeifolia	B		0.00007 x-OH[3]	0.00008					(Chen *et al.*, 1996) (Chen *et al.*, 1997)
Hernandia ovigera	Sd			X			X		(Yamaguchi *et al.*, 1982)
Hernandia sonora	Sd			X	X	X	X	X	(Udino *et al.*, 1999)
Piperales									
Fam. Piperaceae									
Piper cubeba	F		X						(Badheka *et al.*, 1986)
Ranunculales									
Fam. Berberidaceae									
Diphylleia cymosa	L				0.54				(Broomhead and Dewick, 1990b)
Diphylleia grayi	Rt				1.27	0.126			(Broomhead and Dewick, 1990b; Murakami and Matsushima, 1961)
Podophyllum hexandrum	Rt/Rh			0.017	4.27	0.01			(Broomhead and Dewick, 1990b)
Podophyllum peltatum	Rt/Rh			0.023	0.25	0.33			(Broomhead and Dewick, 1990b)
Podophyllum pleianthum	Rt/Rh			0.01	0.135				(Broomhead and Dewick, 1990b)

(Contd.)

Table 11.2: (Contd.)

Plant	*Plant part*[1]	POL[2]	YAT[2]	DOP[2]	PT[2]	ß-PEL[2]	PAM[2]	6MPT[2]	*Reference*
Podophyllum / Dysosma versipellis	Rt/Rh			X	0.32	1.35			(Broomhead and Dewick, 1990b; Yu *et al.*, 1991)
Fam. Rananculaceae									
Pulsatilla chinensis	WP					X			(Mimaki *et al.*, 1999)
Ord. **Rosidae**									
Linales									
Fam. Linaceae									
Sect. *Syllinum*									
Linum album	Rt				0.02	0.04		0.86	(Weiss *et al.*, 1975)
Linum arboreum	Rt				0.02	0.04		0.86	(Konuklugil, 1996)
Linum campanulatum	Rt				0.12	0.16		0.82	(Konuklugil, 1996)
Linum capitatum	Rt				0.02	0.10		0.03	(Broomhead and Dewick, 1990a; Konuklugil, 1996)
Linum cariense	Rt				0.10	0.08		0.12	(Konuklugil, 1996)
Linum elegans	Rt				0.10	0.04		2.30	(Konuklugil, 1996)
Linum flavum	Rt				0.02	0.04		0.65	(Broomhead and Dewick, 1990a; Konuklugil, 1996)
Linum flavum var *compactum*	Rt				0.02	0.04		2.38	(Konuklugil, 1996)
Linum flavum spp *flavum*	Rt				0.06	0.04		0.38	(Konuklugil, 1996)
Linum flavum spp *scabrinerve*	Rt				X	X		0.11	(Konuklugil, 1996)
Linum mucronatum spp *armenum*	Rt				0.02	0.04		0.04	(Konuklugil, 1996)

(Contd.)

Table 11.2: (Cont.)

Plant	*Plant part*[1]	POL[2]	YAT[2]	DOP[2]	PT[2]	ß-PEL[2]	PAM[2]	6MPT[2]	*Reference*
Linum nodiflorum	L				X	0.01		0.25	(Konuklugil, 1996)
Linum pamphylicum	Rt				0.19	0.02		0.06	(Konuklugil, 1996)
Linum tauricum	Rt				0.06			1.32	(Konuklugil, 1996)
Linum thracicum	Rt				0.06	0.01		0.20	(Konuklugil, 1996)
Sect. *Linum*									
Linum austriacum	Rt				X	0.01		0.05	(Konuklugil, 1996)
Linum lewisii	Rt				X	X		X	(Konuklugil, 1996)
Linum meletonis	Rt	X							(Koulman and Konuklugil, 2003)
Linum monogynum	Rt				X	X		0.01	(Konuklugil, 1996)
Linum sibiricum	Rt				X	X		0.01	(Konuklugil, 1996)
Sect. *Dasylinum*									
Linum hirsutum	Rt				X	X		X	(Konuklugil, 1996)
Linum viscosum	Rt				X	X		X	(Konuklugil, 1996)
Sect. *Linopsis*									
Linum corymbulosum	Rt				X	X		X	(Konuklugil, 1996)
Polygalales									
Fam. Polygalaceae									
Polygala polygama	WP	X		X	0.076				(Hokanson, 1978) (Hokanson, 1979)
Sapindales									
Fam. Burseraceae									
Bursera morelensis	E			0.31					(Jolad *et al.*, 1977)
Bursera fagaroides	WP						X		(Bianchi *et al.*, 1969)
Bursera permollis	S/B			0.004			0.006		(Wickramaratne *et al.*, 1995)
Bursera simaruba	Res	0.05							(Peraza-Sanchez and

(Contd.)

Table 11.2: (Cont.)

Plant	*Plant part*[1]	POL[2]	YAT[2]	DOP[2]	PT[2]	ß-PEL[2]	PAM[2]	6MPT[2]	*Reference*
									Pena-Rodriguez, 1992)
Bursera microphylla	S			0.09					(Jolad *et al.*, 1977)
Commiphora incisa	Res	0.5							(Provan and Waterman, 1985)
Fam. Rutaceae									
Haplophyllum cappadocicum	WP	X							(Gözler *et al.*, 1996)
Apiales									
Fam. Apiaceae									
Anthriscus sylvestris	Rh		0.0126	0.007	X				(Ikeda *et al.*, 1998; Kozawa *et al.*, 1978)
Chaerophyllum aureum	L			X					(Rollinger *et al.*, 2003)
Cicuta maculata	Rh			?[4]					(Hartwell *et al.*, 1953b)
Ord. **Asteridae**									
Lamiales									
Fam. Lamiaceae									
Eriope macrostachya	S.					X			(Raffauf *et al.*, 1987)
Hyptis verticillata	W.P				0.25	X			(Kuhnt *et al.*, 1994)

[1]: B= bark, F= fruit, E= exudate, L= leaf, Rh= rhizome, Rt= root, S= stem, Sd= seeds, Res=resin or exudtae, WP= whole plant.
[2]: The concentration in weight percentage based on dry weight. POL: polygamain; YAT: yatein; DOP: deoxypodophyllotoxin; PT: podophyllotoxin; ß-PEL: ß-peltatin; PAM: ß-peltatin-A-methylether; 6MPT: 6-methoxypodophyllotoxin. X: detectable amounts.
[3]: x-OH= is a hydroxylated derivative of YAT.
[4]: Spectroscopic data hint to the presence of deoxypodophyllotoxin but this is not yet confirmed.

maculata and *Pastinaca sativa*. From morphological and phylogenetical evidence it is clear that *A. sylvestris* and *A. cerifolium* are closely related species (Downie et al., 2000; Plunkett and Downie, 1999). Since lignans are not present in each species belonging to the Apiaceae family, this class of secondary metabolites is unsuitable as a chemotaxonomical marker.

BIOSYNTHESIS OF PODOPHYLLOTOXIN AND RELATED LIGNANS

Background

The biosynthesis of PT and related lignans is still a matter of debate. Due to the large number of lignan-producing plant species, this research field is scattered and not concentrated on one specific model species. Different groups focus on different plants and there is no certainty that there is any similarity in the biosynthesis of PT in the different species. Therefore, general conclusions cannot be drawn as yet and it is only possible to speculate about major parts of lignan biosynthesis.

Presently, one cannot conclude as to how many genes are involved in the biosynthesis of any secondary metabolite. As pointed out earlier, PT is found all over the plant kingdom. Plants with limited phylogenetic relationship produce PT. Still, plant secondary metabolism is thought of as one of the fast-evolving aspects of the plant. The occurrence of convergent evolution or better, repeated evolution can only explain the presence of PT in the different families of the plant kingdom. The complexity of the biosynthesis of secondary metabolites is, however, very poorly understood. For the formation of PT from matairesinol, at least 6 enzymatic steps are necessary (see Fig. 11.8). Moreover, the glycosyl-transferases are involved as well as regulatory genes on factors. Whether such a complex system is able to evolve fast and easy as some authors claim to be so for plant secondary metabolism (Pichersky and Gang, 2000) seems to be a paradox.

Unambiguous results obtained for and conclusions drawn from other biosynthetic pathways should be cautiously extrapolated to the lignan biosynthesis.

Towards lignans

All lignans are ββ-dimers of phenylalanine dervatives. The shikimic acid route produces the aromatic amino acid phenylalanine. The biosynthesis from phenylalanine to coniferyl alcohol is not a straight route but a complex network, which is depicted in Fig. 11.5 (Humphreys et al., 1999; Whetten et al., 1998). Coniferyl alcohol is stored in the cell as coniferin (its β-D-glycoside). *Linum flavum* cell suspension cultures are able to accumulate over 12% of coniferin on dry weight basis, which shows that the coniferin pool can be an important factor in the lignan biosynthesis (Van Uden et al., 1990b). Coniferyl

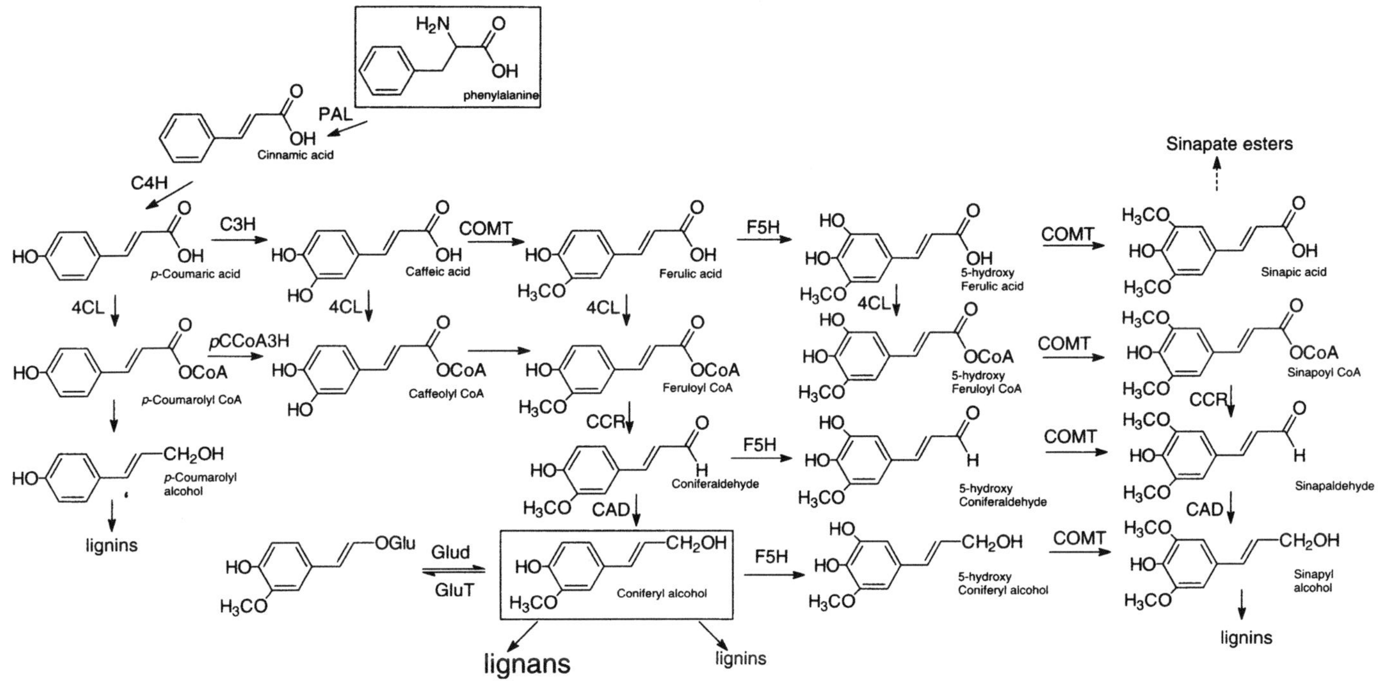

Fig. 11.5 The phenylpropanoid biosynthesis. The enzymes involved are caffeoyl CoA *O*-methyltransferase (CCoOMT), cinnamate 4-hydroxylase (C4H), cinnamoyl alcohol dehydrogenase (CAD), cinnamoyl CoA reductase (CCR), *p*-coumarate 3-hydroxylase (C3H), 4-(hydroxyl)cinnamoyl CoA ligase (4CL), ferulate 5-hydroxylase (F5H), *p*-coumaroyl CoA 3 - hydroxylase (*p*CCoA3H), phenylalanine ammonia lyase (PAL), caffeic acid/5-hydroxylferulic acid/5-hydroxyconiferyl alcohol and aldehyde *O*-methyltransferase (COMT), Glucosyl transferase (Glut), Glucosidase (Glud).

alcohol can also be converted into 5-hydroxy coniferyl alcohol by ferulate 5-hydroxylase (Humphreys et al., 1999). This compound can be further converted into sinapyl alcohol by the action of an O-methyl transferase (Li et al., 2000). Like coniferyl alcohol and *p*-coumaryl alcohol, sinapyl alcohol is also used as a building block for the lignin biosynthesis. Different oxidases are able to generate radicals from these building blocks, allowing polymerization with each other. To what extend this polymerization is controlled is still a matter of discussion (Lewis, 1999).

The dirigent protein

For the formation of lignans, plant oxidases oxidize coniferyl alcohol into its free radical form. In vitro the free radicals will dimerize into a racemic mixture. However, in the plant, the dimerization is controlled and the formation of one stereoisomer is favored. This observation gave rise to discovery of the dirigent protein (Davin et al., 1997). The dirigent protein does not catalyse the reaction in a standard way. There is no active site in the protein and it does not chemically bind (covalent or non-covalent) to the substrates but directs the coupling of two monomer radicals in a stereospecific way. Davin and Lewis claim that homologues genes are present in all the plants they examined, a.o.: *Forsythia* species, *Thuja plicata, Populus tremuloides, Oryza sativa, Linum usitatissimum,* and *Nicotiana tabacum* (Davin and Lewis, 2000). The function of each of these dirigent proteins might differ between the species. The coniferyl alcohol monomers can couple not only through an 8-8′ linkage (Fig. 11.6), but also through an 8-5′ or 8-4′ linkage, which leads to the biosynthesis of neolignans like in *Piper aequale* or *Persea obovatifolia* (Maxwell et al., 1999; Tsai et al., 1998). This biosynthesis also happens in a stereo-controlled way, suggesting the involvement of a dirigent protein. *Eucommis ulmoides* produces a derivative of pinoresinol, which is probably not a dimer formed form two coniferyl alcohols but from two sinapyl alcohol monomers. Again, this happens in a stereospecific manner and from this plant a homologous gene of the dirigent protein has been cloned (Davin and Lewis, 2000).

A recent study with ^{13}C labelled phenylpropanoids fed to cell suspension cultures of *Linum album* showed that ferulic acid was converted into podophyllotoxin (Seidel *et al.*, 2002). The presence of a ^{13}C atom in ferulic acid yielded podophyllotoxin with a mass of M+2. Doubly labelled ferulic acid yielded M+4. Additional experiments with labelled dimethoxycinnamic acid and sinapic acid also yielded podophyllotoxin with an M+1 and M+2. The feeding experiments with labelled methylenedioxycinnamic acid yielded only podophyllotoxin M+1 (see fig. 7; (Seidel et al., 2002). It is not clear, whether, the fed substrates are converted back to ferulic acid before they enter the podophyllotoxin biosynthetic pathway.

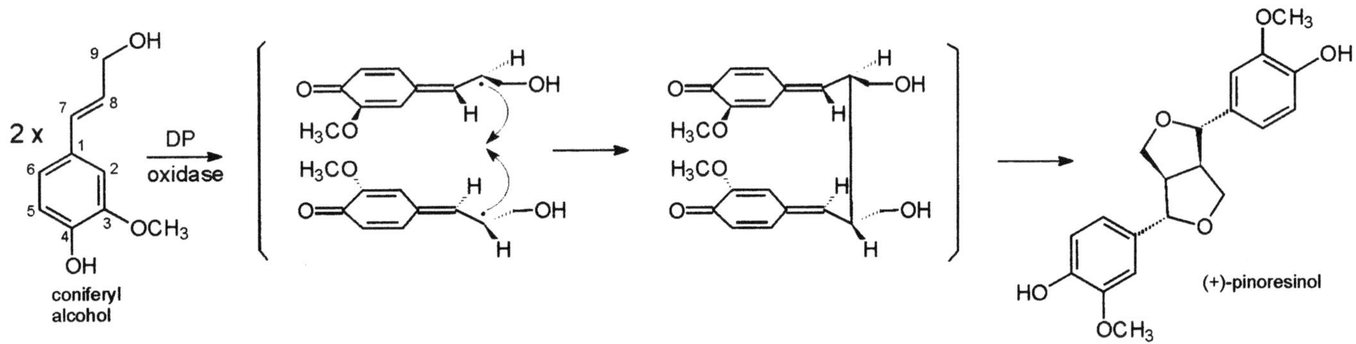

Fig. 11.6 Dirigent-mediated formation of (+) pinoresinol. Dirigent protein (DP) in *Forsythia intermedia* according to Davin and Lewis (Davin *et al.*, 1997).

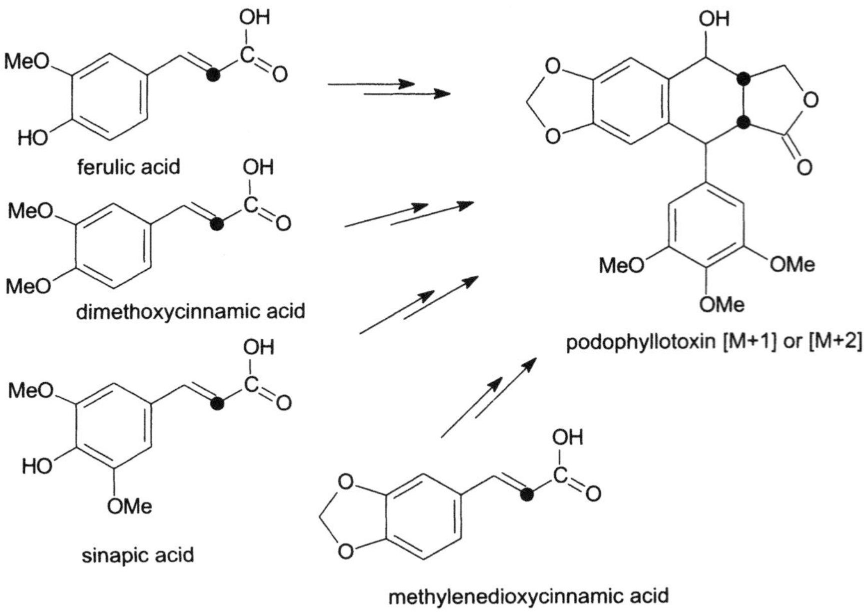

Fig. 11.7 Feeding of ^{13}C labeled phenyl-propanoids to cell suspension cultures of *Linum album*. The black dot indicates the presence of the ^{13}C atom. The label can end up.on the indicated positions in the product depending on the substrate (Seidel *et al.*, 2002).

From pinoresinol to matairesinol

From pinoresinol, the lignan biosynthesis will split up into different directions, leading to a great diversity of lignans. For the biosynthesis leading towards PT, the first number of steps have been proven to result in matairesinol (see Fig. 11.7). These steps were first elucidated in *Forsythia intermedia* (Katayama et al., 1993; Rahman et al., 1990b; Umezawa et al., 1991). Pinoresinol is enantiospecifically reduced in two steps, leading to secoisolariciresinol. The same steps have been confirmed to occur in *Linum flavum* (Xia et al., 2000).

From matairesinol to podophyllotoxin

Two theories have been published about the biosynthesis of PT starting from matairesinol. The first theory dates back to halfway onto the 1980s and was published by Dewick and co-workers. It was proven for *Podophyllum* that matairesinol, YAT and DOP can be converted into PT. Labelled possible precursors of PT were fed to the roots of *Podophyllum hexandrum,* followed by the monitoring of the metabolites. It was shown that *Podophyllum* could convert matairesinol, YAT and DOP into PT. The oxidized analogue of yatein, anhydropodorhizol was not converted into

PT. Figure 11.8 summarizes the precursor feeding experiments (Jackson and Dewick, 1984; Kamil and Dewick, 1986). It can be concluded that at least 6 steps are necessary to convert matairesinol into PT: Four steps are required for the formation of YAT from matairesinol, while one step is needed for the conversion of YAT to DOP and 5 DOP into PT as well.

Van Uden and co-workers delivered further evidence for this theory by feeding DOP to *Linum* cells, this resulted in 6-MPT (Van Uden et al., 1995) and much smaller amounts of PT (unpublished results). Cell suspension cultures of *Linum flavum* are able to convert large amounts of DOP into the glycoside of 6-MPT (Van Uden et al., 1997). Recent studies have shown that the first step from DOP towards 6-MPT is carried out by deoxypodophyllotoxin 6-hydroxylase, a cytochrome P450 monooxygenase, in *Linum album* (unpublished results). This enzyme converts DOP into β-PEL (β-peltatin).

On the other hand, DOP is hydroxylated at the 7 position yielding PT and it glycoside. Enzyme assays showed that there is an S-adenosyl-L-methionine (SAM)-dependent *O*-methyltransferase that converts β-PEL into PAM (β-peltatin-A-methylether). Cell cultures are able to convert this intermediate very rapidly into 6-methoxypodophyllotoxin (unpublished results). This reaction is probably also carried out by a cytochrome P450 monooxygenase, which we will call the putative lignan 7-hydroxylase because it seems to posses a very low substrate specificity. The biosynthesis of 6-MPT from DOP is shown in Fig. 11.9.

Recently, the feeding of matairesinol to the roots of *L. flavum* by Xia and co-workers resulted in 7-hydroxymatairesinol. This compound was not yet known from *Linum* species. The feeding of synthetic labelled 7'-hydroxymatairesinol to *Linum flavum* roots yielded labelled 6-MPT, as is shown in Fig. 11.10 (Xia et al., 2000). This is in contradiction with the results discussed previously and it seems very unlikely that this is the normal biosynthetic pathway in *Linum flavum*. Under standard conditions, 7'hydroxymatairesinol cannot be found in *Linum flavum* and only after the feeding of an 'overdoses' matairesinol to the roots 7'-hydroxylmatairesinol is accumulated. One could argue that if 7'-hydroxymatairesinol was really an intermediate, it would be very difficult to detect, since it is then rapidly converted in the biosynthesis of 6-MPT. This is in analogue with the rapid conversion of DOP to 6-MPT by *L. flavum* cells where DOP cannot be measured under standard conditions.

In our experiments with *Linum flavum* cells we could show that after feeding of AHP also, hydroxylation occurred at the 7-position (unpublished results). Apparently, the 7-hydroxylation of intermediates is quite a common feature in the lignan pathway. An enzyme, which has the capacity to hydroxylate feeded AHP, DOP and matairesinol, is probably a less-specifically acting cytochrome P450 monooxygenase. This yields

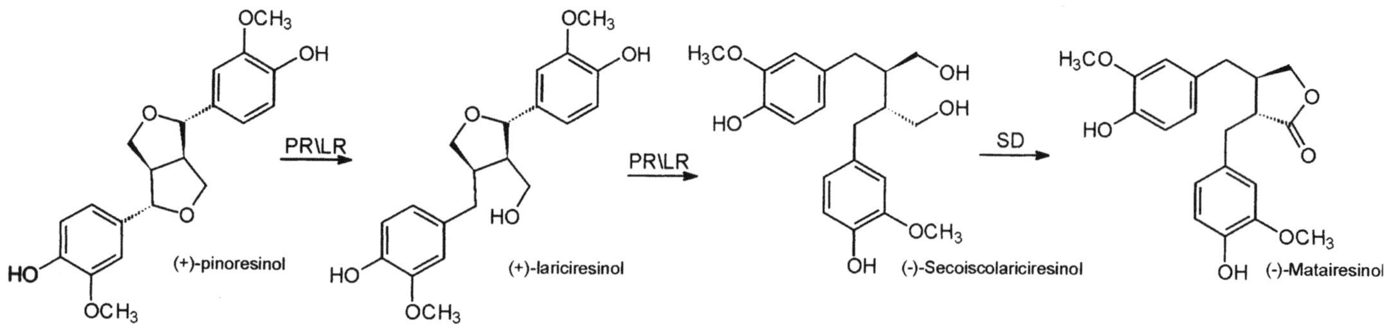

Fig. 11.8 The biosynthesis of matairesinol starting with pinoresinol. The enzymes involved are pinoresinol-lariciresinol reductase (PR\LR) and secoisolariciresinol dehydrogenase (SD) (Rahman et al.,1990b).

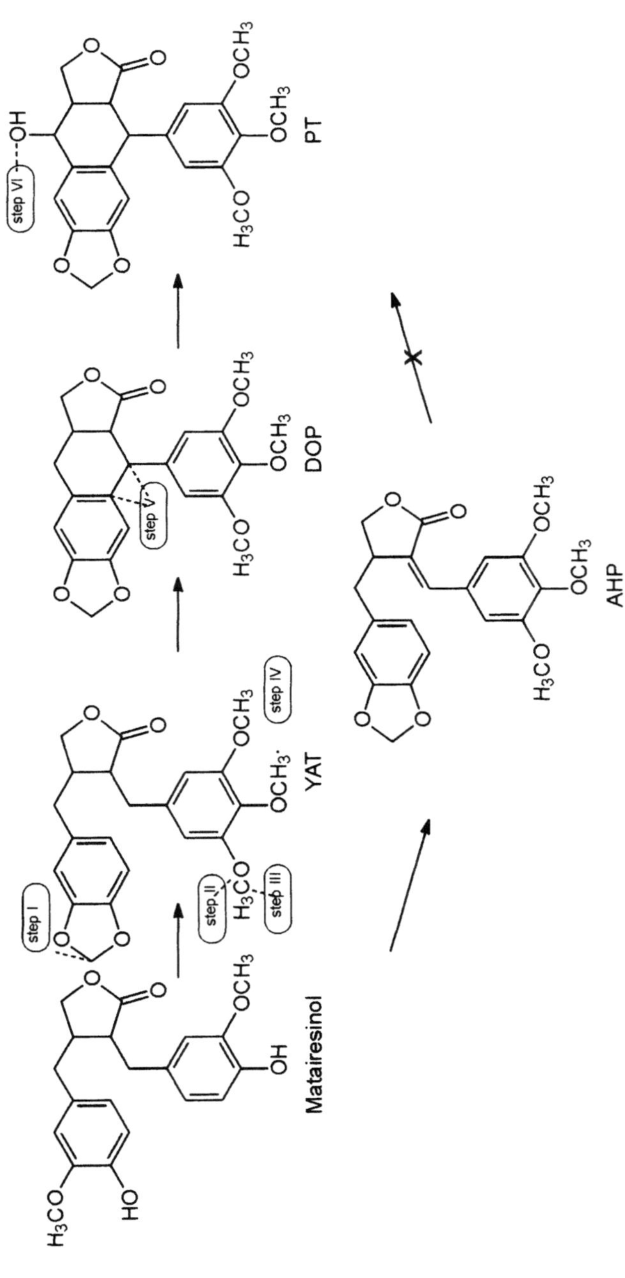

Fig. 11.9 The biosynthesis of Podophyllotoxin (Jackson and Dewick,1984) and the minimal number of enzymatic steps necessary.

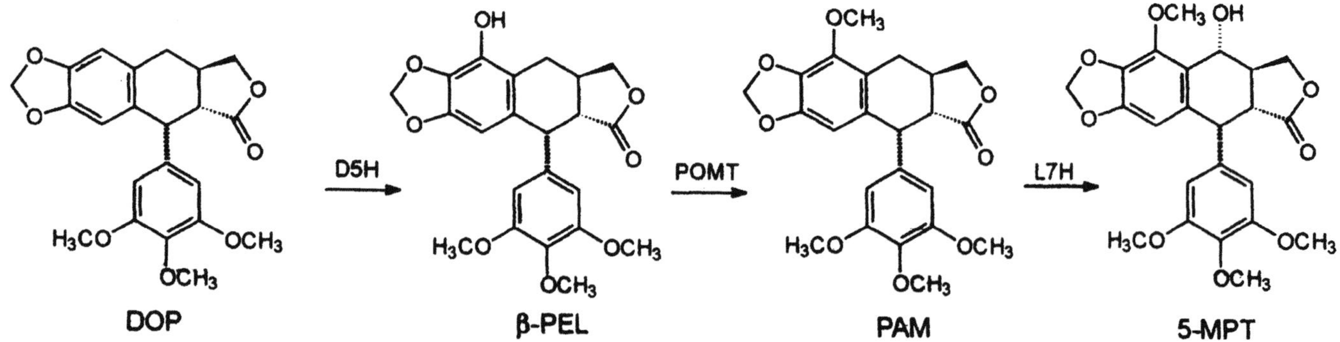

Fig. 11.10 The biosynthesis of 6-MPT (6-methoxypodophyllotoxin) according to van Uden and Petersen (Van Uden et al., 1997). The enzyme abbreviations are deoxypodophylotoxin 6-hydroxylase (D6H), peltatin SAM-dependent *O*-methyltransferase (POMT), putative lignan 7-hydroxylase (L7H).

compounds which are not all normally accumulated in the plant or present as biosynthetic intermediates. The presence of this putative lignan 7′-hydroxylase can explain the occurrence of different 7′-hydroxylignans but does not explain why a compound like 7′-hydroxymatairesinol, which has not been found previously in *Linum flavum*, can be converted into 6-MPT.

The answer to this question might be found in a parallel with the biosynthesis from phenylalanine to coniferyl and sinapyl alcohol (see Fig. 11.5). The biosynthesis of these compounds is not a straight route but more of a network. It might be possible that the biosynthesis of lignans, starting with matairesinol, is a complex network as depicted in Fig. 11.11 Almost all lignan-accumulating plants produce different lignans that are not formed via one single straight route. For instance, *Callitris drumondiii* produces hinokinin PT as well. *Linum flavum* produces PT, 6-MPT and β-peltatin. We developed a GC/MS method for lignan profiling and identified a number of lignans in *Anthriscus sylvestris* (Koulman et al., 2001). We also combined this data with a literature survey of all the lignans in *Anthriscus sylvestris.* If we try to fit all this data in a schematic figure and then fill the gaps with results from literature obtained for *Podophyllum, Linum,* and *Forsythia,* a general but hypothetical, biosynthetic network appears (see Fig. 11.11). This hypothetical scheme is meant to give a glance of the complexity of the lignan biosynthesis. This scheme allows an individual to speculate about the more general features of a pathway instead of a stepwise way of thinking. First of all, there is not one straight biosynthetic pathway leading to one end product. *A. sylvestris* mainly produces DOP, YAT, and AHP. These compounds are not end products; this data confirms that plants accumulate intermediates from their biosynthesis. The network also shows that there are different pathways that can lead to the same end product. Many enzymatic steps belong to perform the same type of chemical reactions such as methylations and hydroxylations.

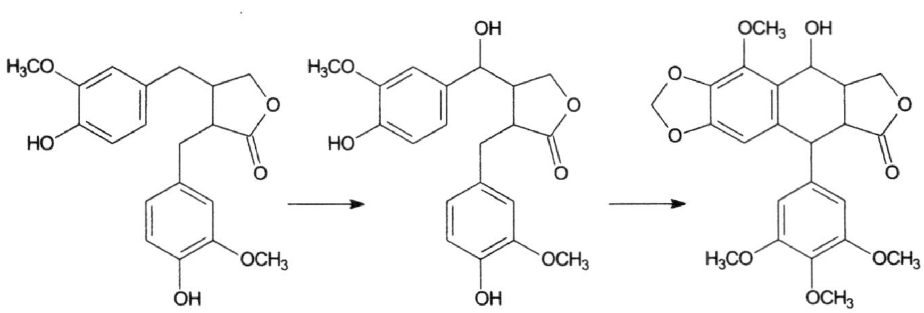

Fig.11.11 The proposed biosynthesis of 5-methoxypodophyllotoxin according to Xia et al., (2000).

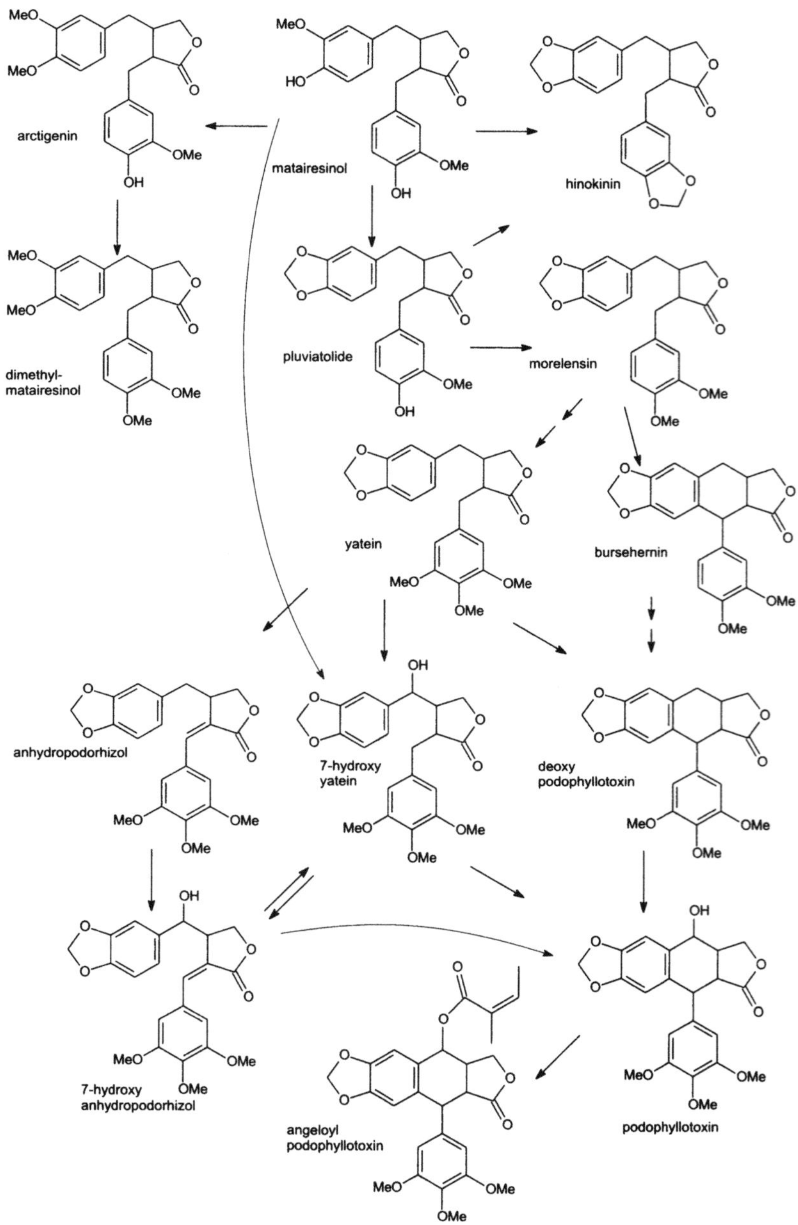

Fig. 11.12 The hypothetical biosynthetic network in *Anthriscus sylvestris*, based upon compounds and conversions found in particularly *A. sylvestris* (*A.s.*) and in *Linum flavum* (*L.f.*) and *Podophyllum hexandrum* (*P.h.*). The glycosides are not included. The network is based on literature and unpublished results (Ikeda *et al.*, 1998; Koulman *et al.*, 2003)

As with the biosynthesis of coniferyl alcohol and its analogues, it might be very well possible that there is a limited number of enzyme classes that are able to produce these ranges of compounds. The minimum number of enzyme classes required are a hydroxylase (cytochrome P-450), an O-methyltransferase, an oxidizing enzyme to form the double bond formation, a dioxymethylene bridge-forming enzyme (probably also a cytochrome P-450), and a ring-closing enzyme. It is difficult to believe that in this system, only one very specific hydroxylase is active. Most probably there are two—one for the aromatic hydroxylations and one for the other aliphatic hydroxylations. This would denote the need for a total number of 6 enzymes with broad substrate specificity so as to enable the lignan biosynthesis. Differences in affinity for the different substrates could explain the variations in product concentration. The levels of gene-expression, regulation and uptake in cell compartments will also play a role in the accumulation of these compounds. Our hypothetical network may be of support for further studies on the biosynthesis of lignans but also to develop new strategies to genetically manipulate and improve the lignan production of plants.

ECOLOGICAL FUNCTION OF PODOPHYLLOTOXIN AND RELATED LIGNANS IN PLANTS

Background

The ecological or biological function of PT and related lignans has been neglected to a great extent. The high cytotoxicity of PT and related lignans has led to the generally well-accepted theory that these compounds protect the plant against herbivores.

Insecticidal and other activities

Japanese research dating from the early 1980s clearly shows that lignans present in roots of *Anthriscus sylvestris* have insecticidal activity. DOP is a primarily responsible compound for this lethal activity on a number of different insect larvae, e.g. *Culex pipiens, Epilachna spara* and adult insects such as *Blatella germanica* (Kozawa et al., 1982). A further study showed that DOP has a slow effect on larvae of the Silkworm *Bombys mori* (Inamori et al., 1983).

In an other study, it was shown that PT was the constituent of *Podophyllum hexandrum* that was responsible for its insecticidal activity against *Drosophila melanogaster*. This was found during a bioguided isolation based on the insecticidal activity (Miyazawa et al., 1999).

Further, in vitro and in vivo studies showed that PT and related lignans are active against viruses (Charlton, 1998) and different kinds of tumour cells (Damayanthi and Lown, 1998). In case of other lignans, germinating inhibitory, antimicrobial and antifungal activity has been reported, but there is no evidence for PT or related lignans concerning these activities (MacRea and Towers, 1984). From these results, it is impossible to define the exact ecological role of these compounds. The insects used in the above-mentioned studies do probably not interact with the plants in situ and there are no field studies available involving PT or DOP producing plants.

Induction and elicitation

Plants like *Anthriscus sylvestris* and *Podophyllum hexandrum* do not produce just one lignan but usually accumulate a whole range of different lignans. For instance, *A. sylvestris* accumulates next to DOP also high amounts of YAT and AHP, and much smaller amounts of PT, hinokin, arctigenin en phylligenin (see Fig. 11.11). There is also a large variation in the concentration of these compounds within this plant species (Koulman et al., 2001). There are different theories explaining the reason why plants produce such a diverse spectrum of compounds instead of just one main compound. On the one hand, a combination of compounds could have a synergistic effect; on the other, a large variation in secondary metabolite pattern will probably make it more difficult for predators to adapt to the bioactivity of these compounds (Shelton, 2000).

The ecological role of secondary metabolites is often characterized by the induction of the biosynthesis of these compounds in the plant after contact with natural enemies like microorganisms, predators or stress conditions (heat, dryness) (Abraham et al., 1999). Thus far, there is no proof that lignan biosynthesis is induced by stress or predation. Also, the infection of a virus in *A. sylvestris* does not change the levels of lignans in the plant (unpublished results).

The induction by ecological factors can be compared with elicitation in plant cell cultures. Elicitation is an increase of secondary metabolism caused by stress factors like exposure to microorganisms. Cell cultures derived from PT-producing plants did not increase lignan production after elicitation. Only *Juniperus chinensis* callus cultures had an elevated PT production after the addition of chito-oligosaccharides which mimic fungal cell wall constituents (Muranaka et al., 1998). There are examples of secondary metabolites that are not induced but still prevent predation, like the furanocoumarins in *Pastinaca sativa* (Zangerl and Berenbaum, 1997).

This chemical-ecological research has become increasingly important for plant sciences. It is of great help for molecular biological research when a so-called on-off situation is found. By comparison of a production situation with a non-production situation of a plant species that accumulates interesting secondary metabolites, it is easier to find the genes that play a role in the biosynthesis. These genes can be detected and identified with

DNA-chips or by subtractive analysis. The availability of these genes is of utmost importance for further study of the regulation of biosynthetic pathways or the biotechnological production of secondary metabolites using recombinant DNA technology.

BIOTECHNOLOGICAL PRODUCTION OF PODOPHYLLO-TOXIN AND RELATED LIGNANS

Background

Different research groups use a biotechnological approach to solve the supply problem of PT for the production of Etoposide and related drugs. Experiments with *Podophyllum* plants in India did not result in a cultivation system with viable agricultural prospects (Choudhary et al., 1998).

A biotechnological production system might, therefore, be a useful alternative. There are several options in this field, most of which have been studied. The most commonly followed approach is the use of plant cell cultures, other closely-related cell systems are certain strains of fungi. Presently, no prokaryotic cells are able to produce lignans. The different productions systems will be further discussed below.

Plant cells

The production of fine chemicals or drugs with in vitro cultured plant cell systems has been the subject of study for many years. Given the totipotency of the plant cell, implying that all the genetic information present in the plant is essentially available, it seems feasible to produce most, if not all, secondary metabolites using in vitro-grown cultures. The successful production of antibiotics by fungi and bacteria has been a stimulating example for researchers in the early days of plant cell cultures. Since then, it has become clear that plant cell cultures do not always accumulate (either qualitatively or quantitatively) the same compounds found in the parent plant from which they were established.

However, different products can be produced with plant cells, like paclitaxel, L-DOPA (Pras et al., 1993), rosmarinic acid (De-Eknamul and Ellis, 1985) or coniferin (Van Uden et al., 1990b). Rarely are these systems applied for the industrial production of these compounds. At the moment, there are no in vitro production systems for PT that are of commercial interest.

The most direct approach for the biotechnological production of PT is the use of a cell suspension culture in a large-scale fermentor. This system would produce PT directly (Table 11.3). A suitable candidate for fermentation could be a *Linum album* suspension culture. Experiments are underway using large-scale fermentors (500 L) with these cells, but it is questionable if an economically-feasible system will be developed, due to the low production and the instability of these cells (< 0.5% PT based on dryweight, pers. com.).

Bioconversion systems

Plant cells form an attractive biocatalytic system, since precursors can be supplied directly to the culture. Several reports dealing with the production of valuable compounds by adding precursors to various culture species have been published. Often, the precursor undergoes more than one bioconversion, resulting in complex mixtures of (unknown) products, or the substrate is metabolised via unknown mechanisms (Kawaguchi et al., 1988). Nevertheless, a number of one-step bioconversons by freely-suspended cells have been described; for instance, the production of codeine with a *Papaver somniferum* culture fed with codeinone (Furuya et al., 1984).

The same approach is possible for the production of 6-MPT, the 6-methoxy derivative of PT. Cell suspension cultures of *Linum flavum* are able to convert DOP (see Table 11.3) into the glycoside of 5MPT (Van Uden et al., 1997). For several reasons, this system has not yet reached the stage of commercial application. Firstly, Etoposide and other analogues cannot be synthesized from 6-MPT. Secondly, the isolation of DOP is a rather expensive and time-consuming process. DOP can be found in the roots of *Anthriscus sylvestris*. This plant is common weed in Northwest Europe and its roots contain upto 2.5% DOP. The selection of a high DOP-producing strain could facilitate a cheaper isolation procedure.

Fungi can also be used for the bioconversion of DOP in to PT-analogues; *Aspergillus niger* cultures produced a racemic mixture of PT (Kondo et al., 1989; Kondo et al., 2000). The use of fungal systems for the production of plant secondary metabolites is poorly developed, but might become of interest in the future.

Recently, Pras and Woerdenbag have reviewed the production of secondary metabolites by bioconversion (Pras and Woerdenbag, 1999).

GENERAL DISCUSSION AND CONCLUSIONS

The research of podophyllotoxin and related lignans is still a very actual topic in natural product chemistry. Several features like the occurrence in the plant kingdom and the biosynthesis makes the research of this group of natural products worthwhile.

There are very reliable quantitative and qualitative analytical methods for most lignans and the future will lie in the further development of a LC-MS system. LC-MS is indispensable for drug research and its role in the study of secondary metabolism is already, and will be of great importance.

Table 11.3 Cell cultures producing podophyllotoxin or 5-MPT

Species	Culture	Medium[1]	Lignans[2]	Reference
Callitris drummondi	Suspension	MS + 4 mg/l NAA	0.1% PT	(Van Uden et al., 1990c)
Juniperus chinensis	Callus	SH 3 mg/l NAA + Kinetin 0.2 mg/l	0.005% PT	(Muranaka et al., 1998)
Podophyllum hexandrum	Callus	B5 + 4 mg/l NAA	0.3% PT	(Van Uden et al., 1989)
	Suspension	B5 + 4 mg/l NAA	0.1% PT	(Woerdenbag et al., 1990)
Podophyllum peltatum	Callus	MS + 1 ppm 2,4D + 0.2 ppm kinetin	0.71% PT	(Kadkade, 1982)
	Suspension	½ MS + hormones	0.32-0.36% PT	(Kutney et al., 1991)
Linum flavum	Roots	MS + 2 µM 2,4-D	0.7-1.3% 6-MPT	(Berlin et al., 1988)
	Suspension	MS + 3 mg NAA	0.004 % 6-MPT	(Van Uden et al., 1990a)
	Suspension with DOP	MS + 4 mg/l NAA + 4 mg/l BAP + 0.8 g /l DOP	4.41% 6-MPT-glu	(Van Uden et al., 1997)
Linum nodiflorum	Suspension	MS + 0.4 mg/l NAA	0.6% 6-MPT + 0.2%PT	(Konuklugil et al., 1999)
Linum album	Suspension	MS + 0.4 mg/l NAA	0.2-0.5% PT and 6-MPT	(Smollny et al., 1998)

[1] Medium used for culturing. The kind of culture media (MS: Murashige and Skoog (Murashige and Skoog, 1962), SH: Schenk and Hildebrand (Schenk and Hildebrandt, 1972), B5: Gamborg's B5 (Gamborg et al., 1968)), and used phytohormones (NAA: naphthalene acetic acid, 2,4D: 2,4-dichlorophenoxy acetic acid, BAP: 6-benzylaminopurine).

[2] The concentration in weight percentage based on the dry weight.

LC-MS offers the only possibility to qualitatively and quantitatively study the lignan production in plants without disregarding the glycosylated intermediates. At the moment, there is no clear picture about the role of glycosylations in the lignan biosynthesis. But most specialists will agree that this role could be of a far greater impact than we anticipate at the moment.

Not only with LC-MS it will also be useful to screen plants for lignans. The biosynthesis of PT and related lignans widely occurs in the plant kingdom. There is no reason to assume that all PT-producing plants have been discovered. One could follow two different approaches. First, there are different plant families that contain PT-producing species. To establish the presence of more PT-producing species in some of those families, far greater research is needed: the Apiaceae family being a good example. Secondly, there probably exist plants producing PT from families that have not yet been studied. The study of these plants might not only reveal high-producing plants but also plants that produce high levels of intermediates. Such efforts will probably also yield new lignans with cytotoxic activities that could be of pharmaceutical interest and this justifies further in-depth research of the lignan biosynthesis.

There are a lot of theories and assumptions about this biosynthesis and the role and origin of plant secondary metabolism in general. A close survey of the lignan biosynthesis makes that some question marks spring into mind about the current views on secondary metabolism. Even then, some definitions about secondary metabolism are clearly not applicable for lignans, e.g. lignans are not taxonomically restricted in their distribution (Berenbaum, 1995). This can be explained by repeated evolution (Pichersky and Gang, 2000), but the general complexity of the biosynthesis of secondary metabolites like that of lignans, makes it difficult to believe such a system can evolve fast, which is said to be the important condition for repeated evolution. This phenomena can be more easily understood for a less complex biosynthesis leading to monoterpenes (Cseke et al., 1998). Even when the minimum number of 6 enzyme types is responsible for the complete lignan biosynthesis, as postulated in this chapter, this pathway is much more complex than the monoterpene biosynthesis.

In any hand- or textbook about biochemistry, you can find the complex 'metabolic maze' which shows the complex interacting primary metabolism. A clear example is the formation of lignin where synapoyl alcohol can be formed through different routes. The secondary metabolism that is based on this primary metabolism is usually depicted as straight cascade of enzymatic reactions, leading to the end-product. There might be side-branches, but these lead to their own end-product (Alberts et al., 1994). The presented data for lignans formation show that such a biosynthetic maze is also possible in secondary metabolism.

One of the main questions concerning lignan biosynthesis—which is usually not addressed in current research—is the similarity of the biosynthesis between the different plant species. The omnipresent dirigent protein does not mean that lignan biosynthesis is similar all over the plant kingdom, as Xia stated (Xia et al., 2000). A possible lignan formed from one molecule coniferyl alcohol and one molecule synapoyl alcohol is still not excluded. In this reaction, the dirigent protein could also play an important role.

The most logical approach in lignan research is to study these possible similarities and divergences in the biosynthesis of different plant species. This will also contribute to the understanding of the evolutionary background of lignans. For these kind of studies, it will be necessary for genes of enzymes in the lignan biosynthesis to be cloned and functionally analyzed. This will enable the screening of other plant species for homological genes.

It might be an interesting option to genetically modify *A. sylvestris*. This plant is able to produce PT but only in very low amounts. The over-expression of the DOP-hydroxylating enzyme should give a higher PT production (Fig. 11.8). Currently, research groups are trying to pick up the gene responsible for the hydroxylation of DOP in *Linum flavum* or *L. album*. This cytochrome P-450 gene can be transferred to *A. sylvestris,* leading to a transgenic *A. sylvestris* that is able to produce PT.

Another biotechnological approach that is currently investigated is based on the use of heterologous expression of plant genes. The organic synthesis of YAT is relatively easy and probably suitable for large scale production (Medarde et al., 1995a; Medarde et al., 1995b). Further synthesis towards PT is not possible due to difficulty of the stereoselective ring closure. It might be possible to use the enzyme responsible for this ring closure outside the plant for a large-scale production of DOP. It is, therefore, necessary to identify the gene of this enzyme and clone it in a suitable expression system.

Finally, it can be concluded that the fundamentals of lignan biosynthesis are very poorly understood. The coming years will reveal if this biosynthesis is a straight route or a proposed network. A biotechnological production of PT will not possible as long as this question will be circumvented.

REFERENCES

Abraham, K.J., Pierce, M.L. and Essenberg, M., 1999, The phytoalexins desoxyhemigossypol and hemigossypol are elicited by *Xanthomonas* in *Gossypium* cotyledons. Phytochem., 52:829-836.

Adlercreutz, H., 1999, Phytoestrogens. State of the Art. Environ.Toxicol. Pharmacol., 7: 201-207.

Alberts, B., Bray, D., Lewis, J., Raff, M., Roberts, K. and Watson, J.D., 1994, Molecular Biology of the Cell, 41-88. Garland Publishing, Inc., New York & London.

Arimoto, M., Yamaguchi, H. and Nishibe, S., 1996, Structure elucidation and synthesis of the lignans from the seeds of *Hernandia ovigera* L. In: Stereoselective Synthesis (Part K), 551-606. Attu-ur-Rahman H.E.J. (ed.). Vol. 18, Elsevier, Amsterdam.

Arturo, S., Miguel del Corral, J.M., Gordaliza, M. and Castro, A. 1990, Lignans from *Juniperus sabina*. *Phytochem.*, 29: 1335-1338.

Aynehchi, Y., 1971, Desoxypodophyllotoxin, the cytotoxic principle of *Callitris columellaris* F. Muell. J. Pharm. Sci., 60:121-122.

Ayres, D.C., and Loike, J.D., 1990, Lignans: Chemical, Biological and Clinical Properties, University Press, Cambridge.

Badheka, L.P., Prabhu, B.R. and Mulchandani, N.B., 1986, Dibenzylbutyrolactone lignans from *Piper cubeba*. Phytochem., 25:487-489.

Berenbaum, M.R., 1995, The chemistry of defense: Theory and practice. Proc. Natl. Acad. Sci. USA., 92:2-8.

Berlin, J., Bedorf, N., Mollenschott, C., Wray, V., Sasse, F. and Hofle, G., 1988, On the podophyllotoxins of root cultures of *Linum flavum*. Planta Med., 54:204-206.

Bernasconi, C., 1989, Etoposide: fifteen years experience. Bone Marrow Transplant., 4 Suppl. 1:52-55.

Bianchi, E., Sheth, K. and Cole, J.R., 1969, Antitumor agents from *Bursera fagaroides* (Burseraceae) (b-peltatin-A-methylether and 5'-desmethoxy-b-peltatin-A-methylether). Tetrahedron Lett., 32:2759-2762.

Broomhead, A.J. and Dewick, P.M., 1990a, Aryltetralin lignans from *Linum flavum* and *Linum capitatum*. Phytochem., 29:3839-3844.

Broomhead, A.J. and Dewick, P.M., 1990b, Tumour-inhibitory aryltetralin lignans in *Podophyllum versipelle*, *Diphylleia cymosa* and *Diphylleia grayi*. Phytochem., 29:3831-3837.

Cairnes, D.A., Ekundayo, O. and Kingston, D.G.I., 1980, Plant anticancer agents. X. Lignans from *Juniperus phoenicea*. J. Nat. Prod, 43:495-497.

Carlstrom, K., Hedin, P.J., Jonsson, L., Lerndal, T., Lien, J., Weitoft, T. and Axelson, M., 2000, Endocrine effects of the podophyllotoxin derivative drug CPH 82 Reumacon (R) in patients with rheumatoid arthritis. Scand. Rheumatol., 29:89-94.

Chang, L.C., Song, L.L., Park, E.J., Luyengi, L., Lee, K.J., Farnsworth, N.R., Pezzuto, J.M. and Kinghorn, A.D., 2000, Bioactive constituents of *Thuja occidentalis*. J. Nat. Prod, 2000:1235-1238.

Charlton, J.L., 1998, Antiviral activity of lignans. J. Nat. Prod., 61:1447-1451.

Chen, C.L. and Uckun, F.M., 2000, Highly sensitive liquid chromatography-electrospray mass spectrometry (LC-MS) method for the determination of etoposide levels in human serum and plasma. J. Chromatogr. B, 744:91-98.

Chen, I.-S., Chen, J.-J., Duh, C.-Y. and Tsai, I.-L., 1997, Cytotoxic Lignans from Formosan *Hernandia nymphaeifolia*. Phytochem., 45:991-996.

Chen, J.-J., Ishikawa, T., Duh, C.-Y., Tsai, I.-L. and Chen, I.-S., 1996, New Dimeric Aporphine Alkaloids and Cytotoxic Constituents of *Hernandia nymphaeifolia*. Planta Med., 62:528-533.

Cho, J.Y., Kim, A.E., Yoo, E.S., Baik, K.U. and Park, M.H., 1999, Immunomodulatory effect of arctigenin, a lignan compound, on tumour necrosis factor-alpha and nitric oxide production, and lymphocyte proliferation. J. Pharm. Pharmacol., 51:1267-1273.

Choudhary, D.K., Kaul, B.L. and Khan, S., 1998, Cultivation and conservation of *Podophyllum hexandrum* — An overview. J. Med. Arom. Plant. Sci., 20:1071-1073.

Cockayne, T.O., 1961, Leechdoms, Wortcunning and Starcraft of Early England, pp.312-313. The Holland Press, London.

Cronquist, A., 1988, The Evolution and Classification of Flowering Plants, New York Botanical Garden Scientific Publ. Department, Bronx, N.Y.

Cseke, L., Dudareva, N. and Pichersky, E., 1998, Structure and evolution of linalool synthase. Mol. Biol. Evol., 15:1491-1498.

Damayanthi, Y. and Lown, J.W., 1998, Podophyllotoxins: Currrent Status and Recent Developments. Curr. Med. Chem., 5:205-252.

Davin, L.B. and Lewis, N.G., 2000, Dirigent Proteins and Dirigent Sites Explain the Mystery of Specificity of Radical Precursor Coupling in Lignan and Lignin Biosynthesis. Plant. Physiol., 123:453-461.

Davin, L.B., Wang, H.-B., Crowell, A.L., Bedgar, B.L., Martin, D.M., Sarkanen, S. and Lewis, N.G., 1997, Stereoselective Bimolecular Phenoxy Radical Coupling by an Auxiliary (Dirigent) Protein without an Active Center. Science, 275:362-366.

De-Eknamul, W. and Ellis, B.E., 1985, Effects of auxins and cytokinins on growth and rosmarinic acid formation in cell suspension cultures of *Anchusa officinalis*. Plant Cell. Rep., 4:50-53.

Downie, S.R., Katz-Downie, D.S. and Spalik, K., 2000, A phylogeny of Apiaceae tribe Scandiceae: evidence from nuclear ribosomal DNA internal transcribed spacer sequences. Am. J. Bot., 87:76-95.

Facino, R.M., Carini, M. and De Angelis, L., 1997, A rapid screening by liquid chromatography/ mass spectrometry and fast-atom bombardment tandem mass spectrometry of phenolic constituents with radical scavenging activity, from *Krameria triandra* roots. Rapid Commun. Mass Spectrum., 11:1303-1308.

Fang, J.-M., Hsu, K.-C. and Cheng, Y.-S., 1989, Lignans from Leaves of *Calocedrus formosana*. Phytochem., 28:3553-3555.

Fang, X., Nanayakkara, N.P.D., Phoebe, C.H., Pezzuto, J.M., Kinghorn, A.D. and Farnsworth, N.R., 1985, Plant Anticancer Agents XXXVII. Constituents of *Amanoa oblongifolia*. Planta Med., 346-347.

Flechon, A., Culine, S. and Droz, J.P., 2001, Intensive and timely chemotherapy, the key of success in testicular cancer. Crit. Rev. Onco. Hema., 37:35-46.

Frega, A., Stentella, P., DiRenzi, F., DelleChiaie, L., Cipriano, L. and Pachi, A., 1997, Assessment of self-application of four topical agents on genital warts in women. J. Eur. Acad. Dermatol.Venereol., 8:112-115.

Furuya, T., Yoshikawa, T. and Taira, M., 1984, Biotransformation of codeinone to codeine by immobilized cells of *Papaver somniferum*. Phytochem., 23:999-1001.

Gamborg, O.L., Miller, R.A. and Oijama, V., 1968, Nutrient requirements of suspension cultures of soybean root cells. Exp. Cell. Res., 50:151-158.

Ganzera, M., Moraes-Cerdeira, R.M. and Khan, I.A., 1999, Separation of *Podophyllum* lignans by micellar electrokinetic capillary chromatography (MECC). Chromatographia, 49:552-555.

Gözler, B., Gozler, T., Saglam, H. and Hesse, M. 1996, Minor lignans from *Haplophyllum cappadocicum*. *Phytochem.*, 42: 689-693.

Gross, G., 2001, Clinical diagnosis and management of anogenital warts and papillomavirus-associated lesions. Hautarzt, 52:6-17.

Gunther, R.T., 1959, The Greek Herbal of Discorides. 54-57. Hafner Publishing Co., New York pp. 54-57.

Hadimani, S.B., Tanpure, R.P. and Bhat, S.V., 1996, Asymmetric total synthesis of (-) Podophyllotoxin. Tetrahedron Lett., 37:4791-4794.

Hanauske, A.R., Wuster, K.C., Lehmer, A., Rotter, M., Schneider, P., Kaeser, F.A., Rastetter, J. and Depenbrock, H., 1995, Activity of NK 611, a new epipodophyllotoxin derivative, against colony forming units from freshly explanted human tumours in vitro. Eur. J. Cancer, 31A:1677-1681.

Hande, K.R., 1998, Etoposide: Four decades of development of a topoisomerase II inhibitor. Eur. J. Cancer, 34:1514-1521.

Hartwell, J.L. and Schrecker, A.W., 1958, Lignans of *Podophyllum*. In: 83-166. Zechmeister L. (ed.). Vol. 15, Springer-Verlag, Wien.

Hartwell, J.L., Johnson, J.M., Fitzgerald, D.B. and Belkin, M., 1953a, Podophyllotoxin from *Juniperis* Species; Savinin. J. Am. Chem. Soc., 75:235-236.

Hartwell, J.L., Schrecker, A.W. and Johnson, J.M., 1953b, The Structure of Silicicolin. J. Am. Chem. Soc., 75:2140.

Hoffmann, J.J., Wiedhopf, R.M. and Cole, J.R., 1977, Cytotoxic and Tumor Inibitory agent from *Polygala macradeniai* Gray (Polygalaceae): 4'-Demethyldeoxypodophyllotoxin. J. Pharm. Sci., 66:586-587.

Hokanson, G.C., 1979, The lignans of *Polygala polygama* (Polycalaceae): deoxypodophyllotoxin and three new lignan lactones. *J.Nat.Prod.*, 42: 378-384.

Hokanson, G.C., 1978, Podophyllotoxin and 4'-demethylpodophyllotoxin from *Polygala polygama* (Polygalaceae). J. Nat. Prod., 41:497-498.

Horwitz, S.B. and Loike, J.D., 1977, A comparison of the mechanism of action of VP-16-213 and Podophyllotoxin. J. Nat. Prod. (Lloydia), 40:82-89.

Humphreys, J.M., Hemm, M.R. and Chapple, C., 1999, New Routes for Lignin Biosynthesis Defined by Biochemical Characterization of Recombinant Ferulate 5-hydroxylase, a Multifunctional Cytochrome P450-Dependent Monooxygenase. Proc. Natl. Acad. Sci. USA., 96:10045-10050.

Ikeda, R., Nagao, T., Okabe, H., Nakano, Y., Matsunaga, H., Katano, M. and Mori, M., 1998, Antiproliferative constituents in Umbelliferae plants. III. Constituents in the root and the ground part of *Anthriscus sylvestris* Hoffm.. Chem. Pharm. Bull., 46:871-874.

Imbert, T.F., 1998, Discovery of Podophyllotoxins. Biochimie, 90:207-222.

Inamori, Y., Kato, Y., Kubo, M., Baba, K., Matsuyama, Y., Sakai, M. and Kozawa, M., 1983, Mechanisms of Insecticidal Action of Deoxypodophyllotoxin (Anthricin). I. Distribution of Deoxypodophyllotoxin in Tissues of the 5th Instar Larvae of Silkworm, *Bombyx mori* LINE. Chem. Pharm. Bull., 31:4464-4468.

Jackson, D.E. and Dewick, P.M., 1984, Biosynthesis of *Podophyllum* Lignans-II. Interconversions of artyltetralin lignans in *Podophyllum hexandrum*. Phytochem., 23:1037-1042.

Jiao, Y., Davin, L.B. and Lewis, N.G., 1998, Furanofuran lignan metabolism as a function of seed maturation in *Sesam indicum* : Methylenedioxy bridge formation. Phytochem., 49:387-394.

Jolad, S.D., Wiedhopf, R.M. and Cole, J.R., 1977, Cytotoxic Agents from *Bursera morelensis* (Burseraceae) Deoxypodophyllotoxin and a New Ligan, 5'-desmethoxydeoxy-podophyllotoxin. J. Pharm. Sci., 66:892-893.

Kadkade, P.G., 1982, Growth and podophyllotoxin production in callus tissues of *Podophyllum peltatum*. Plant Sci. Lett., 25:107-115.

Kamil, W.M. and Dewick, P.M., 1986, Biosynthesis of the lignans a- and b-peltatin. Phytochem., 25:2089-2092.

Katayama, T., Davin, L.B., Chu, A. and Lewis, N.G., 1993, Novel benzylic ether reductions in lignan biogenesis in *Forsythia intermedia*. Phytochem., 33:581-591.

Kawaguchi, K., Hirotani, M. and Furuya, T., 1988, Biotransformation of digitoxigenin by cell suspension cultures of *Strophantus amboensis*. Phytochem., 27:3475-3479.

Kier, L.B., Fitzgerald, D.B. and Burgett, S., 1963, Isolation of Podophyllotoxin from *Callitris drummondii*. J. Pharm. Sci., 52:502-503.

Kilkkinen, A., Virtamo, J., Vartiainen, E., Sankila, R., Virtanen, M.J., Adlercreutz, H. and Pietinen, P., 2004, Serum enterolactone concentration is not associated with breast cancer risk in a nested case-control study. *International Journal of Cancer*, 108: 277-280.

Kondo, K., Ogura, M., Midorikawa, Y., Kozawa, M., Tsuijbo, H. and Inamori, Y., 1990, The conversion of deoxypodophyllotoxin to epipodophyllotoxin by *Penicillium* F-0543. Agric. Biol. Chem., 52:2673-2674.

Kondo, K., Ogura, M., Midorikawa, Y., Kozawa, M., Tsuijbo, H., Baba, K. and Inamori, Y., 1989, Conversion of deoxypodophyllotoxin to podophyllotoxin-related compounds by microbes. Agric. Biol. Chem., 53:777-782.

König, R. and Hopp, J., 1993, C. Plinius secundi, Naturalis Historiae, Libri XXXVII. In: Artemis & Winkler Verlag, München.

Konuklugil, B., 1996, Aryltetralin Lignans from Genus *Linum*. Fitoterapia, 67:379-381.

Konuklugil, B., Schmidt, T.J. and Alfermann, A.W., 1999, Accumulation of Aryltetralin Lactone Lignans in Cell Suspension Cultures of *Linum nodiflorum*. Planta Med., 65:587-588.

Koulman, A. and Konuklugil, B., 2003, Lignan profile of *Linum meletonis*. *Biochem.Sys.Ecol.*, 32: 91-93.

Koulman, A., Bos, R., Medarde, M., Pras, N. and Quax, W.J., 2001, A fast and simple GC MS method for lignan profiling in *Anthriscus sylvestris* and biosynthetically related plant species. Planta Med., in press.

Koulman, A., Kubbinga, M.E., Batterman, S., Woerdenbag, H.J., Pras, N., Woolley, J.G. and Quax, W.J., 2003, A phytochemical study of lignans in whole plants and cell suspension cultures of *Anthriscus sylvestris*. *Planta Med.*, 69: 733-739.

Kozawa, M., Baba, K., Matsuyama, Y., Kido, T., Sakai, M. and Takemoto, M., 1982, Components of the Root of *Anthriscus sylvestris* Hoffm. II. Insecticidal Activity. Chem. Pharm. Bull., 30:2885-2888.

Kozawa, M., Morita, N. and Hata, K., 1978, Structure of Anthriscusin, a New Phenylpropanoid Ester from the Roots of *Anthriscus sylvestris* Hoffm. Chem. Pharm. Bull., 26:1337-1338.

Kuhnt, M., Rimpler, H. and Heinrich, M., 1994, Lignans and other Compounds from the Mixe Indian Medicinal plant *Hyptis verticillata*. Phytochem., 36:485-489.

Kuo, Y.-H., Li, S.-Y., Wu, M.-D., Huang, R.-Y., Yang Kuo, R.-L. and Chen, C.-F., 1999, A New Anti-HBaAg Lignan, Kudsumarin A, from *Kadsura matsudai* and *Schizandra arisanensis*. Chem. Pharm. Bull., 47:1047-1048.

Kupchan, S.M., Hemingway, R.J. and Hemingway, J.C., 1967, Tumor Inhibitors XIX. Desoxypodophyllotoxin, the Cytotoxic Principle of *Libocedrus decurrens*. J. Pharm. Sci., 56:408-409.

Kutney, J.P., Arimoto, M., Hewitt, G.M., Jarvis, T.C. and Sakata, K., 1991, Studies with plant cell cultures of *Podophyllum peltatum* L. I. Production of podophyllotoxin, deoxypodophyllotoxin, podophyllotoxone, and 4'-demethylpodophyllotoxin. Heterocycles, 32:2305-2309.

Lewis, N.G., 1999, A 20th century roller coaster ride: a short account of lignification. Curr. Opin. Plant Biol., 2:153-162.

Li, L., Popko, J.L., Umezawa, T. and Chiang, V.L., 2000, 5-Hydroxyconiferyl Aldehyde Modulates Enzymatic Methylation for Syringyl Monolignol Formation, a New View of Monolignol Biosynthesis in Angiosperms. J. Biol. Chem, 275:6537-6545.

MacRea, W.D. and Towers, G.H.N., 1984, Biological activities of Lignans. Phytochem., 23:1207-1220.

MacRea, W.D., Hudson, J.B. and Towers, G.H.N., 1988, a-(-)-Peltatin, an antiviral constituent of *Amanoa aff.oblongifolia*. J. Ethnopharm., 22:223-226.

Mascaux, C., Paesmans, M., Berghmans, T., Branle, F., Lafitte, J.J., Lemaitre, F., Meert, A.P., Vermylen, P. and Sculier, J.P., 2000, A systematic review of the role of etoposide and cisplatin in the chemotherapy of small cell lung cancer with methodology assessment and meta-analysis. Lung Cancer, 30:23-36.

Maxwell, A., Dabideen, D., Reynolds, W.F. and McLean, S., 1999, Neolignans from *Piper aequale*. Phytochem., 50:499-504.

Medarde, M., Pelaez-Lamamie de Clairac, R., Lopez, J.L., Ga Gravalos, D. and San Feliciano, A., 1995a, Synthesis, antitumor and antiviral evaluation of halo- and demethyl-yatein derivatives. Arch.pharm., 328:640-644.

Medarde, M., Pelaez-Lamamie de Clairac, R., Tome, F., Lopez, J.L. and San Feliciano, A., 1995b, Heterolignanolides: Antitumor activity of furyl-, thienyl-, and pyridyl analogs of lignanolides. Arch.pharm., 328:403-407.

Mimaki, Y., Kuroda, M., Asano, T. and Sashida, Y., 1999, Triterpene Saponins and Lignans from the Roots of *Pulsatilla chinensis* and their Cytotoxic Activity against HL-60 Cells. J. Nat. Prod, 62:1279-1283.

Miyazawa, M., Fukuyama, M., Yoshi, K., Kato, T. and Ishikawa, Y., 1999, Biological Active Components against *Drosophila melanogaster* from *Podophyllum hexandrum*. J. Agric. Food Chem., 47:5108-5110.

Morton, M., Arisaka, O., Miyake, A. and Evans, B., 1999, Analysis of phyto-oestrogens by gas chromatography-mass spectrometry. Environ. Toxicol. Pharmacol., 7:221-225.

Mross, K., Huttmann, A., Herbst, K., Hanauske, A.R., Schilling, T., Manegold, C., Burk, K. and Hossfeld, D.K., 1996, Pharmacokinetics and pharmacodynamics of the new podophyllotoxin derivative NK 611. A study by the AIO groups PHASE-I and APOH. Cancer Chemother. Pharmacol., 38:217-224.

Muhammad, I., Mossa, J.S., Al-Yahya, M.A., Ramadan, A.F. and El-Feraly, F.S., 1995, Further antibacterial diterpenes from the bark and leaves of *Juniperus procera* Hochst. ex Endl. Phytother. Res., 9:584-588.

Murakami, T. and Matsushima, A., 1961, Studies on the Constituents of Japanese Podophyllaceae Plants. I. On the Constituents of the Root of *Diphylleia Grayi*. Yakugaku Zasshi, 81:1596-1600.

Muranaka, T., Miyata, M., Ito, K. and Tachibana, S., 1998, Production of Podophyllotoxin in *Juniperus chinensis* Callus Cultures treated with Oligosaccharides and a Biogenetic Precursor. Phytochem., 49:491-496.

Murashige, T. and Skoog, F., 1962, A revised medium for rapid growth and bioassay with tabacco tissue cultures. Physiol. Plant, 15:473-497.

Peraza-Sanchez, S.R. and Pena-Rodriguez, L.M., 1992, Isolation of picropolygamain from the resin of *Bursera simaruba*. *J.Nat.Prod.*, 55: 1768-1771.

Pichersky, E. and Gang, D.R., 2000, Genetics and biochemistry of secondary metabolites in plants: an evolutionary perspective. Trends Plant Sci., 5:439-445.

Plunkett, G.M. and Downie, S.R., 1999, Major lineages within Apiaceae subfamily Apioideae: A Comparison of Chloroplast Restriction Site and DNA Sequence Data. Am. J. Bot., 86:1014-1026.

Pras, N. and Woerdenbag, H.J., 1999, Production of Secondary Metabolites by Bioconversion. In: Biotechnology; Secondary Metabolites. pp.265-303. Ramawat, K.G. and Mérillon, J.M., (eds). Science Publishers, Inc., USA.

Pras, N., Woerdenbag, H.J., Batterman, S., Visser, J.F. and Malingré, Th.M., 1993, *Mucuna pruriens*: improvement of the biotechnological production of the anti-Parkinson drug L-DOPA by plant cell selection. Pharm. World Sci., 15:263-268.

Provan, G.J. and Waterman, P.G., 1985, Picropolygamain: a new lignan from *Commiphora incisa* resin. *Planta Med.*, 51: 271-272.

Raffauf, R.F., Kelley, C.J., Ahmad, Y. and Le Quesne, P.W., 1987, a- and b-peltatin from *Eriope macrostachya*. J. Nat. Prod, 50:772-773.

Rahman, M.A., Dewick, P.M., Jackson, D.E. and Lucas, J.A., 1990a, Lignans of *Forsythia intermedia*. Phytochem., 29:1971-1980.

Rahman, M.A., Dewick, P.M., Jackson, D.E. and Lucas, J.A., 1990b, Biosynthesis of lignans in *Forsythia intermedia*. Phytochem., 29:1841-1846.

Rai, L.K., Prasad, P. and Sharma, E., 2000, Conservation threats to some important medicinal plants of the Sikkim Himalaya. Biol. Cons., 93:27-33.

Rollinger, J.M., Zidorn, C., Dobner, M.J., Ellmerer, E.P. and Stuppner, H., 2003, Lignans, phenylpropanoids and polyacetylenes from *Chaerophyllum aureum* L. (Apiaceae). *Zeit. f Naturfors,*, 58: 553-557.

San Feliciano, A., del Corral, J.M.M., Lopez, J.L. and de Pascual, T.B., 1992, Lignans from Polar Extracts of *Juniperus thurifera*. Phytochem., 31:267-270.

San Feliciano, A., Miguel del Corral, J.M., Gordaliza, M. and Castro, M.A., 1989, Acetylated lignans from *Juniperus sabina*. *Phytochem.*, 28: 659-660.

Schenk, R.V. and Hildebrandt, A.C., 1972, Medium and techniques for induction and growth of monocotyledonous and dicotyledonous plant cell cultures. Can. J. Bot., 50:199-204.

Sconocchia, S., 1983, Scribonivs largvs compositiones. In: Hansen G.C. (ed.). B.G. Teubner Verlagsgeselffschaft, Leipzig.

Seidel, V., Windhövel, J., Eaton, G., Alfermann, A.W., Arroo, R.R.J., Medarde, M., Petersen, M. and Woolley, J.G., 2002, Biosynthesis of podophyllotoxin in *Linum album* cell cultures. *Planta*, 215: 1031-1039.

Shaari, K. and Waterman, P.G., 1994, Podophyllotoxin-type lignans as major constituents of the stems and leaves of *Casearia clarkei*. J. Nat. Prod, 57:720-724.

Shelton, A.L., 2000, Variable chemical defences in plants and their effects on herbivore behaviour. Evo. Ecol. Res., 2:231-249.

Smollny, T., Wichers, H.J., Kalenberg, S., Shahsavari, A., Petersen, M. and Alfermann, A.W., 1998, Accumulation of podophyllotoxin and related lignans in cell suspension cultures of *Linum album*. Phytochem., 48:975-979.

Stähelin, H.F. and Von Wartburg, A., 1991, The chemical and biological route from podophyllotoxin glucoside to etoposide: Ninth cain memorial award lecture. Cancer Res., 51:5-15.

Stähelin, H.F., 1973, Activity of a new glycosidic lignan derivative (VP 16-213) related to podophyllotoxin in experimental tumors. Eur. J. Cancer, 9:215-221.

Tsai, I.-L., Hsieh, C.-F. and Duh, C.-Y., 1998, Additional cytotoxic neolignans from *Persea obovatifolia*. Phytochem., 48:1371-1375.

Udino, L., Abaul, J., Bourgeois, P., Corrichon, L., Duran, H. and Zedde, C., 1999, Lignans from the seeds of *Hernandia sonora*. Planta Med., 65:279-281.

Umezawa, T., Davin, L.B. and Lewis, N.G., 1991, Formation of lignans (-)-secoisolariciresinol and (-)-matairesinol with *Forsythia intermedia* cell-free extracts. J. Biol. Chem, 266:10210-10217.

Van Uden, W., Bos, J.A., Boeke, G.M., Woerdenbag, H.J. and Pras, N., 1997, The large scale isolation of deoxypodophyllotoxin from rhizomes of *Anthriscus sylvestris* followed by its bioconversion into 5-methoxypodophyllotoxin b-D-glucoside by cell cultures of *Linum flavum*. J. Nat. Prod, 60:401-403.

Van Uden, W., Bouma, A.S., Bracht Waker, J.F., Middel, O., Wichers, H.J., De Waard, P., Woerdenbag, H.J., Kellogg, R.M. and Pras, N., 1995, The production of podophyllotoxin and its 5-methoxy derivative through bioconversion of cyclodextrin-complexed desoxypodophyllotoxin by plant cell cultures. Plant Cell Tiss. Org. Cult., 42:73-79.

Van Uden, W., Homan, B., Woerdenbag, H.J., Pras, N., Malingré, Th.M., Wichers, H.J. and Harkes, M., 1992, Isolation, purification, and cytotoxicity of 5- methoxypodophyllotoxin, a lignan from a root culture of *Linum flavum*. J. Nat. Prod, 55:102-110.

Van Uden, W., Pras, N. and Woerdenbag, H.J., 1994,. *Linum* Species (Flax): In Vivo and In Vitro Accumulation of Lignans and Other Metabolites. In: Medicinal and Aromatic Plants VI. pp.219-244. Bajaj Y.P.S. (ed.). Springer-Verlag, Berlin Heidelberg.

Van Uden, W., Pras, N., Batterman, S., Visser, J.F. and Malingré, Th.M., 1990b, The accumulation and isolation of coniferin from a high-producing cell suspension of *Linum flavum* L. Planta, 183:25-30.

Van Uden, W., Pras, N., Batterman, S., Visser, J.F. and Malingré, Th.M., 1990c, The accumulation of podophyllotoxin-b-D-glucoside by cell suspension cultures derived from the conifer *Callitris drummondii*. Plant Cell Rep., 9:257-260.

Van Uden, W., Pras, N., Visser, J.F. and Malingré, Th.M., 1989, Detection and identification of podophyllotoxin produced by cell cultures derived from *Podophyllum hexandrum* Royle. Plant Cell Rep., 8:165-168.

Van Uden, W., Pras, N., Vossebeld, E.M., Mol, J.N.M. and Malingré, Th.M., 1990a, Production of 5-methoxypodophyllotoxin in cell suspension cultures of *Linum flavum* L. Plant Cell Tiss. Org. Cult., 20:81-88.

Weiss, S.G., Tin-Wa, M., Perdue, R.E. and Farnsworth, N.R., 1975, Potential Anticancer Agents II: Antitumor and Cytotoxic Lignans from *Linum album* (Linaceae). *J.Pharm.Sci.*, 64: 95-98.

Whetten, R.W., MacKay, J.J. and Sederoff, R.R., 1998, Recent Advances in Understanding Lignin Biosynthesis. Annu. Rev. Plant Physiol. Plant Mol. Biol., 49:585-609.

Wichers, H.J., Harkes, M. and Arroo, R.R.J., 1990, Occurrence of 5-methoxypodophyllotoxin in plants, cell cultures and regenerated plants of *Linum flavum*. Plant Cell Tiss. Org. Cult., 23:93-100.

Wichers, H.J., Versluis-De haan, G.G., Marsman, J.W. and Harkes, M., 1991, Podophyllotoxin related lignans in plants and cell cultures of *Linum flavum*. Phytochem., 30:3601-3604.

Wickramaratne, B.D.M., Mar, W., Chai, H., Castillo, J.J., Farnsworth, N.R., Soejarto, D.D., Cordell, G.A., Pezzuto, J.M. and Kinghorn, A.D., 1995, Cytotoxic Constituents of *Bursera permollis*. Planta Med., 61:80-81.

Witterland, A.H., Koks, C.H. and Beijnen, J.H., 1996, Etoposide phosphate, the water soluble prodrug of etoposide. Pharm. World Sci., 18:163-170.

Woerdenbag, H.J., Van Uden, W., Frijlink, H.W., Lerk, C.F., Pras, N. and Malingré, Th.M., 1990, Increased podophyllotoxin production in *Podophyllum hexandrum* cell suspension cultures after feeding coniferyl alcohol as a beta-cyclodextrin complex. Plant Cell Rep., 9:97-100.

Wong, S.K., Tsui, S.K., Kwan, S.Y., Su, X.L. and Lin, R.C., 2000, Identification and characterization of *Podophyllum emodi* by API-LC/MS/MS. J. Mass Spectrom., 35:1246-1251.

Xia, Z.-Q., Costa, M.A., Proctor, J., Davin, L.B. and Lewis, N.G., 2000, Dirigent-mediated podophyllotoxin biosynthesis in *Linum flavum* and *Podophyllum peltatum*. Phytochem., 55:537-549.

Yamaguchi, H., Arimoto, M., Tanoguchi, M., Ishida, T. and Inoue, K., 1982, Studies on the Constituents of the Seeds of *Hernandia ovigera* L. III. Structures of Two New Lignans. Chem. Pharm. Bull., 30:3212-3218.

Yu, P.-Z., Wang, L.-P. and Chen, Z.-N., 1991, A New Podophyllotoxin-type lignan from *Dysosma versipellis* var. *tomentosa*. J. Nat. Prod, 54:1422-1424.

Zangerl, A.R. and Berenbaum, M.R., 1997, Cost of chemically defending seeds: Furanocoumarins and *Pastinaca sativa*. Am. Nat., 150:491-504.

Zhao, Y., Nookandeh, A., Schneider, B., Sun, X., Schmitt, B. and Stöckigt, J., 1999, Lignans from *Torreya jackii* identified by stopped-flow high-performance liquid chromatography-nuclear magnetic resonance spectroscopy. J. Chromatogr.A, 837:63-71.

CHAPTER 12

Accumulation of Biologically Active Furanocoumarins within In vitro Cultures of Medicinal Plants

Halina Ekiert

LINEAR FURANOCOUMARINS— GENERAL INFORMATION

One of the most therapeutically interesting subgroups of coumarin compounds are linear furanocoumarins, whose structure is based on a prototypic compound referred to as psoralen. Thus, the entire group of the derivatives is termed psoralens.

The simplest and the most common in nature, linear structures are methoxy derivatives of psoralen: xanthotoxin (8-methoxypsoralen, 8-MOP), bergapten (5-methoxypsoralen, 5-MOP) and isopimpinellin (5,8-dimethoxypsoralen). Imperatorin and isoimperatorin are also frequently encountered (Fig. 12.1) (Soine 1964; Nielsen 1970, 1971; Tandon and Rastogi 1979; Beyrich 1981; Kohlmünzer 1998). Furanocoumarins occur in only a few botanical families such as *Apiaceae, Fabaceae, Moraceae, Rutaceae and Orchidaceae* (Nielsen 1970, 1971; Hegnauer 1973; Hoppe 1975; Kohlmünzer 1998).

All furanocoumarins, linear as well as angular, derive from 7-hydroxycoumarin, umbelliferone, their biogenetic precursor (Fig.1). Till date, all the stages of biogenetic pathway leading to methoxy derivatives of psoralen, i.e. xanthotoxin, bergapten and isopimpinellin are yet to be established (Dewick 1997). Among other centers, research groups from university in Freiburg and in Marburg/Lahn (Germany) have focussed interest on this issue. These studies are usually carried out on plant in vitro cultures (Matern et al., 1988, 1999; Ekiert 1994; Hehmann et al., 2004).

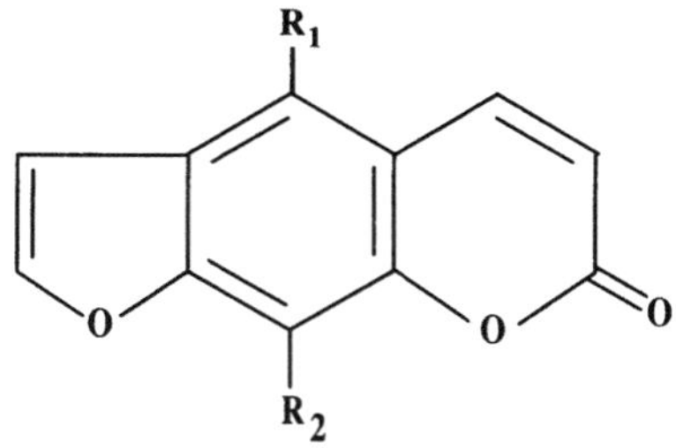

	R_1	R_2
Psoralen	H	H
Xanthotoxin	H	OCH_3
Bergapten	OCH_3	H
Isopimpinellin	OCH_3	OCH_3
Imperatorin	H	$OCH_2-CH=C(CH_3)_2$

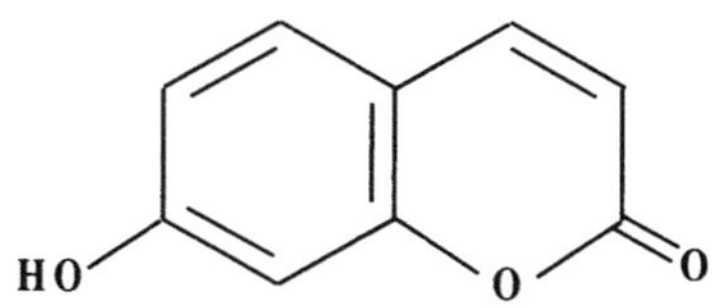

Umbelliferone

Fig.12.1. Chemical structure of the analyzed linear furanocoumarins and their precursor, umbelliferone.

BIOLOGICAL PROPERTIES OF LIENAR FURANO-COUMARINS AND THEIR THERAPEUTIC SIGNIFICANCE

Linear furanocoumarins are capable of sensitizing the cells of the human skin to UV radiation and inducing repigmentation of discoloured skin. They also exhibit antiproliferative properties. Some of the compounds are known for their fungicidal, molluscicidal and insecticidal actions. Studies conducted in recent years indicate that certain compounds within this group are calcium and/or potassium channel blockers.

The biological properties of linear furanocoumarins have been used in photochemotherapy (PUVA – therapy) of a number of dermatological disorders, characterized by disarrayed skin pigmentation and excessive cell proliferation. Diseases treated with psoralens include vitiligo, psoriasis, mycosis fungoides, atopic eczema, lichen planus, pityriasis lichenoides, parapsoriasis en plaques, urticaria pigmentosa, alopecia areata and others (Soine 1964; Pathak et al., 1981; Ben Hur and Pill-Soon 1984; Rodigiero et al., 1984; Roenigh 1984; Rodigiero 1985; Turjanmaa et al., 1985; Plewig et al., 1986; Rodigiero and Dall' Acqua 1986; Wolff and Tessa 1986; Rimpler 1990; Bethea et al., 1999; Ojala 2001).

The recently discovered capacity of the compounds to block Ca^{2+} and K^+ channels opens new prospects of therapeutic application of this group of metabolites in fields such as cardiology and neurology (e.g. successful trials of efficacy of these compounds in treatment of sclerosis multiplex) (Vuorela et al., 1988; Sanmann et al., 1994; Ditzen et al., 1995; Ojala 2001).

PLANT IN VITRO CULTURES AS A SOURCE OF LINEAR FURANOCOUMARINS

Due to their therapeutic potential, linear furanocoumarins are a frequent topic of plant biotechnological studies. These studies concentrate most often on *Ammi majus* L., a species considered to be the richest natural source of linear furanocoumarins. *A. majus* L. is a focus of interest of Asian (mainly in India) as well as European (e.g. in Poland, Germany) research centres (Sehgal 1972; Grewal et al., 1976; Hamerski and Matern 1988a, 1988b; Hamerski et al., 1990a, 1990b; Ekiert 1990, 1993; Koul and Koul 1993; Purohit et al., 1995; Królicka et al., 2001a, 2001b; Pande et al., 2002; Sidwa-Gorycka et al., 2003). Attempts are on to investigate the biosynthetic potential of other representatives of the family *Apiaceae*, namely, *Petroselinum sp.*, *Pastinaca sativa* L., *Heracleum sphondylium* ssp. *sphondylium* L., *Daucus carota* L., *Pimpinella anisum* L. or *Anethum graveolens* L. (Ekiert 2000). Furthermore, examination of different types of in vitro cultures of *Ruta graveolens* L. (*Rutaceae*) are currently in progress (Table 12.1) (Reinhard et al., 1968, 1971; Steck et al., 1971; Petit-Paly et al., 1989; Massot et al., 2000).

Studies into the accumulation of psoralens and others coumarins compounds within *in vitro* cultures are often centred on their endogenous biosynthesis and optimization of culture conditions. Numerous studies report on the possibility of using both biotic and abiotic elicitors in order to stimulate biosynthesis of this group of metabolites. These experiments commonly utilize fungal elicitors, Pmg and Ac (Table 12.2) (Tietjen et al., 1983; Hauffe et al., 1986; Scheel et al., 1986; Hamerski and Matern 1988a, 1988b; Hamerski et al., 1990a, 1990b; Koul and Koul 1993; Bohlmann et al., 1995; Kitamura et al., 1998; Sharan et al., 1998; Siatka and Reichling 2000; Królicka et al., 2001a; Smith et al., 2001).

Stimulation of furanocoumarin biosynthesis facilitates attempts to elucidate the biosynthetic pathway of methoxy derivatives of psoralen, i.e. xanthotoxin, bergapten and isopimpinellin.

Transformation attemps with Agrobacterium rhizogenes have also been undertaken with *Ammi majus* L. in order to obtain hairy roots culture and high yield of furanocoumarins (Królicka et al., 2001b).

Table 12.1 Examples of accumulation of linear furanocoumarins and umbelliferone in cell, tissue and organ cultures of medicinal plants.

Metabolite	Plant species	Type of culture	Yield (max.)	Authors
Bergapten	*Ammi majus* L.	Suspension	8.43 mg% d.w.	Ekiert 1988
	Ammi majus L.	Callus	25.16 mg% d.w.	Ekiert 1988, 1993
	Ammi majus L.	Callus	no data	Koul and Koul 1993
	Anethum graveolens L.	Callus	no data	Ekiert 1997 (unpublished)
	Daucus carota L.*	Callus	0.023**/0.11*** μg/g f.w.	Zobel and Brown 1992
	Heracleum sphondylium ssp. *sphondylium* L.	Callus	2.5-3.6 μg/g d.w.	Tirillini and Ricci 1998
	Pastinaca sativa L.*	Callus	0.25**/0.93*** μg/g f.w.	Zobel and Brown 1992
	Petroselinum sativum L.	Callus	no data	Reinhard 1967
	Pimpinella anisum L.	Callus	5 mg% d.w.	Kartnig et al. 1975
	Ruta graveolens L.	Callus	160 mg/100g f.w.	Reinhard et al., 1968
	Ruta graveolens L.	Suspension	0,28 – 1.27 mg/l	Reinhard et al., 1968
	Ruta graveolens L.	Callus	ab. 0.08g/100g d.w.	Massot et al., 2000
	Ruta graveolens L.	Suspension	ab. 0.01g/100g d.w.	Massot et al., 2000
	Ruta graveolens L.	Shoots	ab. 0.3g/100g d.w.	Massot et al., 2000
Xanthotoxin	*Ammi majus* L.	Suspension	4.08 mg% d.w.	Ekiert 1988
	Ammi majus L.	Callus	6.67 mg% d.w.	Ekiert 1988, 1993
	Ammi majus L.	Shoot-regenerating callus (4 weeks old)	10.6 mg/g d.w.	Purohit et al., 1995
	Ammi majus L.	In vitro bearing immature green fruits	11.9 mg/g d.w.	Pande et al., 2002
	Daucus carota L.	Callus	0.051**/0.034*** μg/g f.w.	Zobel and Brown 1992
	Heracleum sphondylium ssp. *sphondylium* L.	Callus	0.7 – 1.2 μg/g d.w.	Tirillini and Ricci 1998
	Pastinaca sativa L.	Callus	0.32**/0.35*** μg/g f.w.	Zobel and Brown 1992
	Ruta graveolens L.	Callus	220mg/100g f.w.	Reinhard et al., 1968
	Ruta graveolens L.	Suspension	2.15-6.0 mg/l	Reinhard et al., 1968
	Ruta graveolens L.	Callus	ab. 0.05g/100g d.w.	Massot et. al., 2000
	Ruta graveolens L.	Suspension	ab. 0.01g/100g d.w.	Massot et. al., 2000
	Ruta graveolens L.	Shoots	ab. 0.49g/100g d.w.	Massot et. al., 2000

Contd.

Table 12.1 *Contd.*

Metabolite	Plant species	Type of culture	Yield (max.)	Authors
Isopimpinellin	*Ammi majus* L.	Suspension	3.55 mg% d.w.	Ekiert 1988
	Ammi majus L.	Callus	28.57 mg% d.w.	Ekiert 1988, 1993
	Heracleum sphondylium ssp. *sphondylium* L.	Callus	3.8 – 5.1 µg/g d.w.	Tirillini and Ricci 1998
	Ruta graveolens L.	Callus	ab.0.1g/100g d.w.	Massot et al., 2000
	Ruta graveolens L.	Suspension	ab. 0.02g/100g d.w.	Massot et al., 2000
	Ruta graveolens L.	Shoots	ab.0.28g/100g d.w.	Massot et al., 2000
	Thamnosma montana Torr. and Frem.	Callus	no data	Kutney et al., 1973
Imperatorin	*Ammi majus* L.	Suspension	2.89 mg% d.w.	Ekiert 1988
	Ammi majus L.	Callus	22.11 mg% d.w.	Ekiert 1988, 1993
	Heracleum sphondylium ssp. *sphondylium* L.	Callus	0.7 – 1.5 µg/g d.w.	Tirillini and Ricci 1998
Psoralen	*Daucus carota* L.	Callus	-**/ trace***	Zobel and Brown 1992
	Pastinaca sativa L.	Callus	-**/ trace***	Zobel and Brown 1992
	Ruta graveolens L.	Callus	ab. 0.02g/100g d.w.	Massot et al., 2000
	Ruta graveolens L.	Shoots	ab. 0.23 g/100g d.w.	Massot et al., 2000
Umbelliferone	*Ammi majus* L.	Suspension	23.88 mg% d.w.	Ekiert 1988
	Ammi majus L.	Callus	35.83 mg% d.w.	Ekiert 1988, 1993
	Ammi majus L.	Callus	no data	Koul and Koul 1993
	Anethum graveolens L.	Callus	ca. 10 mg% d.w.	Kartnig et al., 1975
	Anethum graveolens L.	Callus	no data	Ekiert 1997 (unpublished)
	Haplophyllum patavinum	Callus	no data	Cappelletti et al., 1998
	(L.) G. Don	Suspension	no data	Cappelletti et al., 1998
	Petroselinum sativum L.	Callus	no data	Reinhard 1967
	Pimpinella anisum L.	Callus	5 mg% d.w.	Kartnig et al., 1975
	Ruta graveolens L.	Suspension	0.001mg/l	Reinhard et al., 1968

*bergapten + isopimpinellin
** surface concentration, *** remainder concentration

Table 12.2 Examples of elicitors used in order to stimulate the synthesis of furanocoumarins and umbelliferone in cultures in vitro.

Plant species	Metabolite	Elicitor	Author
Ammi majus L.	Umbelliferone, isopimpinellin, marmesin, ammirin	Pmg-elicitor, Ac-elicitor and scleroglucan	Hamerski et al., 1990 a, b, Hamerski and Matern 1988 a, b
Ammi majus L.	Bergapten	Yeast extract	Koul and Koul 1993
Ammi majus L.	Umbelliferone Marmesinin	Silicon dioxide, jasmonic acid, lysates of bacteria (*Enterobacter sakazaki, Erwinia chrysanthemi*), scleroglucan	Królicka et al., 2001a
Ipomoea batatas (L.) Lam. cv Centennial	Umbelliferone	Mercuric chloride	Smith et al., 2001
Petroselinum crispum L.	Xanthotoxin, bergapten, isopimpinellin, psoralen	Pmg-elicitor and Ac-elicitor	Hauffe et al., 1986
Petroselinum hortense L.	Psoralen, xanthotoxin, bergapten, isopimpinellin	Pmg-elicitor and Ac-elicitor	Tietjen et al., 1983
Ruta graveolens L.	Xanthotoxin	*Rhodotorula rubra*	Bohlmann et al., 1995

Pmg-elicitor – elicitor from fungus *Phytophthora megasperma* f. sp. *glycinea*
Ac-elicitor – elicitor from fungus *Althernaria carthami* Chowdhury
Scleroglucan – β-D-glucan (1→3), (1→6) from fungus *Sclerotium sclerotiosum*

The influence of different factors on the accumulation of linear furanocoumarins in various types of in vitro cultures of 4 medicinal plant species has been extensively studied in our laboratory, and the results are summarized below.

The objects of these studies are: bergapten, xanthotoxin, isopimpinellin, imperatorin, psoralen and their biogenetic precursor, umbelliferone (Fig.12.1).

DISTRIBUTION, IMPORTANCE AND CHEMISTRY OF THE INVESTIGATED PLANT SPECIES

1. *Ammi majus* L. (bishop's weed), *Apiaceae*

A. majus L. is a species occurring mostly in the Mediterranean area, as well as in southern Europe and African countries (Morocco, Egypt, Ethiopia). This species can also be found in other regions with similar climate such as

Argentine, southern USA, more rarely in Australia and New Zealand (Hegi 1965; Tutin et al., 1968; Hoppe 1975).

A. majus L. has a special position in the history of phytotherapy. It was known and used by ancient Egyptians to treat vitiligo. Its therapeutic use was first mentioned in the Ebers' papyrus dated from c1500 B.C. (Fahmy and Abu-Shady 1947; Ramadan 1982).

A. majus L. is considered to be the richest natural source of linear furanocoumarins. Plants of different origins have been demonstrated to contain more than 20 various psoralen derivatives. The first studies into chemical composition of this plant were carried out in the 1940s in research centers in Cairo (Egypt). The main components isolated from the fruits of this species were xanthotoxin, bergapten and imperatorin, marmesin and its glucoside, marmesinin, isopimpinellin, isoimperatorin, alloimperatorin and ammirin (Hegnauer 1973; Hoppe 1975; Ekiert 1993). Plants of American origin were shown to contain eight other linear furanocoumarins (Ivie 1978).

Representatives of other groups of coumarin compounds are very scarce and include a few angular furanocoumarins and simple coumarins, such as umbelliferone and umbelliprenin (Ekiert 1988). Current Egyptian-German studies have discovered three other new linear furanocoumarins (e.g., apiumetin and 5-methoxy-8-hydroxypsoralen), three pyrano-coumarins (dihydroxy-xanthyletin, lomatin, ammirol) and simple coumarins (e.g., scopoletin) in fruits originating from Egypt. The most splendid achievement was the isolation and identification of three simple coumarins: isoarnottinin and its two glucosides, till then unknown to occur in nature (Elgamal et al., 1993).

Plants originating in Egypt are the most valued material of all with xanthotoxin as its dominating component (about 0.5% d.w.). Plants acclimated for pharmaceutical purposes to the Middle European countries (Germany – East Germany, Poland, Czechoslovakia, Romania) and the Soviet Union in the 1960s usually yielded lower contents of xanthotoxin (Ekiert 1993).

Fruits of this species have served as a source of material used to manufacture dermatological preparations such as Meladinine (Memphis Chemical Co., Basotherm, Pharma-Medica, Promedica, Agpharm), Oxsolaren (Elder, Gerot), Ammifurin (Medexport), Geralen (Gerot), Lamadin (Alkaloid), and Psoraderm-5 (Pharmacal, Begaderm). The main components of these preparations are bergapten, xanthotoxin, isopim-pinellin and imperatorin (Podlewski and Chwalibogowska-Podlewska 1986, 1999; Wolff and Tessa 1986).

2. *Pastinaca sativa* L. (common parsnip, wild parsnip), *Apiaceae*

Pastinaca sativa L. grows wild in temperate climates in all the continents. It is most commonly found in Europe and also frequently encountered in Caucasia and Siberia (Hegi 1965; Tutin et al., 1968; Hoppe 1975; Ruminska and Ozarowski 1990). Known as a vegetable to ancient Greeks and Romans who consumed its edible sweet roots with carroty scent, it is still cultivated and eaten as a vegetable, mostly in Germany. The roots contain 10–12% of sugars, vitamins (C, B_1, B_2) and N, P, Ca and Fe compounds (Hoppe 1975, Ruminska and Ozarowski 1990).

Native to Eurasia, *P. sativa* was introduced into North America in 1609 with the first European settlement at Jamestown (Virginia). By 1818, it had become extensively naturalized throughout eastern North America and today remains one of the most abundant weeds of roadsides and waste places (Berenbaum et al., 1984).

Fruits of *P. sativa* contain many linear furanocoumarins, dominant ones being xanthotoxin, bergapten and imperatorin (contents between 155 – 260 mg/100 g d.w.). They can be a potential source of these pharmaceutically important components (Beyrich 1966, 1981). Due to the not fully successful acclimatization attempts with *Ammi majus* L. in the Middle European countries, in the 1960s, attention was focussed on parsnip, a plant possessing many natural stands in this area. The studies, carried out mostly in Germany (East Germany), were concentrated on 3 species of the genus *Pastinaca*: *P. opaca* Bernh. *P. urens* Req. and *P. sativa* L., of which *P. sativa* was selected as the best and an easy to cultivate source of pharmaceutical material (Beyrich 1965a,b, 1966, 1967, 1981).

In the Soviet Union, *P. sativa* fruits have been used to manufacture a dermatological preparation named Beroxan (Ozarowski 1980), while its roots have served as a source for developing a preparation (Pastinacin) that lowers blood pressure and prevents formation of renal calculi and gallstones (Ruminska and Ozarowski 1990). Studies conducted on different parsnip varieties cultivated in Canada for their culinary values have demonstrated the widely variable contents of furanocoumarins in their fruits (0.06 – 0.19% d.w.) (Ceska et al., 1987). Research conducted in the USA (Berenbaum et al., 1984) has revealed that the content of furano-coumarins in ripe fruits of the wild plants is usually ca. 3 times higher than that in fruits of cultivated plants, and amounts to ca. 1 % d.w. Imperatorin was shown to be the main compound in fruits of seven examined cultivated varieties (Berenbaum et al., 1984); however, other studies have indicated that bergapten or xanthotoxin can also dominate quantitatively in fruits (Ceska et al., 1987).

Psoralens occur mainly in the fruits of *Pastinaca sativa* L. Apart from fruits, other reproductive parts of the plant (flowers and buds) also contain high quantities of these metabolites (Zangerl et al., 1997). Roots contain very low quantities of furanocoumarins (Zogg et al., 1989).

Due to a wide variability in the quantitative composition of furanocoumarins in *P. sativa* fruits, it was proposed to establish its in vitro cultures and also to attempt to control the accumulation of these group of metabolites.

3. *Ruta graveolens* L. (common rue, garden rue), *Rutaceae*

Ruta graveolens L. is also a species characteristic of southern Europe. It grows wild in the Balkan Peninsula and Crimea. It can also be found in southern France and Bulgaria, as also on rocky and limestone slopes of mountains in Greece, Spain, and in southern Alps. Owning to its medicinal and decorative values, it has been cultivated in Europe, North America and SW Asia (Hegi 1965; Tutin et al., 1968; Hoppe 1975). Garden rue also possesses historical importance. The earliest information on its use is dated back to Hippocrates' times (fifth century B.C.) (Becela-Deller 1995).

Until now, more than 200 different components have been isolated from this plant, among which four groups of secondary metabolites can be distinguished: coumarin compounds, alkaloids, flavonoids and volatile oil components. Of coumarin compounds, more than 40 compounds were isolated, which include simple coumarins, and furano- and pyrano-coumarins. Rutarin is the quantitatively dominating metabolite in the overground parts of the plant. The parent compound of furano-coumarins—psoralen and its derivatives: bergapten, xanthotoxin and isopimpinellin are the next most important metabolites.

Two groups, furoquinolin and acridon derivatives, dominate among the alkaloids. Flavonoid compounds are represented *inter alia* by rutin (2 – 5%), which was first isolated from this plant by a pharmacist, August Weiss, in 1842. Concomitant occurrence of volatile oil and alkaloids in this species is unique in the plant kingdom (Hegnauer 1973; Hoppe 1975; Eilert 1994; Kohlmünzer 1998).

Due to the exceptional richness of chemical composition, this species is a frequent focus of research (Smolarz et al., 1997; Milesi et al., 2001; Springob 2001). It was chosen from among 4 species of the genus *Ruta* (*R. montana, R. chalepensis, R. angustifolia, R. graveolens*) as the species containing the most valuable quantitatively composition of linear furanocoumarins. In Western Europe—primarily France—this species has been proposed as a source of furanocoumarins for pharmaceutical and cosmetic applications (Milesi et al., 2001).

4. *Ruta graveolens* ssp. *divaricata* (Tenore) Gams, *Rutaceae*

This plant is considered to be a variety of *R. graveolens* L. by Engler (1964), while other authors, Hegi (1965) and Hoppe (1975), described this plant as a subspecies.* Natural stands of this plant occur in the Mediterranean area—in Italy, on the Adriatic Sea coast (Nanos Mountains), Balkan Peninsula (Hoppe 1975).

The chemical composition of this plant has not been elucidated. It is

*In our laboratory, we have also adopted the latter view. It has been accepted as well by researchers from the Institute for Biosciences, Würzburg University (Germany), who shared with us the plant material and *in vitro* cultures.

known that the dominating metabolites of its volatile oil are different than in the oil from *R. graveolens* L. (terpene hydrocarbons and non-terpene aliphatic compounds, respectively) (Kubeczka K. H. op. cit. Abou-Mandour 1982). Our studies of the coumarin fraction from overground parts of this plant have led to the isolation and, later, identification of psoralen, bergapten, xanthotoxin, isoimperatorin, rutamarin (dihydrofuranocoumarin) and daphnoretin methylether (coumarine dimer) (Ekiert and Kisiel 2002).

EXPERIMENTAL IN VITRO CULTURES

INFLUENCE OF GROWTH SUBSTANCES ON ACCUMULATION OF FURANOCOUMARINS

I. *Ammi majus* L. in vitro cultures

Callus cultures of this species (Fig. 12.2) were maintained on 12 variants of Linsmaier-Skoog (L-S) medium (1965) with differences in the quantitative composition of growth substances (0.1 – 10 mg/l): α-naphthaleneacetic acid (NAA) and 6-benzylaminopurine (BAP). Four-week-old subcultures on the tested media yielded 1.4 – 4.4-fold biomass growth. HPLC analysis of the extracts of callus tissues cultured on the tested variants of the medium have demonstrated the presence of linear furanocoumarins: bergapten, xanthotoxin, isopimpinellin, imperatorin, psoralen and umbelliferone. Imperatorin was the quantitatively dominating metabolite in callus tissues (max. 169.3 mg/100 g d.w.). One of the tested medium facilitated a high accumulation of xanthotoxin (145.3 mg/100 g d.w.). Contents of the remaining furanocoumarins were lower, amounting to less than 15 mg/100 g d.w. for bergapten and less than 7 mg/100 g d.w. for isopimpinellin. Very low contents of psoralen (0.13 – 0.61 mg/100 g d.w.) were detected only in calluses cultured on 4 media under examination. Umbelliferone was accumulated in large quantities (max. 536.3 mg/100 g d.w.) in calluses from the majority of the media.

Total content of the furanocoumarins and umbelliferone changed from 41.0 to 871.1 mg/100 g d.w. (Table 12.3).

Table 12.3 Minimum and maximum contents of the metabolites in callus tissues of *Ammi majus* L. cultured on different variants of L-S medium.

Metabolites	Contents (mg/100 g d.w.)
Psoralen	0.13 – 0.61
Bergapten	1.17 – 14.36
Xanthotoxin	0.37 – 145.33
Isopimpinellin	0.39 – 6.37
Imperatorin	3.10 – 169.27
Umbelliferone	13.28 – 536.29
Total	40.95 – 871.05

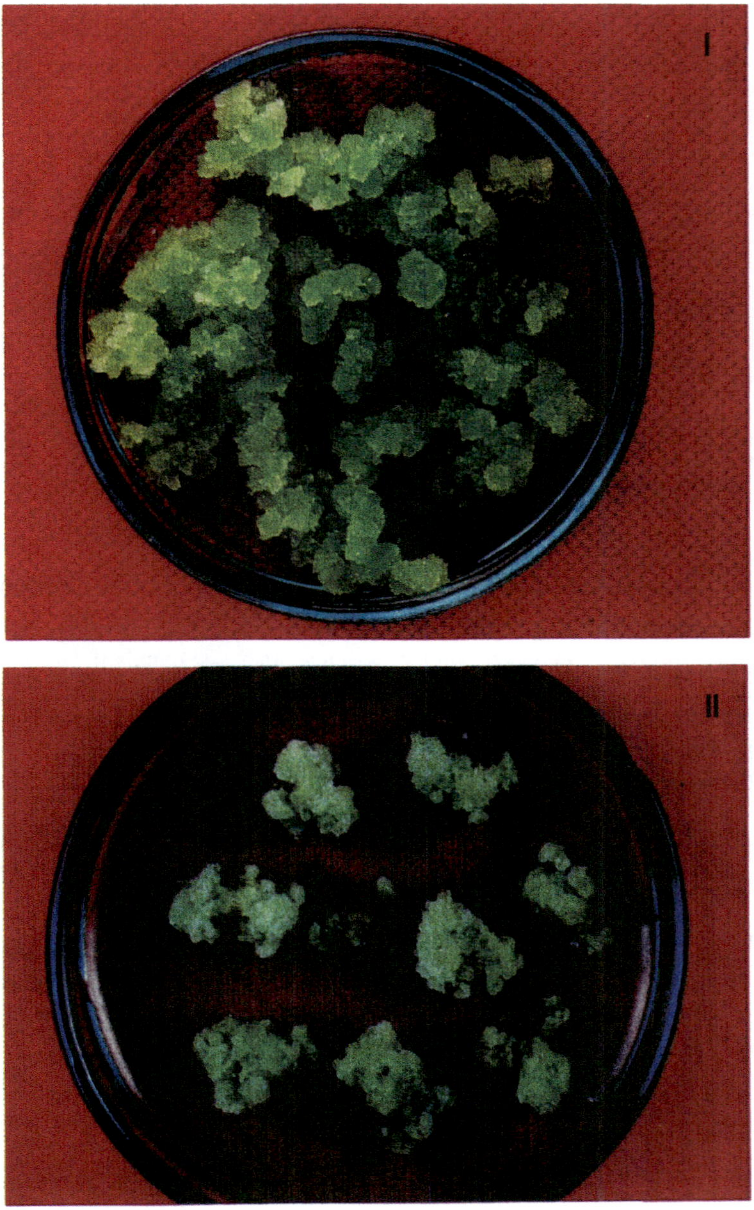

Fig.12.2 I Callus tissues of *Ammi majus* L. Agar culture, L-S medium, NAA - 2mg/l, BAP – 2mg/l.

II Callus tissues of *Pastinaca sativa* L. Agar culture, L-S medium, NAA- 2mg/l, BAP – 2mg/l.

Fig.12.3 I Shoots of *Ruta graveolens* L. Stationary liquid culture, L-S medium, NAA - 2mg/l, BAP – 2mg/l.

II Shoot - differentiating callus of *R.g.* ssp. divaricata (Tenore) Gams. Stationary liquid culture, L-S medium, NAA - 2mg/l, BAP – 2mg/l.

III Agitating shoot culture of *Ruta graveolens* L., L-S medium, NAA – 0.1mg/l, BAP-0.1mg/l.

IV Agitating culture of *R.g.* ssp. divaricata (Tenore) Gams., L-S medium, NAA – 0.1mg/l, BAP-0.1mg/l.

The results of these above-mentioned studies have demonstrated clear influence of growth substances on contents of the metabolites.

The variant containing 0.1 mg/l NAA and 0.1 mg/l BAP was proposed as the best medium facilitating production of the metabolites. The studies proved to be successful since the contents of imperatorin, xanthotoxin and umbelliferone were considerable and interesting from a practical point of view, despite the fact that they were obtained in undifferentiated callus tissues. The culture was proposed as a valuable model for studies of accumulation of furanocoumarins as also a potential rich source of the three aforementioned metabolites (Ekiert and Gomólka 2000b; Ekiert 2003d).

II. *Pastinaca sativa* L. in vitro cultures

Callus cultures of this plant species (Fig.12.2) were maintained on 8 variants of Linsmaier-Skoog medium, containing different amounts (0.1 – 10 mg/l) of NAA and BAP. In addition, 2 variants of L-S medium were supplemented with β-indolylbutyric acid (IBA) and BAP (1 mg/1 mg and 5 mg/5 mg).

Six-week subcultures on the tested media led to 1.5 – 3.0-fold biomass growth.

HPLC analysis of callus extracts showed that tissues cultured in vitro produced psoralen, bergapten, xanthotoxin, isopimpinellin and umbelliferone, but no imperatorin was detected. Isopimpinellin was the quantitatively dominating metabolite (max. 238.9 mg/100 g d.w.). Contents of psoralen (max. 108.8 mg/100 g d.w.) and xanthotoxin (max. 82.6 mg/100 g d.w.) were also significant. The total content of the metabolites was very diverse, ranging from 115.7 to 408.5 mg/100 g d.w. depending on the composition of growth substances in the medium (Table 12.4).

Table 12.4 Minimum and maximum contents of the metabolites in callus tissues of *Pastinaca sativa* L. cultured on different variants of L-S medium.

Metabolites	Contents (mg/100 g d.w.)
Psoralen	18.60 – 108.80
Bergapten	7.40 – 16.20
Xanthotoxin	14.10 – 82.60
Isopimpinellin	90.20 – 238.90
Imperatorin	- *
Umbelliferone	0.50 – 1.80
Total	115.70 – 408.50

*Contents lower than 0.001 mg/100 g d.w.

The variant containing 3 mg/l NAA and 1 mg/l BAP was proposed as the best 'production' medium.

These studies showed that parsnip in vitro culture can be a convenient model for investigations of accumulation of furanocoumarins and a potential abundant source of these metabolites, particularly isopimpinellin (Ekiert and Gomólka 2000a; Ekiert 2003d).

III. *Ruta graveolens* L. in vitro cultures

Stationary liquid shoot cultures of this species (Fig. 12.3) were maintained on 12 variants of L-S medium supplemented with various quantities (0.1 – 10 mg/l) of NAA and BAP. Dry mass of shoots increase during 6-week-old subcultures on the tested media was 6.7 – 9.9-fold.

HPLC analysis of shoot extracts revealed that the contents of furanocoumarins under examination showed marked differences, dependent on the concentrations of growth substances in the medium. The total contents ranged between 254.2 – 939.7 mg/100g d.w., while the variability ranges for individual furanocoumarins were: 9.0 – 228.8 mg/100g d.w. for psoralen, 20.6 – 80.8 mg/100g d.w. for bergapten, 127.6 – 609.3 mg/100 g d.w. for xanthotoxin, 49.8 – 281.5 mg/100g d.w. for isopimpinellin and 0.3 – 8.0 mg/100g d.w. for imperatorin. Xanthotoxin was the dominating metabolite. Umbelliferone was not detected in any of the analyzed shoot extracts. The media containing 0.1 mg/l NAA + 0.1 mg/l BAP, 1 mg/l NAA + 5 mg/l BAP, and 2 mg/l NAA + 2 mg/l BAP were considered to be the best for accumulation of the analyzed metabolites.

The studies demonstrated that stationary liquid shoot culture of *R. graveolens* is a very abundant source of furanocoumarins, and thereby a valuable study object from the biotechnological perspective.

For further detailed studies of dynamics of furanocoumarin accumulation, we chose L-S medium variant containing 2 mg/l NAA and 2 mg/l BAP, since on this medium we obtained high biomass growth and marked contents of the metabolites (Ekiert et al. 2000b).

IV. *Ruta graveolens* ssp. *divaricata* (Tenore) Gams in vitro cultures*

Stationary liquid cultures of shoot-differentiating callus of this subspecies (Fig. 12.3) were maintained on 12 variants of L-S medium containing different amounts (0.1 – 10 mg/l) of NAA and BAP. During 6-week growth cycles on the tested media, dry mass of shoots rose 4.0 – 7.7 times.

HPLC analysis of callus extracts showed a significant influence of the concentration of growth substances on the accumulation of furanocoumarins. Total contents of these metabolites changed from 41.5 to 566.5 mg/100g d.w. The amounts of individual metabolites ranged between 1.6 – 11.8 mg/100g d.w. for psoralen (on five media variants it was not marked), 3.5 – 63.7 mg/100 g d.w. for bergapten, 7.6 – 294.3 mg/100 g d.w. for xanthotoxin, 14.9 – 145.2 mg/100 g d.w. for isopimpinellin and 0.7 – 14.5 mg/100 g d.w. for imperatorin. Xanthotoxin was the dominating metabolite among the

* The culture was obtained in the framework of cooperation with the Institute for Biosciences, Würzburg Univeristy (Germany).

furanocoumarins. Contents of umbelliferone changed from 4.6 to 20.1 mg/100 g d.w. The best medium for accumulation of the analyzed metabolites was the one containing 0.1 mg/l NAA and 0.1 mg/l BAP.

The results indicated that stationary liquid culture of shoot-differentiating callus of *R. graveolens* ssp. *divaricata* (Tenore) Gams was a poorer source of furanocoumarins than stationary liquid shoot culture of *R. graveolens*. Nevertheless, it was regarded as an interesting research object, particularly to investigate the dynamics of accumulation of furanocoumarins. Thorough studies on this issue were conducted on L-S medium supplemented with 2 mg/l NAA and 2 mg/l BAP, analogously as *R. graveolens* cultures (Ekiert et al. 2000a).

EFFECT OF LIGHT ON ACCUMULATION OF FURANOCOUMARINS

Preliminary studies into the effects of light on the accumulation of furanocoumarins in *R. graveolens* stationary liquid shoot cultures have demonstrated that constant artificial light (900 lx) was more conducive to accumulation of the compounds than darkness or constant artificial light following UV-C irradiation (8 hours). Total contents of furanocoumarins and umbelliferone amounted to 1022.2, 437.0 and 851.6 mg/100 g d.w., respectively (Ekiert and Gomólka 1999).

Next, we tested the effect of light of different spectra: white, far-red, red, blue, constant UV irradiation and darkness. Maximal total contents of the metabolites was observed in the shoots cultivated under white and blue light (total contents of 695.8 and 542.7 mg/100 g d.w., respectively). White as well as blue light facilitated the accumulation of xanthotoxin (241.8 and 192.8 mg/100 g d.w., respectively), psoralen (179.1 and 112.3 mg/100 g d.w.) and bergapten (121.8 and 109.0 mg/100 g d.w.). Total contents of the compounds cultured under other light conditions were smaller, e.g. under red and far-red light were 359.2 mg/100 g d.w. and 206.7 mg/100 g d.w., respectively (Ekiert et al. 2003, Ptak et al. 2003).

INFLUENCE OF CULTURE TYPE ON ACCUMULATION OF FURANOCOUMARINS

I. Stationary liquid cultures of *Ruta graveolens* L.

Shoot cultures of this species (Fig.3) were maintained on L-S medium containing 2 mg/l NAA and 2 mg/l BAP. Accumulation of furanocoumarins and umbelliferone, and biomass increments were studied during 6-week growth cycles. Culture on the tested medium led to about 5-fold increase in both fresh and dry mass of shoots. Shoots extracts were shown to contain bergapten, xanthotoxin, psoralen, isopimpinellin, imperatorin and umbelliferone. Xanthotoxin (max. 332.2 mg/100 g d.w.) and bergapten (max. 323.7 mg/100 g d.w.) were the dominating metabolites in the shoots. Isopimpinellin (max. 116.7 mg/100 g d.w.) and psoralen (max. 172.6 mg/100 g d.w.) were also accumulated in marked quantities. Imperatorin contents were very low (< 6 mg/100 g d.w.). Only modest contents of furanocoumarins precursor were detected (max. 27.0 mg/100 g d.w.).

The total content of tested metabolites was very high, reaching almost 1g% d.w. (Table 12.5).

Table 12.5 Changes in contents of the metabolites in *Ruta graveolens* L. shoots during 6-week growth cycles. Stationary liquid culture. L-S medium, 2 mg/l NAA, 2 mg/l BAP.

Metabolites	Contents (mg/100 g d.w.)
Psoralen	55.2 – 172.6
Bergapten	136.1 – 323.7
Xanthotoxin	119.8 – 332.2
Isopimpinellin	59.0 – 116.7
Imperatorin	2.2 – 5.4
Umbelliferone	27.0 – 6.0
Total	407.6 – 996.4

These studies demonstrated the fact that contents of individual compounds changed 2 – 4.5 times during culture growth cycles. Maximum accumulation of the majority of the metabolites was observed on the 28th culture day (the beginning of stationary growth phase); later, their contents significantly declined.

The culture was proposed as a potential high productive source of furanocoumarins, particularly therapeutically important bergapten and xanthotoxin (Ekiert et al., 2001; Ekiert 2002; 2003d). Analogical preliminary studies were performed with LS medium containing 0.1mg/l NAA and 0.1mg/l BAP (Ekiert and Moroniewicz 2003).

II. Stationary liquid cultures of *Ruta graveolens* ssp. *divaricata* (Tenore) Gams

Shoot-differentiating callus cultures of this subspecies (Fig. 12.3) were maintained on L-S medium with identical composition as *R. graveolens* cultures. Six-week subcultures yielded 3.7-fold fresh biomass growth and 3.0-fold rise in dry biomass. Xanthotoxin (max. 112.3 mg/100 g d.w.), isopimpinellin (84.0 mg/100 g d.w.) and bergapten (76.2 mg/100 g d.w.) were the dominating metabolites. The remaining furanocoumarins were accumulated at trace amounts (< 10 mg/100 g d.w.). The presence of umbelliferone was detected only in several tested biomass extracts (max. 0.43 mg/100 g d.w.).

The contents of dominating metabolites transformed 1.8 to 3.2 times within 6-week growth cycles and reached maximum on the 35th culture day.

The total contents of the metabolites changed 2.3-fold, peaking at 283.4 mg/100 g d.w. also within 5 weeks (Table 12.6) (Ekiert 2002; 2003 a,d).

Table 12.6 Changes in contents of the metabolites in *Ruta graveolens* ssp. *divaricata* (Tenore) Gams shoot-differentiating callus during 6-week growth cycles. Stationary liquid culture. L-S medium, 2 mg/l NAA, 2 mg/l BAP.

Metabolites	Contents (mg/100 g d.w.)
Psoralen	1.7 – 5.5
Bergapten	29.4 – 76.2
Xanthotoxin	62.8 – 112.3
Isopimpinellin	26.1 – 84.0
Imperatorin	2.2 – 7.6
Umbelliferone	- 0.43*
Total	123.9 – 283.4

* umbelliferone was detected only in few of the analyzed callus extracts

III. Agitating cultures of *R. graveolens* L.

Shoot cultures (Fig. 12.3) were grown on L-S medium containing 0.1 mg/l NAA and 0.1 mg/l BAP. A medium with such composition was chosen because it facilitated the accumulation of furanocoumarins (Ekiert et al., 2001). Accumulation of furanocoumarins and umbelliferone in shoots and medium and shoots biomass growth were examined during 6-week growth cycles.

Fresh and dry biomass grew 4.4-fold and 3.2-fold, respectively, within 6-week subcultures.

Shoot extracts contained large quantities of bergapten (max. 210.0 mg/100 g d.w.) and xanthotoxin (max. 137.0 mg/100 g d.w.). Contents of isopimpinellin (max. 97.0 mg/100 g d.w.) and psoralen (56.0 mg/100 g d.w.) were also significant. Imperatorin and umbelliferone concentrations were at much lower concentrations (3.2 mg/100 g d.w. and 18.0 mg/100 g d.w., respectively).

The total content of the compounds under examination was high, amounting to 521.0 mg/100 g d.w. (Table 7).

Table 12.7 Changes in contents of the metabolites in *Ruta graveolens* L. shoots during 6-week growth cycles. Agitating culture. L-S medium, 0.1 mg/l NAA, 0.1 mg/l BAP.

Metabolites	Contents (mg/100 g d.w.)
Psoralen	7.2 – 56.0
Bergapten	39.0 – 210.0
Xanthotoxin	38.0 – 137.0
Isopimpinellin	19.0 – 97.0
Imperatorin	0.6 – 3.2
Umbelliferone	4.2 – 18.0
Total	108.5 – 521.0

*content in the shoots (99.4%), medium contained only 0.6% of the metabolites

The contents of individual compounds changed in a wide range in the course of culture growth cycles (3.6 – 7.7 times), maximizing on the 42nd culture day. The total content of the compounds also fluctuated (4.8-fold) substantially during 6-week growth cycles, peaking by the 42nd day. The 99.4 % of the metabolites accumulated in the biomass, while the lyophilized media samples contained only their trace amounts (about 0.6 %) (Ekiert et al., 2001, 2002a; Ekiert 2002, 2003b, c, d, e).

IV. Agitating cultures of *Ruta graveolens* ssp. *divaricata* (Tenore) Gams

Cultures of this subspecies (Fig. 12.3) were maintained on L-S medium with identical composition of growth substances as agitating cultures of *R. graveolens* (0.1 mg/l NAA, 0.1 mg/l BAP). Six-week subcultures led to 5.4-fold fresh biomass growth and 3.9-fold dry biomass growth. This culture type yielded low contents of the metabolites, which were the highest for xanthotoxin (25.0 mg/100 g d.w.) and bergapten (18.4 mg/100 g d.w.). The rest of furanocoumarins were accumulated in slight quantities (below 6.5 mg/100 g d.w.). Maximum contents of umbelliferone amounted to 10.5 mg/100 g d.w. Contents of the dominating metabolites changed 2 and 2.6 times, respectively, during 6-week growth cycles, culminating on the 35th culture day.

Total content of the compounds changed twice, reaching a maximum of 64.1 mg/100 g d.w. on the 35th day of culture. A vast majority of the metabolites were accumulated in the biomass (98.3%), whereas only negligible amount of them were found in the lyophilized media (1.7%) (Table 12.8) (Ekiert et al., 2002b, 2002c; Ekiert 2002, 2003b, c, d, e).

Table 12.8 Changes in contents of the metabolites in biomass from culture of *Ruta graveolens* ssp. *divaricata* (Tenore) Gams during 6-week growth cycles. Agitating culture. L-S medium, 0.1 mg/l NAA, 0.1 mg/l BAP.

Metabolites	Contents (mg/100 g d.w.)*
Psoralen	0.5 – 1.5
Bergapten	7.1 – 18.4
Xanthotoxin	12.6 – 25.0
Isopimpinellin	2.0 – 6.4
Imperatorin	1.5 – 4.4
Umbelliferone	4.4 – 10.4
Total	31.3 – 64.1

* contents in the biomass (98.3 %), medium contained only 1.7 % of the metabolites

V. Suspension cultures

***Ammi majus* L. suspension culture**

Preliminary experiments with suspension cultures have not produced expected results. The culture was maintained on L-S medium (0.1 mg/l

NAA and 0.1 mg/l BAP). Four-week-old subcultures led to about 6-fold growth of both fresh and dry biomass. Total furanocoumarin content ranged from 0.67 to 11.7 mg/100 g d.w., reaching a maximum on the 28th culture day.

***Ruta graveolens* L. suspension culture**

Suspension cultures were cultivated on L-S medium (2 mg/l NAA and 2 mg/l BAP). Within 6-week-old growth cycles, dry biomass increased 4.8 times. The results regarding the biosynthetic capacity of the cells was unsatisfactory. Only three furanocoumarins—bergapten (0.31 mg/100 g d.w.), xanthotoxin (0.72 mg/100 g d.w.) and isopimpinellin (1.89 mg/100 g d.w.)—were detected in the cells along with their aggregates, while the medium contained all the investigated of furanocoumarins, but their quantities were even smaller. Isopimpinellin dominated in the media. Most of the metabolites (99.34%) were accumulated in the cells and cellular aggregates.

STRUCTURAL STUDIES OF METABOLITES

The metabolites occurring in greater quantities in the extracts from in vitro cultured material of the investigated plant species were isolated using prep. TLC technique. The isolated metabolites were then subjected to spectral analysis (EI-MS and ^{1}H-NMR spectra).

The following metabolites were identified:

1. *Callus tissues of A. majus L.*—umbelliferone (Ekiert and Bieda 2003; Ekiert and Kisiel-unpublished). This compound is known as a secondary metabolite of the intact plant, as well as of elicited and non-elicited cell cultures.

2. *Callus tissues of P. sativa L.*—bergapten, xanthotoxin, isopimpinellin and sphondin (angular furanocoumarin) (Ekiert and Kisiel, 2000). The compounds are well-known secondary metabolites of the intact plant. Three linear furanocoumarins were isolated for the first time from callus tissues of this plant species. This was the first report on the isolation of sphondin from in vitro plant cultures.

3. *Shoots from stationary liquid culture of R. graveolens L.*—The compounds isolated were: psoralen, bergapten, xanthotoxin, isopimpinellin, rutamarin (dihydrofuranocoumarin) and two furanoquinoline alkaloids (kokusaginine and skimmianine) (Ekiert and Kisiel 1997).

These isolated metabolites are also known as secondary metabolites of rue.

The ability of callus and cell suspension cultures of the plant to produce the isolated coumarins and alkaloids was reported previously. This was the first report on the isolation of these metabolites from the shoots of garden rue cultured in a stationary liquid phase.

4. *Shoot-differentiating callus of R.g. ssp. divaricata (Tenore) Gams*—xanthotoxin, bergapten, rutacultin and dimethylether 3-(1′, 1′-dimethylallyl) - daphnetin (simple coumarins) (Ekiert and Kisiel 2001). To our knowledge, this was the first report on the isolation of these metabolites from the in vitro culture of the plant. Hitherto-existing literature data on in vitro culture of this subspecies focussed on the growth of cells and tissues under different culture conditions, and on the ability of their differentiation and regeneration (Abou-Mandour 1982, 1994, 1997; Hartung and Abou-Mandour 1996).

The composition of furanocoumarins of the native plant had not been studied earlier. We isolated the following coumarin compounds: xanthotoxin, bergapten, psoralen, isoimperatorin, rutamarin and daphnoretin methylether (coumarine dimer) from the overground parts of plants and identified with spectral methods (Ekiert and Kisiel 2002).

All the isolated coumarins are known as secondary metabolites of *R. graveolens* (Eilert 1994).

CONCLUSIONS

The main findings of the present studies can be summarized as follows:

1. The studies demonstrated a marked effect of concentrations of growth substances on the accumulation of furanocoumarins in all in vitro tested cultures of four plant spiecies.
2. Quality of light was shown to have a significant effect on the accumulation of the metabolites in stationary liquid culture of *R. graveolens*. White and blue light proved to be the most beneficial.
3. Culture type (stationary liquid, agitating, suspension) was evidenced to substantially influence the accumulation of the metabolites.
4. Stationary liquid cultures of *R. graveolens* and *R. graveolens* ssp. *divaricata* (Tenore) Gams were a richer source of furanocoumarins rather than agitating cultures of both the plants.
5. Dynamics of accumulation of the metabolites in stationary liquid cultures differed from agitating cultures, which was evident particularly in *R. graveolens* cultures.
6. In agitating cultures, metabolites were accumulated mostly in the biomass, whereas the medium contained only their trace amounts.
7. *Ammi majus* L. and *Pastinaca sativa* L. callus cultures were a rich source of furanocoumarins, despite their low degree of differentiation. *A. majus* cultures can be a potential source of imperatorin and xanthotoxin, while *Pastinaca sativa* L. culture can be a productive source of isopimpinellin. In addition, large amounts of umbelliferone can be obtained from *A. majus* callus (Table 12.9).
8. Stationary liquid culture of *R. graveolens* shoots was found to be the richest source of xanthotoxin, bergapten and psoralen (Table 12.9).

9. The contents of the metabolites observed in the cultures under examination were high and interesting from a practical point of view (Tables 9 and 10). These cultures can be proposed as their potential copious alternative source for pharmaceutical applications.

Table 12.9 Maximum contents of furanocoumarins and umbelliferone obtained in the tested in vitro cultures.

Metabolite	Content (mg/100 g d.w.)	Plant species	Type of culture
Psoralen	173.0	*R. graveolens*	stationary liquid
Bergapten	324.0	*R. graveolens*	stationary liquid
Xanthotoxin	332.0	*R. graveolens*	stationary liquid
Isopimpinellin	239.0	*P. sativa*	agar culture
Imperatorin	169.0	*A. majus*	agar culture
Umbelliferone	536.0	*A. majus*	agar culture

Table 12.10 Maximum total content of furanocoumarins and umbelliferone in the tested in vitro cultures.

Plant species	Type of culture	Contents (mg/100 g d.w.)
A. majus	agar callus culture	871.1
A. majus	suspension culture	11.7
P. sativa	agar callus culture	408.5
R. graveolens	stationary liquid shoot culture	966.4
R. graveolens	agitating shoot culture	521.0
R. graveolens	suspension culture	4.5
R. graveolens ssp. *divaricata*	stationary liquid shoot-differentiating callus culture	283.4
R. graveolens ssp. *divaricata*	agitating callus culture (aggregates)	64.1

EXPERIMENTAL

1. In vitro cultures

Ammi majus L.

The callus culture was established from hypocotyls of sterile seedlings. Fruits were obtained from Botanischer Garten der Technischen Hochschule, Aachen (Germany).

Details for callus initiation were reported earlier (Ekiert and Gomólka 2000b).

Callus cultures were maintained on variants of L-S agar medium (Linsmaier and Skoog 1965) containing various concentration of growth substances, NAA/BAP (mg/l):0.1/0.1; 1.0/1.0; 2.0/2.0; 3.0/1.0; 5.0/5.0; 10.0/10.0; 1.0/5.0; 5.0/1.0; 5.0/10.0; 10.0/5.0; 1.0/10.0; 10.0/1.0 – 12 variants.

Cultures were cultivated under constant artificial light (900 lx), at a temperature of 25±2^{0}C. Callus tissues were subcultured every 4 weeks.

Suspension cultures were maintained on a liquid L-S medium containing 0,1 mg/l NAA and 0,1 mg/l BAP. These cultures were cultivated under identical light and temperature conditions as callus cultures. Growth cycles – 4 weeks.

Pastinaca sativa **L.**

The callus culture was derived from hypocotyls parts of sterile seedlings. Fruits were obtained from the München – Nymphenburg Botanical Garden (Germany). Details for callus initiation have been reported earlier (Ekiert and Gomólka 2000a).

Callus cultures were cultivated on variants of agar L-S medium, containing varying quantities of NAA and BAP (mg/l): 0,1/0,1; 1,0/1,0; 2,0/2,0; 3,0/1,0; 5,0/5,0; 5,0/10,0; 10,0/5,0; 10,0/10,0 – 8 variants, and IBA and BAP (mg/l): 1,0/1,0 and 5,0/5,0 – 2 variants. Cultures were conducted under constant artificial light (900 lx), at a temp. of 25±2^{0}C. Tissues were subcultured every 6 weeks.

Ruta graveolens **L.**

The shoot culture was initiated from hypocotyl segments of sterile seedlings. Seeds were derived from Hortus Centralis Cultura Herbarum Medicarum Facultas Medica, Universitas Purkyniana, Brno (Czech).

**Stationary liquid culture—initial*

The shoot culture was maintained on Petri dishes with U-shaped glass tubes covered with filter paper partly immersed in the medium.

The cultures were cultured on variants of L-S medium.

**The tested concentration of growth substances*

12 variants of L-S medium (the same as for *A. majus* callus cultures), differing in concentration of NAA and BAP, were tested. Cultures were cultivated under constant artificial light (900 lux) at 25±2°C. Growth cycles- 6 weeks.

**The tested light conditions—experiment I*

Shoots were cultured in stationary liquid phase on LS medium (NAA – 2 mg/l, BAP – 2 mg/l) at a temp. of 25±2^{0}C.

The following light conditions were tested:

- constant artificial light (900 lux), LF – 40 W lamp, daylight (Pila)
- constant darkness
- constant artificial light (900 lux) following irradiation with UV – C light for 8 hours (Philips lamp)

Shoots were subcultured every 6 weeks.

**The tested light conditions – experiment II*

Shoots were cultured in a stationary liquid phase on LS medium (NAA – 2 mg/l, BAP – 2 mg/l) at a temp. of 25±2^{0}C under the following light conditions:

white light (390 – 760 nm, Tungsram lamp 40 W F33), UV irradiation (360 – 450 nm, Philips TLD 36 W), blue light (450 – 492 nm, Philips TLD 36 W), red light (647 – 770 nm, Philips TLD 36 W), far-red light (770 – 800 nm, 100 W incandescent light with standard filter no. 405 orange + standard filter no. 420 deep blue, Compact Light B. V. Amsterdam), and in darkness. Radiometric measurements were taken in the horizontal plane 35 cm above the cultures. The photosynthetically active radiation (PAR) was 60 $\mu mol\ m^{-2}s^{-1}$ except for red and far-red light (20 $\mu mol\ m^{-2}s^{-1}$).

Growth cycles : 6 weeks.

**Stationary liquid shoot cultures—dynamics of accumulation of metabolites*

Stationary liquid shoot culture were maintained on LS medium (2 mg/l NAA, 2 mg/l BAP), under artificial constant light (900 lux) at 25±2^{0}C. Shoots were collected every week during 6 weeks' growth cycles.

**Agitating shoot cultures—dynamics of accumulation of metabolites*

Cultures were maintained in 500 ml Erlenmayer flasks (4 g f.w. inoculum (4g f.w. inoculum/125 ml) on L-S medium (0,1 mg/l NAA and 0,1 mg/l BAP), under the same light and temperature conditions as for stationary liquid cultures. The shoots and the media were collected every week during 6 weeks, growth cycles. The media were frozen and lyophilized.

**Suspension cultures*

Cultures were maintained on L-S medium (2mg/l NAA and 2 mg/l BAP). Light and temperature conditions remained – the same as for stationary liquid culture. Growth cycles : 6 weeks.

Ruta graveolens ssp. divaricata (Tenore) Gams

In vitro culture of *R. graveolens* ssp. divaricata (Tenore) Gams was established at the Institute for Biosciences, Würzburg University (Germany). Starting material for establishment of the culture originated from the plants harvested in natural sites in the Nanos Mountains in Jugoslavia in 1973, and was grown subsequently at the Botanical Garden of the above mentioned institute. Callus cultures were derived from fragments of young leaves and stems.

**Stationary liquid culture—initial*

The shoot-differentiating callus were cultured on Petri dishes on L-S medium (NAA – 2 mg/l, BAP- 2 mg/l).

**Effect of growth substances*

12 variants of L-S medium (the same as for *A. majus* callus cultures and *R. graveolens* shoot cultures) differing in concentration of NAA and BAP were tested. See conditions for *R. gravelones* shoot culture.

**Stationary liquid cultures—dynamics of accumulation of metabolites.*

See conditions for *R. graveolens* shoot culture.

**Agitating cultures—dynamics of accumulation of metabolites*

See conditions for *R. graveolens* shoot culture.

2. Extraction

The dried and ground callus tissues and shoots (app. 1 g) were extracted with 2 portions (50 ml) of boiling 96% ethanol in a Soxhlet apparatus for 10 hours. The extracts were combined, thickened and evaporated to dryness. The residue was quantitatively dissolved in 10 ml of 96% ethanol and analysed by HPLC method (details, see Ekiert and Gomólka 1999).

3. HPLC

Chromatographic quantification of psoralen, xanthotoxin, bergapten, isopimpinellin, imperatorin and umbelliferone was performed according to the procedure, developed in our laboratory (see, Ekiert and Gomólka 1999). The conditions were as follows:

HPLC apparatus:	Ati Unicam, Cambridge
Pump:	Crystal 200 (Ati Unicam, Cambridge)
Column:	Supelcosil LC-8 (4.6 mm/25 cm)
Solvent system:	methanol – water (1:1.2 V/V); in case of imperatorin: methanol – water (2:1 V/V)
Flow rate:	1 ml/min.
Detector UV:	λ = 310 nm
Standards:	manufactured by Carl Roth

4. Isolation, purification and identification of metabolites

- *Material under investigations:*
- callus tissues of *A. majus* L.
- callus tissues of *P. sativa* L.
- shoots from stationary liquid cultures of *R. graveolens* L.
- shoot-differentiating callus from stationary liquid culture of R. g. ssp. divaricata (Tenore) Gams

- *Extraction:*

ethanolic extracts, details see (Ekiert and Gomólka 1999)

- *Isolation:*

prep. TLC plates (Merck, Art. 11844)

- *Purification:*

HP-TLC (Merck, Art. 5633)

- *Spectral analysis:*

EI-MS (70 eV, 15 eV), LKB-2091 mass spectrometer (Sweden)

^{1}H-NMR (300 MHz, $CDCl_3$, TMS as internal standard), NMR spectrometer MSL – 300 (Bruker).

ACKNOWLEDGEMENT

I would like to express my sincere thanks to Dr. Radosława Wróbel for her help with the translating this chapter into English.

REFERENCES

Abou- Mandour, A. A., 1982, Untersuchungen an *Ruta graveolens* ssp. divaricata. Planta Med, 46: 105-109.

Abou-Mandour, A. A., 1994, Gewebekulturen und daraus regenerierte Arzneipflanzen. DAZ, 134:19-27.

Abou-Mandour A.A., 1997, Ein Standardnährmedium für die Anzucht von Kalluskulturen einiger Arzneipflanzem. Z. Pflanzenphysiol., 85: 273 – 277.

Becela-Deller, Ch., 1995, *Ruta graveolens* L. – Weinraute - Kulturhistorisches Portrait einer traditionellen Heilpflanze. Z. Phytother. 16:275-281.

Ben- Hur, E. Pill- Soon, S., 1984, The photochemistry and photobiology of furocoumarins (psoralens). Adv. Radiat. Biol., 11: 131-171.

Berenbaum, M.R., Zangerl, A.R., Nitao, J.K., 1984, Furanocoumarins in seeds of wild and cultivated parsnip. Phytochemistry, 23:1809-1810.

Beyrich, Th., 1965a, Die Furocumarine von *Pastinaca urens* Req. Pharmazie, 20:655-656.

Beyrich, Th., 1965b, Trennung und Identifizierung der Furocumarine von *Pastinaca opaca* Bernh. Planta Med., 13:439-444.

Beyrich, Th., 1966, Die Furocumarine von *Pastinaca sativa* L. Pharmazie, 21:365-373.

Beyrich, Th., 1967, Die Bildungsstätte der Furocumarine von *Pastinaca sativa* L. Planta Med. 15:306-310.

Beyrich, Th., 1981, Furocoumarine., Wiss Z.E.M. Arndt- Univ. Greifswald, 30: 25- 28.

Bethea, D., Fullmer, B., Syed, S., Seltzer, G., Tiano, J., Rischko, C., Gillespie, L., Brown, D., Gasparro, F. P., 1999, Psoralen photobiology and photochemotherapy: 50 years of science and medicine. J. Dermat. Sci., 19: 78- 88.

Bohlmann, E., Gibraltarskaya, Eilert, U., 1995, Elicitor induction of furanocoumarin biosynthetic pathway in cell cultures of *Ruta graveolens*. Plant Cell, Tissue and Organ Culture, 43: 155 – 161.

Cappelletti, E.M., Innocenti, G., Caniato, R., Filippini, R., Piovan, A., 1998, *Haplophyllum patavinum* (L.) G. Don fil. (Paduan rue): In Vitro Regeneration and the Production of Coumarin Compounds. In: Biotechnology in Agriculture and Forestry 41, Bajaj Y. P. S. (ed.) Springer Verlag, Berlin- Heidelberg- New York: 238- 248.

Ceska, O., Chaudhary, S.K., Warrington, P.J., Aswood-Smith, M.J., 1987, Photoactive furocoumarins in fruits of some umbellifers. Phytochemistry, 26:165-169.

Dewick, P. M., 1997, Medicinal Natural Products, Biosynthetic Approach. John Wiley and Sons Ltd., Chichester.

Ditzen, G., Gerst, F., Hänsel, W., Koppenhöfer, E., 1995, Kaliumkanal-Blocker-ein neuer Ansatz zur symptomatischen Therapie der multiplen Sklerose. DMW, 120:1061-1062.

Eilert, U., 1994, Ruta. In: Hagers Handbuch der pharmazeutischen Praxis. Drogen P-Z, Hänsel, R., Keller, K., Rimpler, H., Schneider, G. (ed.), Springer Verlag Berlin-Heidelberg, Bd.6.

Ekiert, H., 1988, Doctor's Thesis, Medical Academy, Kraków, Poland.

Ekiert, H., 1990, Furanocoumarins in tissue culture of *Ammi majus*. Planta Med, 56: 572- 572.

Ekiert, H., 1993, *Ammi majus* L. Bishop' s Weed: In vitro culture and the production of coumarin compounds. In: Biotechnology in Agriculture and Forestry 21, Bajaj Y. P. S. (ed.), Springer Verlag, Berlin- Heidelberg- New York: 1- 17.

Ekiert, H., 1994, Biosynteza aktywnych biologicznie furanokumaryn w roslinnych kulturach tkankowych. Biotechnologia, 24:121-131.

Ekiert, H., 2000, Medicinal plant biotechnology: the Apiaceae family as the example of rapid development. Pharmazie, 55:561-567.

Ekiert, H., 2002, In vitro cultures of *Ruta graveolens* and *Ruta graveolens ssp. divaricata* as biotechnological source of biologically active furanocoumarins. Int. Symposium on Promotion of International Cooperation in Eastern and Southern Europe in the Field of Medicinal Biotechnology, Lódz (Poland), abstracts: 28.

Ekiert, H., 2003a, Accumulation of biologically active furanocoumarins in *Ruta graveolens ssp. divaricata* (Tenore) Gams *in vitro* culture. Pharmazie - submitted for publication

Ekiert, H., 2003b, Accumulation of coumarin compounds in agitating cultures of *Ruta graveolens* L. and *Ruta graveolens ssp. divaricata* (Tenore) Gams. Acta Soc. Bot. Polon.- submitted for publication.

Ekiert, H., 2003c, Accumulation of coumarins in different types of plant in vitro cultures. Biotechnologia- submitted for publication.

Ekiert., H., 2003d, Roslinne hodowle in vitro jako biotechnologiczne zródlo furanokumaryn. II Kongres Biotechnologii, Lódz (Poland), Materialy Zjazdowe:115.

Ekiert, H., 2003e, Accumulation of coumarins in different types of plant in vitro cultures. Biotechnologia roslinna w biologii, farmacji i rolnictwie. X Ogólnopolska Konferencja Kultur In Vitro i Biotechnologii Roslin, Bydgoszcz (Poland), Materialy Zjazdowe: 91.

Ekiert, H., Bieda, J., 2003, Chromatographic methods in isolation and purification of coumarins from *Ammi majus* L. callus tissues. Chromatograficzne Metody Badania Zwiazków Organicznych. XXVII Sympozjum, Szczyrk (Poland), Materialy Zjazdowe: 82.

Ekiert, H., Choloniewska, M., Gomólka, E., 2001, Accumulation of furanocoumarins in *Ruta graveolens* L. shoot culture. Biotechnology Letters, 23:543 – 545.

Ekiert, H., Gomólka, E., 1999, Effect of light on contents of coumarin compounds in shoots of *Ruta graveolens* L. cultivated in vitro. Acta Soc. Bot. Polon., 68:197-200.

Ekiert, H., Gomólka, E., 2000a, Furanocoumarins in *Pastinaca sativa* L. in vitro culture. Pharmazie, 55:618 - 620.

Ekiert, H., Gomólka, E., 2000b, Coumarin compounds in *Ammi majus* L. callus culture. Pharmazie, 55:684 - 687.

Ekiert, H., Kisiel, W., 1997, Coumarins and alkaloids in shoot culture of *Ruta graveolens* L. Acta Soc. Bot. Polon., 66: 329 - 332.

Ekiert, H., Kisiel, W., 2000, Isolation of furanocoumarins from *Pastinaca sativa* L. callus culture. Acta Soc. Bot. Polon., 69:193 - 195.

Ekiert, H., Kisiel, W., 2001, Kumaryny z hodowli in vitro *Ruta graveolens ssp. divaricata* (Tenore) Gams. Farmacja XXI wieku., XVII Naukowy Zjazd Polskiego Towarzystwa Farmaceutycznego, Poznan (Poland), Materialy Zjazdowe: 223.

Ekiert, H., Kisiel, W., 2002, Coumarins from *Ruta graveolens ssp. divaricata*. The Application of Chromatographic Methods in Phytochemical & Biomedical Analysis, 3rd International Symposium on Chromatography of Natural Products, Lublin- Kazimierz Dolny (Poland), abstracts: 70.

Ekiert, H., Fracz, S., Ptasinska, M., Gomólka, E., 2000a, Accumulation of furanocoumarins in shoot – differentiating callus culture of *Ruta graveolens ssp. divaricata* (Tenore) Gams. Zastosowanie kultur in vitro w fizjologii roslin., IV Ogólnopolska Konferencja, Kraków (Poland), Materialy Zjazdowe: 27.

Ekiert, H., Hennig, M., Gomólka, E., 2002a, Application of HPLC to the Studies into Metabolite Accumulation in Agitating Shoot Culture of *Ruta graveolens* L., 8th Int. Symposium on Separation Sciences, Torun (Poland), abstracts: 4- 3.

Ekiert, H., Kamecka, A.., Gomólka, E., 2001, Optimisation of conditions for furanocoumarins production in *Ruta graveolens* L. shoot culture. Traits, Tracks and Traces, 49th Annual Meeting of Society for Medicinal Plant Research, Erlangen (Germany), abstracts: 248.

Ekiert, H., Kuliga- Jamróz J., Gomółka, E., 2000b, Effect of phytohormone on contents of furanocoumarins in shoots of *Ruta graveolens* L. cultivated in vitro. Zastosowanie kultur in vitro w fizjologii roslin, IV Ogólnopolska Konferencja, Kraków, Materialy Zjazdowe: 28.

Ekiert, H., Macurova, L., Gomółka, E., 2002b, Optymalizacja warunków produkcji furanokumaryn w hodowli in vitro *Ruta graveolens ssp. divaricata* (Tenore) Gams. Zastosowanie kultur in vitro w fizjologii roslin, V Ogólnopolska Konferencja, Kraków (Poland), Materialy Zjazdowe: 78.

Ekiert, H., Moroniewicz, U., 2003, Stationary liquid culture of *Ruta graveolens* L. shoots as a potential biotechnological source of coumarins. Polish- Austrian- German- Hungarian-Italian Joint Meeting on Medicinal Chemistry, Kraków (Poland), abstract: 222.

Ekiert, H., Ptak, A., Swiderski, A., Bach A., 2003, Production of coumarins and pigments in *Ruta graveolens* L. shoots cultured in vitro under wide range of light conditions. Acta Soc. Bot. Polon.- submitted for publication.

Ekiert, H. , Sopata, K., Gomółka, E., 2002c, Dynamics of accumulation of biologically active furanocoumarins in Ruta graveolens ssp. divaricata agitating culture. Int. Symposium on Promotion of International Cooperation in Eastern and Southern Europe in the Field of Medicinal Biotechnology, Lódz (Poland), abstracts: 27.

Elgamal, M., H., A., Shalaby, N., M., M., Duddeck, H., Hiegemann, M., 1993, Coumarins and coumarin glucosides from the fruits of *Ammi majus*. Phytochemistry, 34:819-823.

Engler, A., 1964, Syllabus der Pflanzenfamilien., Gebrüder Borntraeger Verlag, Berlin-Nikolassee, Bd.2.

Fahmy, I.R., Abu-Shady, H., 1947, *Ammi majus* Linn. – pharmacognostical study and isolation of a crystalline constituent ammoidin. Q.J. Pharm. Pharmocol. 20:281-281.

Grewal, S., Sachdeva, U., Atal, C.K., 1976, Regeneration of plants by embryogenesis from hypocotyl cultures of *Ammi majus* L. Indian J. Exp. Biol., 14:716-717.

Hamerski, D., Matern, U., 1988a, Biosynthesis of psoralens. Psoralen 5-monooxygenase activity from elicitor-treated *Ammi majus* cells. FEBS Lett., 239:263-265.

Hamerski, D., Matern, U., 1988b, Elicitor-induced biosynthesis of psoralens in *Ammi majus* L. suspension cultures. Microsomal conversion of demethylsuberosin into (+) marmesin and psoralen. Eur. J. Biochem., 171:369-375.

Hamerski, D., Beier, R.C., Kneusel R.E., Matern, U., Himmelspach, K., 1990a, Accumulation of coumarins in elicitor-treated cell suspension cultures of *Ammi majus*. Phytochemistry, 29:1137-1142.

Hamerski, D., Schmitt D., Matern, U., 1990b, Induction of two prenyltransferases for the accumulation of coumarin phytoalexins in elicitor-treated *Ammi majus* cell suspension cultures. Phytochemistry, 29:1131-1135.

Hartung, W., Abou-Mandour, A. A. 1996, A Beneficial Role of Abscisic Acid for Regenerates of *Ruta graveolens* ssp. *divaricata* (Tenore) Gams. Suffering from Transplant Shock. Angew. Bot., 70:221-223.

Hauffe, D.K., Hahlbrock, K., Scheel, D., 1986, Elicitor-Stimulated Furanocoumarin Biosynthesis in Cultured Parsley Cells: S-Adenosyl-L-Methionine:Bergaptol and S-Adenosyl-L-Methionine:Xanthotoxol O-Methyltransferases. Z. Naturforsch., 41:228-239.

Hegi, G., 1965, Illustrierte Flora von Mittel-Europa. Carl Hanser Verlag, München.

Hegnauer, R., 1973, Chemotaxonomie der Pflanzen, Birkhäuser Verlag, Basel und Stuttgart.

Hehmann, M., Lukacin, R., Ekiert, H., Matern U., 2004, Furanocoumarin biosynthesis in *Ammi majus* L., Cloning of bergaptol-O-methyltransferase. Eur. J. Biochem.-submitted for publication.

Hoppe, H., A., 1975, Drogenkunde, Walter de Gruyter, Berlin - New York., Bd. 1.

Ivie, W.G., 1978, Linear furocoumarins (psoralens) from the seeds of Texas *Ammi majus* L. (Bishop's weed). J. Agric. Food Chem., 26:1394-1403.

Kartnig, T., Moeckel, H., Maunz, B., 1975, über das Vorkommen von Cumarinen und Sterolen in Gewebelculturen aus Wurzeln von *Anethum graveolens* und *Pimpinella anisum*. Planta Med., 27:1-4.

Kitamura, Y., Ikenaga, T., Ooe, Y., Hiraoka, N., Mizukami, H., 1998, Induction of furanocoumarin biosynthesis in Glehnia littoralis cell suspension cultures by elicitor treatment. Phytochemistry, 48: 113-117.

Kohlmünzer, S., 1998, Farmakognozja (Pharmacognosy) - in polish, PZWL, Warszawa.

Koul, S., Koul, A.K., 1993, Development of media for growth and furanocoumarin production of *Ammi majus* cells. Fitoterapia, 64:415-422.

Królicka, A., Lojkowska, E., Krauze-Baranowska, M., Staniszewska, I., Malinski, E., Szafranek, J., Turkiewicz, M., 2001a, Identification of secondary metabolites in in vitro culture of *Ammi majus* treated with elicitors. Acta Hort., 560:255-258.

Królicka, A., Staniszewska, I., Bielawski, K., Malinski, E., Szafranek, J., Lojkowska, E., 2001b, Establishment of hairy root cultures of *Ammi majus*. Plant Science, 160:259-264.

Kutney, J.P., Salisbury, P.J., Verma, A.K., 1973, Biosynthetic studies in the coumarin series.III. Studies in tissue cultures of *Thamnosma montana* Torr. and Frem. The role of mevalonate. Tetrahedron, 29:2673-2681.

Linsmaier, E.M., Skoog, F., 1965, Organic growth factor requirements of tabacco tissues cultures. Physiol. Plant., 18:100-127.

Massot, B., Milesi, S., Gontier, E., Bourgaud F., Guckert, A., 2000, Optimized culture conditions for production of furanocoumarins by micropropagated shoots of *Ruta graveolens*. Plant Cell, Tissue and Organ Culture, 62:11-19.

Matern, U., Lüer, P., Kreusch, D., 1999, Medicinal Potential and Biosynthesis of Plant Coumarins, In: Phytochemicals in Human Health Protection, Nutrition and Plant Defense, Romeo, J.T. (ed.). Kluwer Academic, Plenum Publishers, New York. 161-183.

Matern, U., Strasser, H., Wendorff, H., Hamerski, D., 1988, Coumarins and Furanocoumarins. In: Cell Culture and Somatic Cell Genetics of Plants, Vol.5. Phytochemicals in Plant Cell Cultures, Constabel, F., Vasil, I.K. (ed.), Academic Press, New York: 3-21.

Milesi, S., Massot, B., Gontier, E., Bourgaud, F., Guckert, A., 2001, *Ruta graveolens* L.: a promising species for the production of furanocoumarins. Plant Science, 161:189-199.

Nielsen, B.E., 1970, Coumarins of umbelliferous plants. Kandrup and Wunsch, Copenhagen.

Nielsen, B.E., 1971, Coumarin patterns in the Umbelliferae. In: Biology and chemistry of Umbelliferae. Heywood V.H. (ed.) Acad. Press, London: 325-336.

Ojala, T., 2001, Biological screening of Plant Coumarins.Academic Dissertation, University of Helsinki, Helsinki.

Ozarowski, A. (ed.), 1980, Ziololecznictwo. PZWL, Warszawa.

Pande, D., Purohit, M., Srivastava, P.S., 2002, Variation in xanthotoxin content in *Ammi majus* L. cultures during in vitro flowering and fruiting. Plant Science, 162:583-587.

Pathak, M.A., Parrish, J.A., Fitzpatrick, T.B., 1981, Psoralens in photochemotherapy of skin diseases. Farmaco, Ed. Sci. 36:479-491.

Petit-Paly, G., Ramawat, K.G., Chenieux, J.C., Rideau, M., 1989, *Ruta graveolens*: In vitro production of alkaloids and medicinal compounds. In: Biotechnology in Agriculture and Forestry. Vol.7. Medicinal and Aromatic Plants II, Bajaj Y.P.S (ed.), Springer Verlag, Berlin-Heidelberg, 488-505.

Plewig, G., Hölzle, E., Lehmann, P., 1986, Phototherapy for photodermatoses. Curr. Probl. Dermatol., 15:254-264.

Podlewski, J.K., Chwalibogowska-Podlewska, A., 1986, Leki wpólczesnej terapii, PZWL, Warszawa, 384-385.

Podlewski, J.K., Chwalibogowska-Podlewska, A., 1999, Leki wpólczesnej terapii, Wyd. Fundacji PB Büchnera, Split Trading, 408-409.

Ptak, A., Ekiert, H., Gwiazda J., Gomólka E., Swiderski, A., Bach A., 2003, Optimization of light conditions for production of coumarins in *Ruta graveolens* L. shoots cultivated in vitro. Polish- Austrian- German- Hungarian- Italian Joint Meeting on Medicinal Chemistry, Kraków (Poland), abstracts: 224.

Purohit, M., Pande, D., Datta, A., Srivastava P.S., 1995, Enhanced Xanthotoxin Content in Regenerating Cultures of *Ammi majus* and micropropagation. Planta Med., 61: 481-482.

Ramadan, S., 1982, The *Ammi majus* plant. Hamdard, 25:32-35.

Reinhard, E., 1967, Probleme der Production von Arzneistoffen durch pflanzliche Gewebekulturen. DAZ, 107:1201-1207.

Reinhard, E., Corduan, G., Brocke, W., 1971, Untersuchungen über das Ätherische Öl und Cumarine in Gewebekulturen von *Ruta graveolens*. Herba Hungarica, 10: 9- 26.

Reinhard, E., Corduan, G., Volk, O. H., 1968, Über Gewebekulturen von *Ruta graveolens*. Planta Med., 16: 8- 16.

Rimpler, H. (ed.), 1990, Pharmazeutische Biologie II. Biogene Arzneistoffe. Georg Thieme Verlag, Stuttgart- New York.

Rodigiero, G., 1985, Hyperpigmentation induced by furanocoumarins. Farmaco, Ed. Prat., 40: 172-186.

Rodigiero, G., Dall' Acqua, F., 1986, Present aspects concerning the molecular mechanisms of photochemotherapy with psoralens. Drugs Exp. Clin. Res., 12: 507- 515.

Rodigiero, G., Dall' Acqua, F., Pathak, M. A., 1984, Photobiological properties of monofunctional furanocoumarin derivatives. In: Topics in photomedicine, Smith, K. C.(ed.), Plenum, New York: 319- 398.

Roenigh, H.H. Jr., 1984, Effectiveness of psoralens in mycosis fungoides. In: Photochemotherapeutic aspects of psoralens. Natl. Cancer Inst. Monogr., 66: 179-183.

Ruminska, A., Ozarowski, A., 1990, Leksykon roslin leczniczych. PWRiL, Warszawa.

Sanmann, K., Gerst, F., Bohuslavizki, K.H., Koppenhöfer, E., Hänsel, W., 1994, Furocoumarins as a new class of potassium chanel blockers and their possible significance in demyelinating diseases, Eur. J. Pharm. Sci. 2, 168-168.

Scheel, D., Hauffe, K.D., Jahnen, W., Hahlbrock, K., 1986, Stimulation of phytoalexin formation in fungus-infected plants and elicitor-treated cell cultures of parsley. In: Recognition in Microbe-Plant Symiotic and Pathogenic Interactions, Lugtenberg, B. (ed.), Springer Verlaq, Berlin: 325-331.

Sehgal, C. B., 1972, In vitro induction of polyembryony in *Ammi majus* L. Curr. Sci., 41: 263-264.

Sharan, M., Taguchi, G., Gonda, K., Jouke, T., Shimosaka, M., Hayashida, N., Okazaki, M., 1998, Effects of methyl jasmonate and elicitor on the activation of phenylalanine ammonia- lyase and the accumulation of scopoletin and scopolin in tabacco cell cultures. Plant Science, 132: 13- 19.

Siatka, T., Reichling, J., 2000, Stimulation of scopoletin accumulation in *Archangelica officinalis* Hoffm. cell suspension cultures by fungal elicitors. Herba Polonica, 46:12-17.

Sidwa- Gorycka, M., Królicka, A., Lojkowska, E., 2003, Elicytacja furanokumaryn w kokulturach *Ruta graveolens* i *Ammi majus*. II Krajowy Kongres Biotechnologii, Lódz (Poland), Materialy Zjazdowe: 125.

Smith, R. A., Drummond, S., Haines, A., Porter, J. R., Hock, R. S., 2001, Induction of umelliferone in sweet potato cell suspension culture using mercuric chloride. Biotechnology Letters, 23: 1397- 1400.

Smolarz, H.D., Sokolowska-Wozniak, A., Zgórka, G., 1997, Phenolic acids from the herb of *Ruta graveolens* L. Acta. Pol. Pharm.-Drug Research. 54: 161-163.

Soine, T. O., 1964, Naturally occuring coumarins and related physiological activities. J. Pharm. Sci., 53: 231- 264.

Springob, K., 2001, Polyketidsynthasen aus *Ruta graveolens* L. Molekulare Charakterisierung und Aspekte der Evolution. Academic Dissertation, Universität Marburg/Lahn, (Germany).

Steck , W., Bailey, B. K., Shyluk, J. P., Gamborg, O. L., 1971, Coumarins and alkaloids from cell cultures of *Ruta graveolens.* Phytochemistry, 10: 191- 194.

Tandon, S., Rastogi, R.P., 1979, Recent advances in naturally occurring coumarins. Ind. J. Sci. Res., 38:428-441.

Tietjen, K.G., Hunkler, D., Matern, U., 1983, Differential Response of Cultured Parsley Cells to Elicitors from two non- pathogenic strains of Fungi. 1. Identification of Induced Products as Coumarin Derivatives. Eur. J. Biochem. 131:401-407.

Tirillini, B., Ricci, A., 1998, Furocoumarin production by callus tissues of *Heracleum sphondylium* ssp. *sphondylium* L. Phytother. Res., 12:S25-S26.

Turjanmaa, K., Salo, H., Reunala, T., 1985, Comparison of trioxsalen bath and oral methoxsalen PUVA in psoriasis. Acta Dermatol. Venerol. (Stokh.)., 65: 86-88.

Tutin, T.G., Heywood, V.H., Burges, N.K., Moore, D.M., Valentine, D.H., Walters, J.H., Webb, D.A. ed., 1968, Flora Europaea, vol II, Cambridge Univ. Press, Cambridge.

Vuorela, H., Törnquist, K., Nyiredy, Sz., Sticher, O., Hiltunen, R., 1988, Calcium antagonistic activity of the main furocoumarins from *Peucedanum palustre*. Yearly Congr. Soc. Med. Plant Res., Freiburg (Germany), Thieme, Stuttgart, Abstracts: 44-45.

Wolff, F.A., Tessa, T.V., 1986, Clinical pharmacokinetics of methoxsalen and other psoralens. Clin. Pharmacokinet., 11:62-75.

Zangerl, A.R., Green, E.S., Lampman, R.L., Berenbaum, M.R., 1997, Phenological changes in primary and secondary chemistry of reproductive parts in wild parsnip. Phytochemistry, 44:825-831.

Zobel, A.M., Brown, S.A., 1992, Furanocoumarins on the surface of callus cultures from species of the Rutaceae and Umbelliferae. Can. J. Bot., 71: 966-969.

Zogg, G.C., Nyiredy, Sz., Sticher, O., 1989, Apiaceenwurzeln. Qualitative und quantitative Furanocumarinbestimmung in Apiaceenwurzeln. DAZ, 129:717-722.

Index

M

N

O

P

Q

R

S

T

For Product Safety Concerns and Information please contact our EU representative GPSR@taylorandfrancis.com Taylor & Francis Verlag GmbH, Kaufingerstraße 24, 80331 München, Germany

Batch number: 10726440

Printed by Printforce, the Netherlands